Sodium-Ion Capacitors

Sodium-Ion Capacitors

Mechanisms, Materials, and Technologies

Edited by Guoqiang Zou, Xiaobo Ji, and Hongshuai Hou

WILEY-VCH

Editors

Prof. Guoqiang Zou
Central South University
College of Chemistry and Chemical Engineering
No. 932, Lushan South Road
YueLu District
Changsha 410083
China

Prof. Xiaobo Ji
Central South University
College of Chemistry and Chemical Engineering
No. 932, Lushan South Road
YueLu District
Changsha 410083
China

Prof. Hongshuai Hou
Central South University
College of Chemistry and Chemical Engineering
No. 932, Lushan South Road
YueLu District
Changsha 410083
China

Cover Image: © Andrei Kuzmik/Shutterstock

Library of Congress Card No.: applied for

British Library Cataloguing-in-Publication Data
A catalogue record for this book is available from the British Library.

Bibliographic information published by the Deutsche Nationalbibliothek
The Deutsche Nationalbibliothek lists this publication in the Deutsche Nationalbibliografie; detailed bibliographic data are available on the Internet at <http://dnb.d-nb.de>.

Print ISBN: 978-3-527-35037-7
ePDF ISBN: 978-3-527-83734-2
ePub ISBN: 978-3-527-83735-9
oBook ISBN: 978-3-527-83736-6

Typesetting: Straive, Chennai, India

Contents

Preface

Standing at the point where global warming and climate change risks are appearing extremely prominent, the transformation of the global energy configuration to renewable energy system is an inevitable trend in order to achieve the goals in the Paris Agreement and accomplish the reduction of carbon emission. The utilization of renewable energy is expected to be wielded in a larger scale to reduce the dependence on traditional fossil fuels (coal, oil, etc.) and become the major source of power generation. Significantly, the maturity of renewable energy technology and the rapid expansion of scale have greatly reduced the relevant costs in recent years, forming a strong competition with traditional fossil energy in terms of price.

Advanced secondary ion batteries, as the representational technology of energy storage, play a pivotal role for decarbonization. Triumphing toward clean energy conversion in the next few decades, lithium-ion batteries (LIBs) have been developed in the direction of high-energy density. However, the limited output performance at high-power density seriously restricts the corresponding wide application. Importantly, well-balanced trade-offs between energy density and power density have been presented by metal-ion capacitors, in which capacitor-type electrode with adsorption/desorption behavior and battery-type electrode with intercalation/conversion/alloying mechanism are selected as cathode and anode, respectively. Owing to the overwhelming advantage of much more abundant content of sodium sources in the crust than lithium, sodium-ion capacitors (SICs) have attracted extensive interest in constructing next-generation energy storage systems.

SICs are just like Rubik's cubes with charming diversities; this book comprehensively and systematically describes technology, relevant materials, and development trends of SICs, including characteristic of SICs devices, fundamental understanding of SICs mechanism, classification of SICs cell configurations, cathode materials for SICs, anode materials for SICs, flexible SICs devices, and pre-sodiation technologies. Meanwhile, this book offers guidelines to construct advanced SICs, which is suitable for scientific and technological workers in the field of energy storage systems, as well as senior undergraduates, graduate students, and teachers of related majors in colleges and universities.

This book is edited by Prof. Zou, Prof. Hou, and Prof. Ji of the Central South University of China. The specific chapter writers divided the work as follows: Chapters 1–4 are written by Peng Cai, Wentao Deng, Hongshuai Hou, Guoqiang Zou, and Xiaobo Ji of Central South University; Chapter 5 is written by Xiong Zhang, Wenjie Liu, Lei Wang, Chen Li, and Yanwei Ma of Institute of Electrical Engineering, Chinese Academy of Sciences; Chapter 6 is written by Kangyu Zou, Wentao Deng, Hongshuai Hou, Guoqiang Zou, and Xiaobo Ji of Central South University; Chapter 7 is written by Taoqiu Zhang and Huanwen Wang of Faculty of Material and Chemistry, China University of Geosciences; Chapter 8 is written by Zirui Song, Chang Liu, and Xiaobo Ji of Central South University; Chapter 9 is written by Kangyu Zou, Wentao Deng, Hongshuai Hou, Guoqiang Zou, and Xiaobo Ji of Central South University.

Due to the wide range of content involved in this book, coupled with the limited time and ability, if there are omissions and inadequacies, please criticize and inform us to correct!

7 November 2022

Guoqiang Zou, Xiaobo Ji, and Hongshuai Hou
Central South University
Changsha, China

1

Introduction

Peng Cai, Wentao Deng, Hongshuai Hou, Guoqiang Zou, and Xiaobo Ji

Central South University, College of Chemistry and Chemical Engineering, No. 932, Lushan South Road, Yuelu District, Changsha 410083, China

1.1 A Brief Development of SICs

Nowadays, with the rapid development of daily household appliances, portable instruments, data storage systems, and aerospace facilities, it is even more necessary to develop new energy storage devices with high energy density, high power density, and good cycle stability. In the past, many researchers have been committed to designing excellent energy storage devices that take into account high energy density and high power density, such as rechargeable batteries and electrochemical supercapacitors (SCs) [1–13]. However, for the world, the configuration of electrochemical energy storage devices that provide both high energy density and high power has become an urgent need [14].

The successful commercialization of lithium-ion batteries (LIBs) in 1991 has received extensive attention from researchers [15–18]. LIBs are characterized by high working voltage, high energy density, and wide working voltage window but poor rate performance [19–22]. In comparison, SCs have higher power density and cycle stability, but their application is limited due to their low energy density defects [23, 24]. Therefore, in response to this defect, the concept of hybrid-ion capacitors was brought up [25–32]. In 2001, Amatucci used activated carbon (AC) to construct the first lithium hybrid capacitor as the cathode and nanostructured $Li_4Ti_5O_{12}$ (LTO) as the anode [17, 33, 34]. The energy density of the container is twice that of traditional carbon-based SCs, and at the same time it presents a bright magnification prospect. Since then, after the application of lithium hybrid capacitors in assembling a variety of devices, many researchers are still exploring and paying attention to the cost and future reserves of lithium, especially related to the application of large-scale energy devices and smart grids, and they have gradually proposed the idea of replacing lithium with sodium [32, 35–47]. In addition, there is another kind of alkali metal ion capacitor–potassium ion capacitor (PIC), which is rich in resources, but its related technology research is still in the preliminary exploration stage. There are five main factors restricting the development of

Sodium-Ion Capacitors: Mechanisms, Materials, and Technologies, First Edition.
Edited by Guoqiang Zou, Xiaobo Ji, and Hongshuai Hou.

PICs: (i) low ion diffusion rate in solid electrodes and poor potassium ion reaction kinetics; (ii) large volume changes during potassium insertion/depotassization; (iii) serious side reactions and electrolyte consumption; (iv) dendrite growth and safety hazards; and (v) limited energy density/power density caused by the relatively high atomic mass of *K*. In addition, some aqueous metal-ion capacitors (zinc-ion capacitors) have also received widespread attention. However, due to their lower energy density and longer cycle life, aqueous metal-ion capacitors are more suitable for applications in biological systems.

From the development of lithium-ion capacitors (LICs) and other hybrid-ion capacitors above, it is not difficult to see that mastering the development of LICs is very meaningful for researchers to explore the development of sodium-ion capacitors (SICs). Therefore, in order to better understand the development features and advantages of SICs, other types of hybrid-ion capacitors will be introduced in Section 1.2. Other hybrid-ion capacitors will be introduced in the categories of monovalent hybrid-ion capacitors and multivalent hybrid-ion capacitors. Taking the monovalent hybrid-ion capacitor as an example, the PICs are firstly introduced. Then, a brief introduction to multivalent hybrid-ion capacitors is given. For example: magnesium-ion hybrid capacitors (MICs), calcium-ion batteries (CIBs), zinc-ion hybrid capacitors (ZICs), and aluminum-ion hybrid capacitors (AICs). Compared with high power density, the core problems of these emerging systems may lie in other aspects. Therefore, these systems may not be fully developed in the field of hybrid-ion capacitors. Some basic thinking, attempts, and explorations of hybrid-ion capacitors in these emerging development fields will also be briefly covered. This chapter hopes to inspire readers to fully understand and master SICs by covering a variety of backgrounds and introductions. Table 1.1 and Figure 1.1 demonstrate the different characteristic aspects of the charge carriers of Li^+, Na^+, K^+, Mg^{2+}, Ca^{2+}, Zn^{2+}, and Al^{3+} for their respective energy storage systems.

As representative of monovalent hybrid-ion capacitors, LICs, their development history and successful commercialization path are worth discussing. In 1984, Dr. Yamabe of Kyoto University cooperated with Dr. Yata of Kanebo Co., Ltd. to synthesize a new type of carbonaceous material, namely polyacene semiconductor (PAS), by pyrolysis of phenolic resin [49]. In 1987, Yata et al. reported the intercalation/deintercalation properties of lithium ions in PAS, and research activities on LICs have received much attention since then [50]. In 2005, Fuji Heavy Industries was the first to commercialize LIC based on a porous carbon cathode and a pre-lithiated PAS anode with a stacked structure [51]. Since then, different companies have attempted to commercialize LICs based on various technologies, and a few promising commercial products are described in Table 1.2. Recently commercialized LICs can achieve gravimetric energy and power densities of $20\,Wh\,kg^{-1}$ and $7.5\,kW\,kg^{-1}$, respectively, with lifetimes ranging from 100 000 to 800 000 [52].

Among the many LICs, the most classic is the pioneering work of Amatucci et al. As mentioned before, Amatucci et al. were the first to report LICs via hybridization of EDLCs and LIBs [33]. In 2001, AC(+)//$Li_4Ti_5O_{12}$(−) LIC using 1 M $LiPF_6$ in ethylene carbonate (EC)/dimethyl carbonate (DMC) (2 : 1, v/v) was first reported

Table 1.1 Characteristic physical properties of Li^+, Na^+, K^+, Mg^{2+}, Ca^{2+}, Zn^{2+}, and Al^{3+} ion carriers for hybrid-ion capacitors.

Properties	Li^+	Na^+	K^+	Mg^{2+}	Ca^{2+}	Zn^{2+}	Al^{3+}
Relative atomic mass	6.94	6.94	9.10	24.31	40.08	65.38	26.58
Mass-to-charge ratio	6.94	6.94	9.10	12.15	20.04	32.69	8.86
Theoretical gravimetric capacity of $ACoO_2$ ($mAh\,g^{-1}$)	274	235	206	260	242	-	268
Theoretical volumetric capacity of $ACoO_2$ ($mAh\,cm^{-3}$)	1378	1193	906	—	—	—	—
E^0 ($A/A^{n+}{}_{aq}$) V vs. SHE	−3.04	−2.71	−2.93	−2.37	−2.87	−0.76	−1.66
E^0 ($A/A^{n+}{}_{PC}$) V vs. $Li/Li^+{}_{PC}$	0	0.23	−0.09	—	—	—	—
Shannon's ionic radii (Å)	0.76	1.02	1.38	0.72	1	0.74	0.535
Stokes radii in H_2O (Å)	2.38	1.84	1.25	3.47	3.10	3.49	4.39
Stokes radii in PC (Å)	4.8	4.6	3.6	—	—	—	—
Molar ionic conductivity in PC ($S\,cm^2\,mol^{-1}$)	8.3	9.1	15.2	—	—	—	—
Molar ionic conductivity in $AClO_4$/PC ($S\,cm^2\,mol^{-1}$)	6.54	6.54	—	—	—	—	—
Polarization strength ($10^4\,pm^{-2}$)	2.16	1.11	—	4.73	—	—	—
Desolvation energy in PC ($kJ\,mol^{-1}$)	215.8	158.2	119.2	569.4	—	—	—
Melting point (°C)	180.5	97.8	63.4	650	842	419	660
Coordination preference (O = octahedral, T = tetrahedral, P = prismatic)	O and T	O and P	—	O and T	—	—	—

Source: Naskar et al. [48]/with permission of John Wiley & Sons.

by Amatucci et al. The LICs displayed sloping charge/discharge curves in the 1.5–3.0 V window with 90% capacity utilization at 10 C rate and 10–15% capacity loss after 5000 cycles. In their reports, the $Li_4Ti_5O_{12}$ anodes exhibit almost zero volume change upon Li^+ ion intercalation/deintercalation and show a terminal lithiation voltage of 1.55 V. In a follow-up publication in 2004, typical devices with 40 Wh kg^{-1} large energy density and 4000 W kg^{-1} high power density were developed by using nanostructured $Li_4Ti_5O_{12}$ anode and AC/$LiCoO_2$ composite cathode in 2 M $LiBF_4$/acetonitrile (AN) electrolyte [53]. High durability is achieved in this device due to the utilization of AC EDL cathode material and $Li_4Ti_5O_{12}$ nanostructured anode material. The capacity loss after 9000 cycles at full depth of discharge is 20%, which is quite superior compared to conventional LIBs.

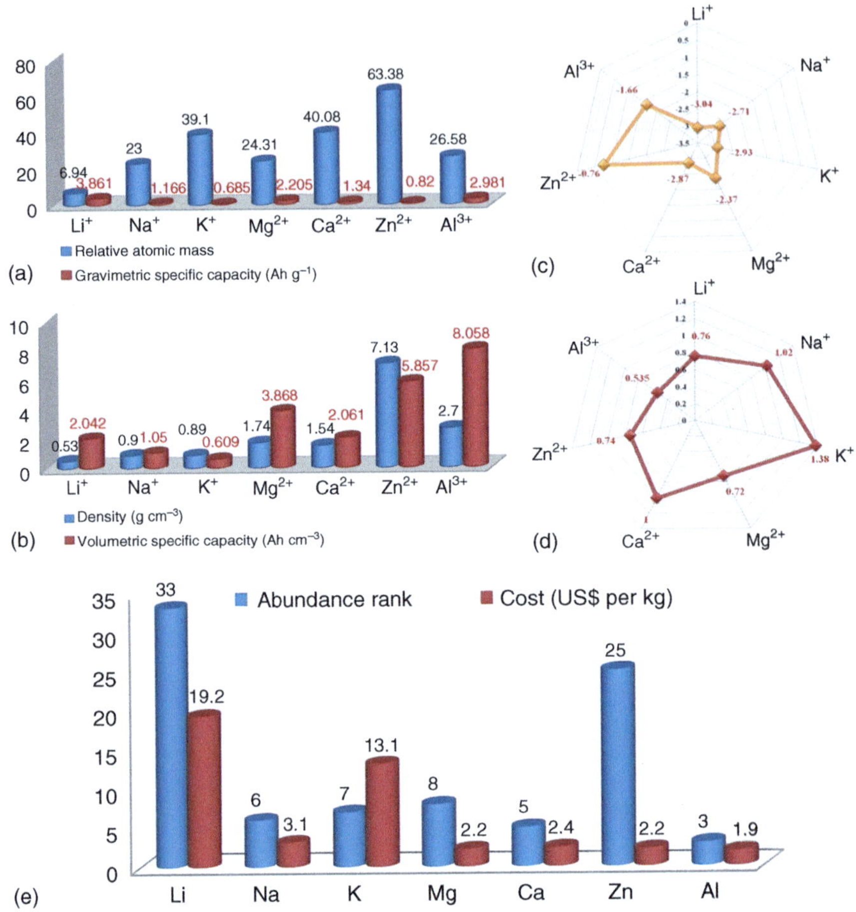

Figure 1.1 (a) Relative atomic mass and gravimetric specific capacity of metal. (b) Density and volumetric specific capacity of metal. (c) Standard electrode potential (V vs. SHE). (d) Shannon's ionic radii (Å) of different charge carriers for rechargeable batteries, and (e) Relative abundance-rank and cost of different charge carriers (Li^+, Na^+, K^+, Mg^{2+}, Ca^{2+}, Zn^{2+}, and Al^{3+}) for rechargeable batteries. Source: Reproduced with permission of Naskar et al. [48] Copyright 2021, Wiley-VCH GmbH.

1.2 Comparison Between Different Hybrid-Ion Capacitors

Similarly, among monovalent hybrid-ion capacitors, PICs have recently drawn attention as promising next-generation energy storage systems due to their relatively abundant potassium reserves and low cost. The development of PICs is accompanied by the development of potassium-ion batteries (PIBs). Since 2015, an increasing number of scientific publications on PIBs and PICs, can be observed. PICs exhibit some advantages [54]. For example, potassium does not thermodynamically form Al–K intermetallics, cheaper aluminum foils can be utilized as anode current

Table 1.2 Product characteristics of some of the most advanced commercial lithium-ion capacitors (LICs).

Company	Device type	Potential (V)	Capacitance (F)	Energy density (Wh kg^{-1})	Cycles
JM Energy Corporation	Prismatic	2.2–3.8	3300	13	300 000
General Capacitor Intl, Inc.	Laminate	2.2–3.8	3000	18	100 000
Taiyo Yuden	Cylinder	2.2–3.8	200	15	100 000
Vina Technology	Cylinder	2.2–3.8	270	—	—
Aowei Technology	Module	2.2–3.8	9000	>20	>30 000
Greenway	21 700	2.0–4.0	333	10.7	50 000
LRNET	Laminate	4	15 000	—	50 000
Asahi Kasei FDK energy	Module	15	600	12 Wh	—
ACT	Laminate	2.0–4.0	5000	15	—
NEC Tokin	Laminate	2.2–3.8	1000	8.0	—
MSR Micro	Laminate/ Prismatic	2.2–3.8	92–825 range	—	—
PuriXel, South Korea	Laminate	2.25–3.00	—	—	100 000

Source: Adapted from Naoi et al. [52].

collectors for PICs. Moreover, the standard electrode potential of K (−2.936 V vs. K^+/K) is lower than that of Na (−2.714 V vs. Na^+/Na), which may result in a wider voltage window, indicating higher energy density than SICs. Furthermore, due to the weak Lewis acidity of K^+, K^+ can form smaller solvated ions (3.6 Å) than Li^+ (4.8 Å) and Na^+ (4.6 Å) in monovalent hybrid-ion capacitors. Thus, PICs achieve fast ionic diffusion rates and high electrical conductivity in propylene carbonate (PC) solvent. Furthermore, unlike SICs, commercially available graphite can be used as anode materials for PICs by forming intercalation compounds (KC_8) with theoretical capacities of 279 mAh g^{-1}. Another significant advantage compared to SICs is that, for most anode materials, the intercalation potential of K ions is about 0.2 V relative to K^+/K, which reduces the possibility of metal potassium plating and effectively avoids the risk of dendrite formation during charging. For example, for hard carbon (HC) in SICs, the sodium potential is about 0.05 V relative to Na^+/Na, while for K^+/K in PICs, the potassium potential is 0.2 V. However, PICs also suffer from more limited cycling stability. Due to the large ionic radius of potassium ions (1.38 Å), among many electrode materials, especially the anodes would show large volume expansion during charge and discharge, resulting in a short lifespan in most PICs (usually <500 cycles). Consequently, more limited cycling stabilities are also great challenges for improving the power densities of PICs. Therefore, PICs are still in the developmental stages.

As discussed before, multivalent hybrid-ion capacitors may have a different focus than the core issues of monovalent hybrid-ion capacitors. To understand the features of MICs, a brief overview of the electrodes and electrolytes of MICs is required. As a divalent cation, energy systems based on Mg ions encounter sluggish kinetics due to the strong electrostatic interactions of Mg ions with anions in the host framework in the cathode. Therefore, the design of the cathode active material is a key factor. Chevrel phases ($Mg_xMo_6T_8$, T = S, Se, Te) and disulfides (MoS_2, WSe_2, etc.) are preferred as promising Mg-ion cathode active materials [55–59]; since in these host materials, the electrostatic interaction between Mg ions and anions is low. In addition, sufficient channel size in the crystal structure also facilitates the facile intercalation/deintercalation of Mg ions. On the other hand, moderately polar anions (S^{2-}, Se^{2-}, Te^{2-}, etc.) in electrode materials can lead to weaker bond strengths between transition metal cations, resulting in relatively low redox potentials for transition metals [60]. Therefore, Chevrel phases and disulfide show redox potentials below 2 V (vs. Mg^{2+}/Mg). This discussion suggests that higher voltages (i.e. higher energies) and fast kinetics (i.e. higher power) are not easily achieved simultaneously in MICs. Besides these materials V_2O_5 [61], MnO_2 [62], MoO_3 [63], $MgCo_2O_4$ [64], TiS_2 [65], TiO_2 [66], sulfur [67], iodine [68], and polyanion-based materials ($MgMnSiO_4$, $MgFeSiO_4$, etc.) [69, 70] have also been reported as cathode active materials for MICs. To better understand the properties of MICs, some characteristics of metallic magnesium anodes in magnesium-ion batteries (MIBs) are also introduced. From the anode point of view, metallic Mg electrode is a good choice for nonaqueous MIBs because of its high negative voltage (−2.37 V vs. SHE) and high theoretical capacity (2205 mAh kg^{-1}). However, complex reactions are prone to occur between the metal magnesium anodes and the electrolytes. Hence, the discussion of anodes and electrolytes needs to be covered. Due to the chemical activity of Mg in conventional electrolytes, it encounters serious problems: Due to the presence of Mg-ion salts in polar aprotic solvents, the Mg surface tends to form a hard passivation layer that inhibits ion pathways and accompanying electrochemical reactions. To circumvent this problem, researchers are turning to alloyed/dealloyed or Mg insertion/deintercalation types of anodes. However, such materials also suffer from slow kinetics and pulverization due to excessive volume changes during charge/discharge. Among alloying/dealloying materials, Bi, Sb, Bi–Sb, Ge, Si, Sn, and Sn-based binary alloys (Cu–Sn, Pb–Sn, and In–Sn) have been reported in the literature [71–74]. In addition, the development of non-Mg metal anodes has promoted the further development of MICs to a certain extent [75, 76]. In 2D materials, the anode of MICs can also use defective graphene and graphene allotrope moieties [77], black phosphorus [78], and $Li_4Ti_5O_{12}$ [79]. In organic systems, the development of MICs is mainly limited by electrolytes. Furthermore, a discussion of electrolytes in MIBs is essential when it comes to electrolytes for MICs. In the work of Aurbach et al., an unconventional electrolyte system based on organohaloaluminate magnesium salts in tetrahydrofuran (THF) and polyethers of the glyme was developed, in which metallic magnesium electrodes work reversibly with relatively fast kinetics [80]. Grignard reagents (RMgX; R: alkyl or aryl, X: Cl or Br) as electrolytes for MIBs have also been reported in the literature for passivation-free metal Mg electrodes, but their strong reducibility limits the oxidative stability of cathodes [81]. Another electrolyte system, namely organoborate (magnesium dibutyldiphenyl $Mg(BPh_2Bu_2)_2$ and tributylphenyl $Mg(BPhBu_3)_2$) in THF (>0.4 mol), can realize anode reversible Mg stripping/plating as well as cathode reversible Mg ion insertion/extraction [82]. In recent years, MIBs of aqueous electrolytes have attracted attention due to their safe and cost-effective properties. For example, in 2019, Zhang et al. constructed devices based on δ-MnO_2@carbon molecular sieve composite as cathode and nanowire VO_2 as anode [83]. In 2017, Zhang et al. reported an aqueous anode with carbon-coated $FeVO_4$ and a todorokite-type magnesium octahedral molecular sieve (Mg-OMS-1) cathode [84]. Nam et al. proposed smart material engineering by introducing crystal H_2O into the layered structure of the brucite MnO_2 cathode to efficiently screen electrostatic interactions between Mg^{2+} and host framework anions [85]. The group also demonstrated lower desolvation energies at the cathode and electrolyte interfaces by adding H_2O to the nonaqueous electrolyte solution. This is because hydrated Mg ions are allowed to intercalate in their hydrated form, thereby minimizing desolvation energy loss. The intercalated hydrated Mg ions in the host framework further minimize the electrostatic interactions between Mg ions and host anions [86]. Therefore, the birnessite MnO_2 cathode in pure aqueous electrolyte exhibits a large reversible capacity of 231.1 mAh g^{-1} at 2.8 V.

However, the above overview of aqueous/nonaqueous electrode materials is used to complement the basic background of MICs. In 2014, Yoo et al. developed the first MICs utilizing AC cloth and magnesium metal as cathodes and anodes, respectively. To prevent hard passivation films and dendrite growth on the metal anodes, 0.25 M organohalide magnesium aluminate complexes ($Mg_2Cl_3^+$–$Ph_2AlCl_2^-$) were utilized in THF electrolytes [87]. In this electrolyte, the Mg electrode works reversibly for thousands of cycles with a CE of approximately 100%. On the other hand, the pores in the cathodes are saturated with large ions (bulky Mg and Al-based ionic complexes consisting of Cl, alkyl or aryl groups, and THF ligands) before the potential limit is reached. Surprisingly, the introduction of 0.5 M LiCl solves this problem, as small ionic substances will be present in the electrolytes. The full-cell device exhibited a specific capacitance of 90 F g^{-1} at 5 mA g^{-1} within 0.9–2.4 V and maintained 79% of the initial capacitance after 4500 charge/discharge cycles. In the following years, several researchers screened suitable materials and electrolyte systems for advanced MICs. Breakthroughs in MICs are more focused on aqueous electrolytes due to the significant challenges in organic systems. Sun et al. and Maitra et al. reported MnO_2 nanowires in $MgSO_4$/$Mg(NO_3)_2$ electrolytes and $MgNiO_2$/$Mg(ClO_4)_2$ electrolytes for low-cost aqueous MICs, respectively [88]. In 2017, Zhang et al. demonstrated aqueous MICs with (+)Mg-OMS-2/graphene//0.5 M $Mg(NO_3)_2$(aq)//AC(−) configuration [89]. The cryptomelane-type manganese oxide octahedral molecular sieves (OMS-2) are a unique electrode with 2×2 and 1×1 tunneling structures of MnO_2 in MICs, which is widely used as an active material in magnesium ion batteries. Besides, the low electronic conductivity of OMS-2 can be mitigated by the preparation of composites containing carbon material. The full cell exhibited a high energy density of 46.9 Wh kg^{-1} (100 mA g^{-1}) and excellent cycling stability

(95.8% capacity retention at 100 mA g^{-1} after 500 cycles) at 0–2 V. Cao et al. reported the utilization of Mn_3O_4 and AC as the active materials in cathode and anode, respectively, using aqueous electrolytes of 2 M $MgSO_4$ [90]. The MICs showed an energy density of 20.2 Wh kg^{-1} (125 W kg^{-1}) and excellent cycling stability (95% capacity retention after 6000 cycles at 0.5 A g^{-1}) with the potential of 0–2 V. In addition, the scaled-up flexible packaging depicts 80% capacity retention after 500 cycles at a current density of 0.5 A g^{-1}. Tian et al. reported Mg^{2+} ion insertion/detachment in neutral aqueous $MgSO_4$ electrolytes for VN anodes [91]. By increasing the scan rate from 1 to 200 mV s^{-1}, the rectangular-like CV and the small polarized redox peaks indicate the fast reaction kinetics due to a surface-controlled process. In order to understand the pseudocapacitance mechanism of VN during charging/discharging, both XPS and CV can confirm that the V(III) to V(II) transition is the key to the VN charge/discharge reaction. Quasi-solid aqueous MICs with a (+)MnO_2@C/$MgSO_4$ gel/VN(−) configuration show a bulk energy density of 13.10 mWh cm^{-3} (72 mW cm^{-3}), and a bulk power density of 440 mW cm^{-3} (10.35 mWh cm^{-3}) and excellent cycling stabilities (5000 cycles at 16 mA cm^{-2}) in the range of 0–2.2 V. In addition, the device was further developed as a flexible solar-charging integrated unit based on screen-printed micro-supercapacitors. However, the number of promising MIC systems is indeed limited from the point of view of laboratory prototypes or practical devices.

Possessing the highest Shannon's ionic radii of all multivalent charge carriers (Table 1.1), Ca^{2+} ions exhibit faster electrochemical reaction kinetics than Mg^{2+}, Zn^{2+}, and Al^{3+} ions, due to low polarization. However, as mentioned above, the promising classes of electrode materials for CICs are similar to MICs. Different experimental and simulation studies have shown that 3D tunneling and layered structures, such as $CaMn_2O_4$, V_2O_5, graphite, etc., are suitable for Ca^{2+} intercalation/de-intercalation. Therefore, these materials can be considered as anode materials for CICs. Inspired by the Mg^{2+} ion system, a similar Chevral phase ($CaMo_6T_8$ [T = S, Se, Te]) is theoretically envisaged for the Ca^{2+} energy storage systems [92]. The operating voltage of $CaMo_6S_8$ is predicted to be 1.4 V (vs. Ca/Ca^{2+}). However, the diffusion of Ca^{2+} is slower in $CaMo_6S_8$ than that of Mg^{2+}. The diffusion energies of Mg^{2+} and Ca^{2+} in $CaMo_6Se_8$ are both lower than those of $CaMo_6S_8$. The diffusion energies of Ca^{2+} in Mo_6S_8 and Mo_6Se_8 are 780 and 520 meV, respectively, which are relatively higher than the diffusion energies of Mg^{2+} (270 and 180 meV for Mo_6S_8 and Mo_6Se_8, respectively). Based on the quantitative diffusion barrier limits for cell operation, the nanostructured Mo_6Se_8 could be promising as a suitable anode for CICs with an average voltage of 1.25 V. However, no experimental data on the insertion/deinsertion of Ca^{2+} ions based on the Chevral phase are available in the literature to date. Cubic framework structures of Prussian blue analogues ($A_xMFe(CN)_6 \cdot yH_2O$ [where A = Li, Na, K, Mg, Ca, etc., and M = Ba, Ti, Mn, Fe, Co, or Ni]) have also been investigated as insertion electrodes for CICs, but their capacities are not up to the mark [93]. While Ca metal anodes may be attractive for CICs due to their high bulk and weight capacities, surface passivation and subsequent hard SEIs formation in conventional electrolytes hinders reversible Ca^{2+} stripping/plating. Therefore, in conventional electrolytes (containing Ca ion salts in polar non-protonic solvents), an efficient Ca ion energy storage system cannot be achieved using metal anodes [94]. On the other hand, alloyed materials such as Ca–Sn and Ca–Si anodes can exhibit reversible alloying/dealloying electrochemical reactions with appreciable capacity by avoiding the formation of hard passivation films [93, 95]. In 1991, Aurbach et al. demonstrated the electrochemical behavior of Ca-metal electrodes in several organic electrolytes, such as $Ca(ClO_4)_2$, $Ca(BF_4)_2$, $LiAsF_6$, and tetrabutylammonium salts (BF_4^- and ClO_4^-) in THF, PC, AN, and γ-butyrolactone solvents [96]. During the electrochemical reduction of electrolyte solutions, $CaCl_2$ (in ClO_4^- salt solutions), $Ca(OH)_2$, $CaCO_3$, and calcium alcohol salts form passivation films which limit the reversible deposition/dissolution of electrodes in Ca-based nonaqueous electrolytes. In 2016, Ponrouch et al. reported that salts in organic electrolytes mixed with carbonate solvents (PC and EC) containing $Ca(ClO_4)_2$, $Ca(BF_4)_2$, and $Ca(TFSI)_2$ exhibited a wide electrochemical stability window at high temperatures (e.g. −0.5 to 3.5 V vs. Ca/Ca^{2+} at 100 °C) [94]. Ultimately, these electrolytes also show irreversibility of Ca^{2+} stripping/plating. In 2017, Wang et al. developed a THF-based electrolyte containing $Ca(BF_4)_2$ salt that can be operated at room temperature [97]. In this electrolyte, the reversible stripping/plating of Ca^{2+} was very satisfactory, but with lower anodic stability (3 V vs. Ca/Ca^{2+}) and lower CE. Unlike the reduction products that form hard SEIs in conventional electrolytes, CaH_2 was identified as the SEIs component in the THF-based electrolytes. Unfortunately, the SEI films were unstable and formed continuously during cycling due to CaH_2 deposition, resulting in a CE lower than that required for practical applications (99.98%). In 2019, Shyamsunder et al. synthesized a new fluorinated alkoxy borate ($Ca(B(Ohfip)_4)_2{\cdot}4DME$), which is based on the hexafluoroisopropoxy ($Ohfip^-$) ligand [98]. It shows reversible stripping/plating of Ca^{2+} in 1,2-dimethoxyethane (DME) solvent at 25 °C with low polarization (170 mV). This salt also exhibits higher anodic stability up to 4.1 and 4.9 V in DME and *N,N*-dimethyltriflamide, respectively. In 2019, Li et al. reported a similar electrolyte system, calcium tetrakis(hexafluoroisopropoxy)borate ($Ca[B(hfip)_4]_2$) in DME, showing reversible stripping/plating of Ca^{2+} at room temperature with high oxidative stability up to 4.5 V vs. Ca/Ca^{2+} and high ionic conductivity (>8 mS cm^{-1}) [99]. Nevertheless, these electrolytes have some limitations and therefore require extensive research on the non-aqueous and aqueous electrolytes. Ta et al. and Lee et al. report computational simulation studies of Ca electrodeposition and species formation processes in nonaqueous electrolytes and modulate the hydration number of Ca^{2+} ions by varying the electrolyte concentration, respectively [100, 101]. Different theoretical and experimental studies on electrodes/electrolytes are essential in order to gain a more comprehensive knowledge on the feasibility of practical CICs. In 2019, Wu et al. reported the first CIC devices based on AC cathodes, Sn foil anodes, and a 0.8 M $Ca(PF_6)_2$ electrolyte solution in a mixed carbonate solvent (EC, PC, DMC, and EMC) [102]. During the charging of CICs, PF_6^- anions were adsorbed to the AC surface and Ca^{2+} were migrated toward the Sn anodes, forming the Ca_7Sn_6 alloy. During the discharge of CICs, the opposite process occurred. Based on the above mechanism, the full device exhibited an operating voltage of

1.5–4.8 V. Moreover, the CV curves at different scan rates (10–100 mV s^{-1}) indicated that the full-cell device delivers good CICs performances. Reversible capacities of 92 mAh g^{-1} (0.1 A g^{-1}) and 82 mAh g^{-1} (0.4 A g^{-1}) and a capacity retention rate of 84% over 1000 cycles at 0.2 A g^{-1} were achieved. To our knowledge, this is the only CIC device reported to date that shows promising electrochemical performances at room temperature.

ZICs are developed from ZIBs. ZIBs have received a lot of attention for their low cost, high safety, and environment friendliness. Typically, ZIBs consist of zinc metal anodes, aqueous electrolytes such as $ZnSO_4$ (aq), and cathodes for Zn^{2+} intercalation/de-intercalation. Unlike the extremely active lithium, sodium, and potassium metal electrodes, zinc metal electrodes are stable in air and can be used directly as the anodes in ZIBs. At the same time, metallic Zn electrodes have a high weight capacity of 823 mAh g^{-1} (corresponding to an ultra-high-volume capacity of 5845 Ah l^{-1}, much higher than 2046 Ah l^{-1} for lithium electrodes and 3833 Ah l^{-1} for Mg metal electrodes) and a low redox potential of −0.76 V compared to standard hydrogen electrodes [103]. In addition, the high ionic conductivity of the aqueous Zn^{2+}-containing electrolyte of the ZIBs facilitates fast charge/discharge rates. Therefore, ZICs inherit the excellent advantages of ZIBs. As mentioned above, the power density of the ZICs is highly competitive compared to other hybrid-ion capacitors. The two electrodes of ZICs (AC and Zn) are also very stable (in this neutral/light system) and ensure excellent cycling stabilities [104]. Most importantly, ZICs have the potential to be a hybrid ion capacitor device with good energy density, power density, and excellent long-term stabilities. ZICs typically utilize porous carbon (or AC) as the cathodes and zinc metal as the anodes. The highly reversible charge storage mechanism of the porous carbon cathode gives ZICs an extremely long service life. Zinc anodes store charge by plating and stripping, thus providing high capacity for ZICs. Given the low redox potential of the zinc anodes, aqueous ZICs with carbon-based cathodes can provide higher voltages (≈1.6–1.8 V) than symmetrical C//C SCs (≈1 V) [105]. Thanks to the hybrid configuration, ZICs can bridge the energy density gap between SCs and rechargeable batteries. In addition, ZICs offer more stable cycling performances and higher power densities than their ZIB counterparts. Based on the low cost of carbon materials and zinc metal, ZICs hold promise for low-cost and large-scale applications, especially for those requiring high power.

Research work on ZICs has only recently begun. In 2016, Wang and coworkers reported the first ZICs assembled from oxidized carbon nanotube cathodes and zinc anodes in aqueous $ZnSO_4$ electrolytes [106]. The ZICs showed a low specific capacitance of 53 F g^{-1}, which was attributed to the low specific surface area (SSA) of the oCNT cathode (211 m^2 g^{-1}). In 2017, Tang and coworkers used porous carbon with a high SSA (3384 m^2 g^{-1}) as ZIC cathodes [107]. Such ZICs provide a capacitance of 170 F g^{-1} at a current density of 0.1 A g^{-1}, corresponding to an energy density of 52.7 Wh kg^{-1} at 1725 W kg^{-1}. In order to increase the energy density of ZICs, research work has attempted to expand the voltage window by optimizing the electrolyte. Typically, aqueous $ZnSO_4$ electrolyte-based ZICs have a voltage window of 0.2–1.8 V. In 2018, Wang et al. developed an ultra-high concentration water-in-salt (WIS) electrolyte [108]. The WIS electrolyte consisted of 20 M (mol kg^{-1}) lithium bis(trifluoromethanesulfonyl)imide (LiTFSI) and 1 M $Zn(TFSI)_2$. The WIS electrolyte provided a wide voltage window with voltages up to 2.1 V supported by a highly stable WIS electrolyte, and achieved a high average CE of 99.7%. In 2019, Lu ad coworkers assembled ZICs with high energy density by employing nitrogen-doped graded porous carbon as cathodes [109]. The nitrogen dopant was shown to obtain pseudocapacitance by lowering the energy barrier for the formation of C—O—Zn bonding. In 2019, Zapien and coworkers fabricated the first flexible ZICs using polyacrylamide hydrogel electrolytes [110]. In 2020, Bimbo and coworkers introduced Ti_3C_2 MXene as an anode material for ZICs [111]. Ti_3C_2-based ZICs exhibited a high capacity of 189 mAh g^{-1} with capacity retention of 96% after 1000 cycles. Zhi and coworkers ZICs were fabricated using phosphene as the cathodes [112]. Phosphene cathodes showed high capacitance of over 300 F g^{-1}. It was worth noting that pseudocapacitances can be used to increase the energy density of ZICs. Qu and coworkers improved the capacitance of ZICs by manipulating the ZICs in an oxygen reduction reaction [113]. Due to the oxygen reduction reaction, the discharge capacities of ZICs were much higher than their charging capacity. Alshareef et al. utilized hydrogen and oxygen pseudocapacitance to increase the capacitances of ZICs. These pseudocapacitance studies provide a new approach to the design of aqueous ZICs with high energy densities [114].

However, the development of practical ZICs requires overcoming multiple challenges [115]. One limitation is the low energy density, which can be improved by increasing the capacitances of the cathode materials and by widening the voltage windows. The voltage windows of ZICs are mainly driven by the electrochemical stabilities of the electrolytes. In alkaline electrolytes, the zinc metal interacts with hydroxide ions, resulting in the accumulation of zinc hydroxide materials on the zinc anodes. Isolated zinc hydroxide materials reduce the overall conductivity and increase the polarization of the zinc anodes. Therefore, alkaline electrolytes should be avoided in ZICs. In neutral and weakly acidic electrolytes (pH above 3), the voltage windows of ZICs are limited by parasitic reactions such as HER, OER, and oxidation of porous carbon cathodes. Another limitation of ZICs is the cycling stabilities of the zinc anodes. Compared to the carbon cathodes, zinc anodes exhibit limited cycling stabilities due to dendrite growth, HER, and corrosion reactions. The formation of zinc dendrites involves a complex process, which is highly dependent on the inhomogeneous electric fields and ion distributions. During charging and discharging, the growth of zinc dendrites increases the SSA of the zinc anodes, thereby accelerating the HER rates. The high HER rate consumes protons in the aqueous electrolytes, leading to the accumulation of hydroxide ions. The hydroxide ions interact with the zinc anode and thus trigger the zinc corrosion reactions. Zinc corrosion results in the accumulation of zinc hydroxide substances on the surfaces of the zinc anodes. These byproducts reduce the overall electrical conductivity and exacerbate the inhomogeneous electric field of the zinc anodes, thereby increasing dendrite growth. Subsequently, the vicious cycles of zinc dendrites, HER, and zinc corrosion continue. Besides metallic zinc anodes, the modification of porous carbon electrodes is also a concern. The capacitances of porous carbon cathodes have been improved by porous structure engineering and pseudocapacitive engineering. Advanced capacitive cathode materials, such as MXenes, phosphorene, and TiN, have proved to be promising candidates for cathode materials for ZICs. In addition to electrodes, the design of electrolytes is also a core area. Aqueous $ZnSO_4$ solutions are the most widely used electrolytes in ZICs. ZIC electrolytes containing $ZnSO_4$ aqueous solutions have a limited voltage window of 0.2–1.8 V. Beyond this voltage range, the electrolyte is subjected to a number of different types of electrodes. Beyond this voltage range, parasitic reactions, such as HER, can occur and damage ZIC devices. To suppress parasitic reactions and expand the voltage window, the electrolyte composition, such as substance, concentration, and solvent, can be optimized. In addition, zinc anodes exhibit poor cycling stabilities compared to porous carbon cathodes, which have an extremely long cycle life. In the case of zinc anodes, the growth of zinc dendrites, ion distributors, and artificial SEIs is suppressed by expanding the electrochemical surface area with a 3D architectural design. The use of low-potential intercalation anodes also helps to avoid zinc dendrite growth.

Despite the research results achieved with ZICs, the current results are based on laboratory-scale devices and focus on specific components rather than complete devices. In the reported ZICs, the capacities of the anodes, cathodes, and electrolytes do not match. For studies of ZICs, zinc anodes and electrolytes are often used in excess. The reported energy densities of ZICs are unreliable, as these values are based on the active mass of the porous carbon cathodes. Further research should therefore focus on capacity matching between anodes and cathodes. In other words, the utilization of zinc should be as high as possible. Cycle life, energy density, and power density are only meaningful if there is a limited use of zinc cathodes. From a practical point of view, the resource cost is an important parameter to prove the feasibility of ZIC devices in practical applications. Estimating the actual energy density and the corresponding capital cost of the device is necessary for the further development of ZICs. As the energy density of ZICs can be further increased by various engineering strategies, ZICs are promising energy storage technologies in the future energy storage market. Cost is one of the key parameters determining whether ZICs can be used for practical applications. Although porous carbon and ZICs have a good industrial base, the cost of commercial AC (US$ 57–114 per kg) is much higher than that of ZICs (US$ 2.1 per kg). Based on the above estimated data, the cost of electrode material for ZIC would be US$ 365.6 per kWh (US$ 7.2 per kWh for zinc anodes and US$ 358.4 per kWh for porous carbon cathodes) [116]. The high cost of porous carbon is therefore a bottleneck for the practical application of ZICs. In order to reduce the cost of ZICs, a simple low-cost porous carbon manufacturing process is required. It is believed that more green synthesis processes will be developed for low-cost porous carbon materials. The last few years have seen tremendous progress in the electrochemical performances of ZICs. In future, a great deal of effort should be put into the manufacture of ZICs for practical applications. In the next few decades, there will be numerous potential opportunities to scale up ZICs with high performances.

As three-electron redox systems (Al/Al^{3+}), energy storage devices based on aluminum ions have a higher theoretical capacity and higher energy density. However, due to the higher polarization, the trivalent Al^{3+} suffers from a higher desolvation energy and a higher solid-state diffusion energy barrier. Considering the ease of practical devices based on divalent ions (Mg^{2+}, Ca^{2+}, and Zn^{2+}), trivalent ions (Al^{3+}) are expected to be more difficult. However, the first reported use of aluminum metal as anodes dates back to 1857, and the concept of AIBs for aluminum ion batteries was introduced in 1972 [117]. Since then, researchers have been working to develop practical aluminum-based devices by finding solutions for screening suitable cathode active materials with easy Al^{3+} ion diffusion (solid state), finding suitable cathode active materials for aqueous and nonaqueous media, optimizing the electrolytes for lower desolventization energy, and minimizing the formation of hard SEIs on aluminum metal electrodes. Therefore, the correct choice of electrolyte composition is key to the success of AICs. Aqueous salt electrolytes with $AlCl_3$ or $Al_2(SO_4)_3$ cannot be used for Al anodes because of the formation of Al_2O_3 through surface passivation and the intrinsic hydrogen precipitation reaction. Nonaqueous electrolytes can overcome these limitations. Historically, binary (NaCl–$AlCl_3$) and ternary (KCl–NaCl–$AlCl_3$) molten salts have been used as possible electrolytes; however, Al^{3+} is not present [118]. This chloroaluminate molten salt electrolyte system is divided into three parts, viz. acidic, neutral, and basic species, based on the molar concentration of $AlCl_3$. Molar concentrations of $AlCl_3$ of less than 50% in the electrolyte provide the basic characteristics by containing $AlCl_4^-$ and Cl^- primary anions, while more than 50% provides an acidic electrolyte with $AlCl_4^-$ and $Al_2Cl_7^-$ anions. There is evidence that reversible Al electrochemical stripping/plating occurs only in acidic compositions [119]. However, the high melting point of such molten chloride aluminate electrolytes limits their practical application. On the other hand, fluorinated salts in organic solvents are not suitable for aluminum-based systems, but this is negligible for lithium/sodium/potassium-based systems. For example, aluminum salts containing fluoride anions in the electrolyte (similar to $LiPF_6$ in LICs) form electronically and ionically nonconductive passivated AlF_3 layers on aluminum metal electrodes, which hinders the reversibility of aluminum stripping/plating. At the same time, high desolvation energy losses at the electrode/electrolyte interface increase the polarization of the electrochemical reactions in typical organic systems (e.g. ethers). These problems are responsible for the slow kinetics of AICs using conventional fluorinated electrolytes in organic media. Recent research and developments have shown that ILs with high ionic conductivity, low volatility, and high chemical/electrochemical stability can be used as electrolyte solvents in AICs at room temperature [120]. Common IL-based electrolytes are prepared by mixing $AlCl_3$ with imidazolium chloride in a specific ratio. The most commonly used imidazolium chlorides are 1-butyl-3-methylimidazolium chloride ([BMIM]Cl) and 1-ethyl-3-methylimidazolium chloride ([EMIM]Cl). Herein, the Lewis acidity of the electrolyte is achieved by maintaining the molar ratio of $AlCl_3$/imidazolium salt greater than one. The reaction mechanism for aluminum stripping/plating is similar to that of the molten chloroaluminate electrolyte described above in terms of the main anion species and reversible reactions. Therefore, in the cathodes in AICs devices, graphite-based materials (3D graphite foam, graphene microflower, etc.), metal sulfides (Mo_6S_8, FeS_2, SnS_2, Ni_3S_2, CuS,

Co_9S_8, etc.), metal oxides (TiO_2, V_2O_5, etc.), Prussian blue analogues (CuHCF), MXene (V_2CT_x), chloroaluminate-doped conductive polymers, etc. have been well reported [52]. Due to their high bulk density, high abundance, and environment friendliness, aluminum metal anodes are mostly accepted by conventional AICs. Notably, after removal of the surface oxide films, the thin Al_2O_3 films in commercial aluminum foils show better electrochemical properties compared to pure aluminum foils [52]. The defective sites in the oxide films facilitate efficient penetration of the electrolytes and minimize dendrite growth. In aqueous aluminum-based systems, surface modification of aluminum foils by ionic liquid impregnation introduces artificial SEIs to minimize hard surface passivation on the aluminum anodes. Salt-packed aqueous or solid/quasi-solid electrolytes are also suitable for aqueous AICs, while the overpotential of the hydrogen precipitation reaction is significantly increased [52]. Interestingly, in addition to aluminum metal electrodes, zinc metal, MoO_3@PPy nanotubes, SWCNT/$W_{18}O_{49}$ nanowires, etc. have also been tried as anode materials in aqueous AICs.

In 2016, Yoo et al. filed a US patent for AICs by preferentially using aluminum (including aluminum foil, aluminum powder, aluminum foam, shell particles with an aluminum coating, and aluminum alloys) anodes, and high surface area porous carbon (including AC, CNT, and graphene) cathodes/[EMIM]Cl with $AlCl_3$ electrolytes [121]. According to the patent description, in addition to aluminum metal, graphite, aluminum-doped graphite, carbon, silicon, titanium dioxide, molybdenum sulfide, and other intercalation/de-intercalation materials can also be used as potential anodes. Similarly, conductive polymers, oxides, sulfides, and nitrides are also suitable for potential cathodes. The patent description is not limited to electrolytes in ionic liquids; it also includes organic/aqueous electrolytes and even solid electrolytes. Tian and his colleagues developed low-cost and high-performance aqueous AICs by using nanostructured V_2O_5-impregnated mesoporous carbon microspheres (MCM/V_2O_5) in a 1 M $Al_2(SO_4)_3$ electrolyte. The AICs operated through the intercalation/de-intercalation of Al^{3+} and the reduction/oxidation of V^{5+}/V^{4+} pairs. The CV of the AICs showed typical redox properties due to the pseudocapacitance contribution of the V^{5+}/V^{4+} pair. The *b*-value of the composite electrodes (close to 1) indicated the dominant capacitance/pseudocapacitance contribution in the energy storage. The higher quality loading of V_2O_5 reduced the *b*-value, suggesting that V_2O_5 led to a diffusion-controlled mechanism. As-prepared AICs devices delivered an energy density of 13.2 Wh kg^{-1} (147 W kg^{-1}), a power density of 5840 W kg^{-1} (7 Wh kg^{-1}), and good cycling performances (capacitance retention of 90% after 10 000 cycles at 1 A g^{-1}) in the range of 0–1.6 V. The AICs designed by Lei et al. [122] were composed of nitrogen-doped graphene as the cathode materials, aluminum foil as the anode materials, glass fibers as the separators, and [EMIM]Cl/$AlCl_3$ as the electrolytes. In this device, the capacitive behaviors were derived from the adsorption/desorption of $AlCl_4^-$ ions in the nitrogen-doped graphene. The rectangular CV curves of the hybrid devices at various scan rates indicated that the nitrogen-doped graphene exhibited good properties in AICs. In addition, a weak reduction peak at 1.9 V indicated the presence of $AlCl_4^-$ intercalation/de-intercalation in/from the layer structures. The AICs device showed an initial discharge capacitance of 160 F g^{-1} (95% CE at 0.3 A g^{-1} within 0.3–2.3 V) increasing to 254 F g^{-1} (90% CE) after 1000 cycles. The nitrogen-doped graphene cathode was activated in the first 30 cycles, after which the capacitance showed a slow increase up to 1000 cycles. Wang et al. reported AICs with a (+)AC//0.5 M $Al_2(SO_4)_3$(aq)//PPy@MoO_3(−) structures [123]. CV curves for the full devices of the AICs showed broad peaks associated with the intercalation/de-intercalation of Al^{3+} into/out of MoO_3@PPy. *b* values (0.5) also demonstrated the solid-state diffusion kinetics of MoO_3 in the $Al_2(SO_4)_3$ electrolytes. The device exhibited an energy density of 28 Wh kg^{-1} (460 W kg^{-1}), a power density of 2840 W kg^{-1} (20 Wh kg^{-1}), and good cycling performances (93% capacitance retention after 1800 cycles at 2 A g^{-1}) in the 0–1.5 V. Considerable electrochemical performances are achieved due to the PPy coating and the nanotube structure of MoO_3. PPy provides a conductive network with reduced charge transfer resistance and also protects MoO_3 from acid etching (acid can be produced by hydrolysis of $Al_2(SO_4)_3$). The nanotube structure facilitates the penetration of electrolytes and buffers MoO_3 volume changes due to Al^{3+} intercalation/de-intercalation. Furthermore, the fabricated (+)AC//PPy@MoO_3(−) devices were able to light up red LEDs, which demonstrated the practicality of the devices. Li et al. reported flexible AICs with (+)SWCNT/PANI//1 M $AlCl_3$ (aq)//SWCNT/$W_{18}O_{49}$(−) [124]. The CV curves of SWCNT/$W_{18}O_{49}$ composites revealed broad redox peaks due to the fast pseudocapacitive reactions. This unique network structure contributed to fast ion transport and provided high electrode conductivity (1626 S cm^{-1}). In addition, the $W_{18}O_{49}$ nanowires exhibited a wide lattice spacing, high aspect ratio, and homogeneous laminar structures, which facilitated the efficient intercalation of Al^{3+} ions. On the other hand, the SWCNT/PANI cathode provided a combination of capacitive/pseudocapacitive mechanisms. The hybrid device showed a bulk energy density of 19.0 mWh cm^{-3} (295 mW cm^{-3}), a bulk power density of 1278 mW cm^{-3} (14.5 mWh cm^{-3}), and an excellent cycling stability (95.9% capacitance retention at 14 mA cm^{-2} after 6000 cycles) at 0–1.8 V. Nevertheless, the number of successful AICs reported in the literature is indeed limited.

Now turn our attention back to the SICs. Both sodium and lithium are elements of the first main group, and their chemical properties are similar. Although the radius of sodium ions is larger than that of lithium ion (0.102 vs. 0.076 nm), the resource reserves of lithium ions are far behind those of sodium ions [125–132]. Sodium is one of the most abundant elements in the earth's crust, about 2.74%, ranking sixth. And when sodium ions are dissolved in H_2O, the radius is relatively small (compared with potassium ions), and it diffuses in H_2O faster than other ions, so the conductivity of sodium ions in the solution is higher [133]. Musashi Energy Solutions Co., Ltd. pointed out that SICs can be used in many fields such as construction machinery, photovoltaics, wind farms, and medical machinery. Although the solid-state diffusion in the electrode material lattice often determines the rate of traditional batteries, sodium ion energy storage devices may change this trend and generate better power than LIBs [134–136]. Therefore, in the long run, the development of sodium-ion capacitors is very promising. Therefore, SICs, as an emerging technology, are considered to be a supplement and extension to existing LICs [137, 138].

In 2012, Kuratani and coworkers developed an SIC energy storage device with hard carbon (HC) as the anode and AC as the cathode. Since then, the research of SICs has developed rapidly. Surprisingly, the energy density of SICs is 20–50 times greater than that of traditional capacitors, and the power density of SICs is 20–40 times greater than that of traditional sodium ion batteries (SIBs) [138–146]. Therefore, SICs can bridge the gap between SIBs and SCs [147, 148]. Compared with traditional SCs, the energy density of SICs is up to four times higher, so mechanical devices used for regenerative braking may be significantly beneficial. In the past few decades, different materials have also been used to enhance the energy storage performance of SIC, and these concepts are gradually deviating from the concept of classic SIC. Therefore, it is of great significance to classify SIC and clarify different energy storage mechanisms in the current context [149–153]. Therefore, it is very important to build high-performance SIC devices based on an in-depth understanding of the mechanism.

1.3 SICs Energy Storage Mechanism Introduction

In order to improve the performance of the cathode and anode, scholars will further explore and study pseudocapacitance materials in SIC devices [154]. On the one hand, many studies will focus on how to use pseudocapacitive materials to replace the traditional EDLC-type cathodes and further assemble them with battery materials to realize hybrid devices. SIC devices involving pseudocapacitor cathodes are in the middle position between EDLC cathodes and battery cathodes. Pseudocapacitance can be regarded as a supplementary mechanism of EDLC because it is not a mechanism of electrostatic adsorption, but compared to EDLC, pseudocapacitance has a similar CV curve shape and comparable constant CDC curves [155]. On the other hand, the pseudocapacitance material in the anode is also the main interest of SICs research. With the rapid development of anode nanotechnology and nanoscience, nanomaterial anodes have played an important role in electrochemical energy devices over the years. Nano-scale materials are small in size but large in surface area. In this case, it is impossible to accurately distinguish between surface and volume. Therefore, some anodes based on the Faraday mechanism usually exhibit strong redox reactions in the matrix. When the size is reduced to nanometers, they behave like pseudocapacitive materials, characterized by the disappearance of CV and CDC curves. Therefore, in recent years, the boundary between battery materials and pseudocapacitive materials has become blurred [156]. It should be noted that the misunderstanding of the confusing electrochemical energy storage mechanism will lead to the opaque development of controversial battery configurations (dual-ion battery configuration and dual-capacitor configuration) or SIC devices (flexible devices and prefabricated technology). Therefore, in addition to the chaotic mechanism, the neglected cell configuration and the newly developed flexible equipment and prefabricated areas were completely researched.

1.4 Key Technologies of SICs

SICs mainly consist of cathode/anode materials, collectors, electrolytes, separators, and metal shells. Among other things, the design of flexible devices and the development of pre-sodiation methods are the focus of future research into SIC devices. On the one hand, flexible electronics have revolutionized our lives and permeate every aspect of our daily life. Flexible energy storage devices are receiving increased attention due to their enormous potential in the emerging flexible electronics market, including roll-up displays, flexible mobile phones, skin sensors for human health monitoring, and implantable medical devices. The development of flexible SICs is important for the next generation of SIC devices. On the other hand, pre-sodiation methods are an integral part of electrochemical energy storage systems and can effectively compensate for irreversible capacity losses, increase the operating voltage, and improve the concentration of Na^+ in the electrolytes. Therefore, flexible SICs devices and pre-sodiation methods will be two important areas of key technologies for SICs devices. Both of these areas are important research areas for SICs devices and play a key role in driving the development of SICs devices.

The current energy storage devices are rigid and bulky and cannot meet the basic requirements of flexible energy storage. Therefore, efforts must be made to achieve structural optimization, light weight, and flexibility. A promising research direction is the development of lighter, smaller, and thinner modern flexible devices, including soft electronics, winding displays, and wearable products. This also requires advances in the corresponding energy supply devices. The development of flexible electrodes with low cost, high performance, good stability, safety, and reliability is a major challenge for many practical applications. The rapid development of portable electronic devices has contributed to the development of better electrochemical energy storage technologies. Flexible SIC devices are complex devices whose effective energy densities and cycling stabilities depend on various design factors, including the configuration and choice of electrolytes. Each of these factors plays a key role in improving electrochemical performance. In this chapter, several simple flexible SIC devices are classified according to the different configurations of SICs, which gives the reader a step ahead to familiarize themselves with the significance of the configuration and the simple development of flexible SIC devices. For this chapter, these introductions are important. A more detailed analysis and discussion can be studied in Chapter 7.

The first kind of flexible SIC device presented here consists of an EDLC-type cathode and a battery-type anode. For battery-type anodes, titanium-based composites, in particular $Na_2Ti_3O_7$ and $Na_2Ti_2O_{5-x}$, have been widely explored as attractive flexible materials due to their low potential and good electronic conductivity [157]. In the absence of binders or metal collectors, Zhang and coworkers developed flexible SIC devices based on $Na_2Ti_3O_7$ nanosheet arrays/CT anodes and flexible rGO film cathodes [158]. By a simple hydrothermal method combined with a subsequent calcination process, the $Na_2Ti_3O_7$ nanosheet arrays were grown directly on the CT and showed good adhesion. The synthesized material can be bent at arbitrary angles and has good flexibility. The whole device using nanostructured $Na_2Ti_3O_7$/CT and rGO films as electrodes exhibits excellent electrochemical properties while maintaining good flexibility. The CDC curves of the flexible SICs devices can be recycled at different rates over a potential range of 1–3 V from 0.1 to 1.5 A g^{-1} and show stability at 0.5 A g^{-1} (over 80.3% of capacity after 2500 cycles) as well as higher coulombic efficiency (close to 100%). In another case, Gu and coworkers prepared flexible $Na_2Ti_2O_{5-x}$ electrodes and by introducing oxygen vacancies [159]. They assembled flexible SICs (labeled rGO/AC//$Na_2Ti_2O_{5-x}$@Ti//rGO/AC) using $Na_2Ti_2O_{5-x}$@Ti as the anodes and rGO/AC as the cathodes. The hybrid SIC devices offer high coulombic efficiency (close to 100%) and good cycling stabilities (82.5% capacity retention) at 1 A g^{-1} after 5000 cycles. The flexible SICs devices operate at potentials of 1–3.8 V and have maximum energy and power densities of 15.6 Wh l^{-1} and 120 W l^{-1}, respectively.

The flexible SIC devices presented in this place consist of cell-type cathodes and EDLC-type anodes. As for the EDLC-type anodes, research has mainly focused on AC. In order to provide high power characteristics, Fan et al. designed and fabricated a novel type of N-doped mesoporous carbon nanosheets (mp-CNS) consisting of interconnected Co/Zn metal organic framework (MOF) nanosheets [160]. In order to give the flexible SIC devices high power, the pseudocapacitive VO_2@mp-CNSs/CFC material is synthesized by depositing interconnected VO_2 on top of mp-CNSs/CFC. Flexible SIC devices are assembled from pseudocapacitive VO_2@mp-CNSs/CFC anodes and cellular NVP@mp-CNSs/CFC cathodes. At 1 A g^{-1}, the flexible SIC achieves a capacity retention of 78% after 2000 cycles. Furthermore, no significant structural damage was observed at different bending angles from 0° to 90°.

The above examples of flexible electrodes are innovative or improved in that there are no binders or collectors, but the electrolyte remains a conventional liquid organic electrolyte. In addition, most of the liquid organic electrolytes used in hybrid flexible SIC devices are highly volatile and flammable, posing a safety hazard. Therefore, effective solutions need to be proposed. Due to the low literature coverage in this area, only quasi-solid electrolytes are briefly discussed here. Fan et al. and Yu et al. used P(VDF-HFP) membranes to avoid these problems. Yu et al. used sea urchin-like $Na_2Ti_3O_7$, peanut shell-derived carbon (PSC), and P (VDF-HFP) membranes as anodes, cathodes, and electrolytes to construct flexible quasi-solid SICs [161]. Flexible $Na_2Ti_3O_7$//PSC quasi-solid SICs provided 86% capacitance retention and nearly 100% coulombic efficiency after 3000 cycles. The prepared $Na_2Ti_3O_7$//PSC quasi-solid state SICs devices showed no significant capacity loss under different bending conditions, confirming their superior mechanical strength. The flexible $Na_2Ti_3O_7$//PSC quasi-solid-state SIC devices can power up to 50 red LEDs and show both high power and high energy characteristics.

For flexible SIC devices, the concept of the configuration and the development of the electrolyte have been briefly outlined. The development of the relevant details will be provided in detail in Sections 7.1, 7.2, and 7.3. We believe that these discussions have given the reader a lot of insight. In addition to flexible SIC devices, another topic that cannot be avoided is the field of pre-sodiation methods in SICs. In pursuit of the growing demand for new applications with higher energy densities, researchers have invested in and developed a large number of materials for energy storage electrodes. However, the associated pre-sodiation techniques have not received sufficient attention and have seriously hampered the commercialization of SIC devices. Pre-sodiation methods, i.e. the pre-doping of sodium ions, refer to the addition of a sodium source to the SIC devices to ensure sufficient sodium content. In general, pre-sodiation methods serve to compensate for the initial irreversible capacity loss of SIC devices. In a general sodium ion system, the storage and release of capacity is based on the reversible intercalation and de-intercalation of Na^+ in the electrode materials, while the Na^+ required for the anode is generated from the cathodes. During charging, a certain amount of active sodium is lost, mainly due to the formation of SEIs in the anodes. As a result, the irreversible sodium will decrease in the subsequent cycles. Surprisingly, the loss of active sodium content can be compensated by pre-sodiation methods to further ensure an increase in energy densities. In addition, pre-sodiation methods increase the operating voltage of SIC devices during cycling and reduce the consumption of electrolytes. Fortunately, various pre-sodiation methods have been successfully developed and work has been reported in recent years. Different pre-sodiation methods exhibit different mechanisms, which play a crucial role in the development of SICs. As fewer pre-sodiation methods have been reported for SICs devices, the main effective pre-sedimentation strategies are summarized as follows: (i) Direct contact methods; (ii) electrochemical methods; and (iii) adding additives methods. In different places, these names may be replaced by different expressions in different forms, but the underlying mechanisms are similar. Other similar or different examples are discussed in Chapter 8.

Direct contact methods: Moeez et al. developed a direct contact method to achieve presoaking of the electrodes and metal sodium, demonstrating an improved coulombic efficiency for the first cycle with improved cycle stability and reversible capacity [162]. In addition, a thick passivation layer is formed at the interface during the process; this helps to improve cycling stabilities by preventing dissolution of the active materials and deposition on the anode surfaces. The direct contact method is a simple and economical method of pre-sodiation, which has potential applications in practical production. To achieve contact, sodium metal is connected directly to the electrodes with a small amount of electrolytes (0.5 ml). Constant pressure is applied using, for example, plastic slides and adhesive clips so that the layered sodium metal is in good contact with the electrode and is maintained for 30 minutes. The sodium ions can enter the electrode at the same time and achieve charge neutrality. By bringing the sodium metal into contact with the electrodes, the sodium ions can be inserted into the active material to compensate for the initial loss of sodium. It is important to note that the measurement of the sodium metal and the open circuit voltage (OCV) is crucial to determine the degree of pre□sodium. In addition, the placement of a trace amount of electrolytes between the sodium metal and the electrode is important for the insertion of excess sodium ions in this pre-treatment. The electrolytes not only provide a bridge for sodium ions to migrate from the source to the electrode, but also help to form a stable passivation layer between the electrodes and the electrolytes. The passivation layer protects the electrode materials from undesirable side reactions. In view of the high reactivity of sodium metal, all experiments should be carried out in Ar glove boxes at $H_2O < 0.1$ and $O_2 < 0.1$ levels.

Electrochemical methods: In the case of electrochemical methods, pre-sodiation methods are accomplished by cycling at relatively low current densities, where the sodium and targeted electrodes are isolated from each other. This method can be achieved by performing several charging and discharging cycles to achieve the desired potential. The process is short-circuit free, safe, green, and controllable, and opens up a new avenue of research in pretreatment technology. This method is currently the most widely used method. Zou et al. utilized the working electrodes, sodium metal electrodes, electrolytes, and separators to assemble a CR2016 cell under a high purity argon atmosphere [163]. The half-cell anode was cycled five times at $0.1\,A\,g^{-1}$. The sodium metal electrodes are then removed, and the SICs devices are assembled in a glove box with the cathode (working electrode) remaining. However, while the electrochemical method is technically controllable and easy to perform, the assembly/disassembly step of the cell pre-treatment adds cost and process complexity to commercialization.

Adding additives methods: A novel voltage-induced efficient in situ pre-sodiation method was successfully implemented by Zou et al. [164] through introducing an organic sodium salt ($Na_2C_6O_6$) as a sodium source as extra adding additives in the cathodes. In addition, $Na_2C_6O_6$ possesses a stable sodium–oxygen bond, and the decarbonization product cyclohexanedione in the electrolytes is neutral and soluble, which can ensure the forward decarbonization reaction. It is worth noting that sacrificing organic sodium salts has the advantages of abundant resources, low cost, and environment friendliness, which is conducive to promoting commercialization. In view of these advantages, the $Na_2C_6O_6$-added components exhibit excellent electrochemical properties. Furthermore, as an organic sodium salt, $Na_2C_6O_6$ is electrochemically unstable at high voltage and can provide an adequate source of sodium. To ensure the electrochemical behaviors of the $Na_2C_6O_6$ anode, the assembled half-cells were first evaluated. In the first cycle of the CV curve, there is a low intensity anodic peak at 3.60 V and a distinct anodic peak at 3.98 V with a maximum oxidation current. In addition, there is no peak in the cathodic section, which indicates that the extraction of sodium ions is mostly irreversible. In other words, the extracted sodium source enters the electrode and cannot be returned to the positive material. Also, in the second cycle that followed, the previous anodic peak disappeared completely, further verifying the irreversible reaction of the cathode. Furthermore, a specific capacity of $309.8\,mAh\,g^{-1}$ could be extracted from the $Na_2C_6O_6$ electrode at $25\,mA\,g^{-1}$, indicating that the additional specific capacity was caused by a side reaction. However, in the first subsequent GCD curve, only an exceptional discharge capacity of $24.9\,mAh\,g^{-1}$ can be obtained, which may be related to the side reactions or capacitive behaviors of the conducting carbon. The initial charging process is irreversible, which agrees well with the CV curves. The experimental results show that the $Na_2C_6O_6$ electrodes exhibit high irreversible capacities and can provide abundant sources of sodium, which facilitates the utilization of the anodes. The organic sodium salt exhibits irreversible electrochemical properties at high potential, which can provide an abundant sodium source without any negative impact on the subsequent physical operation. The results also show that both the carbon and TiO_2 anodes have good electrochemical compatibility with $Na_2C_6O_6$. This rational strategy is expected to provide a potential method to greatly simplify the physical assembly process, which may significantly accelerate the commercialization of SICs.

The above results first systematically and comprehensively analyzed and compared the EDLC mechanism, the battery-type mechanism, and the confusion pseudocapacitance mechanism. Subsequently, the SIC battery configuration with different orientation mechanisms for the cathode and anode was discussed. In addition, the characteristics of electrode materials in different SIC battery configurations are further summarized. Finally, the key technologies and future developments in related fields are summarized. The above conclusion summarized the current research progress of SICs from multiple angles, established a basic understanding of SICs, and provided a solid theoretical basis for subsequent research.

References

1 Hou, H., Shao, L., Zhang, Y. et al. (2017). *Adv. Sci.* 4: 1600243.
2 Zou, G., Hou, H., Foster, C.W. et al. (2018). *Adv. Sci.* 5: 1800241.
3 Hou, H., Banks, C.E., Jing, M. et al. (2015). *Adv. Mater.* 27: 7861.
4 Ge, P., Hou, H., Cao, X. et al. (2018). *Adv. Sci.* 5: 1800080.
5 Wu, T., Zhang, C., Zou, G. et al. (2019). *Sci. China Mater.* 62: 1127.
6 Cao, D.P., Yin, C.L., Shi, D.R. et al. (2017). *Adv. Funct. Mater.* 27: 9.
7 Han, Y.L., Hu, J.L., Yin, C.L. et al. (2016). *J. Mater. Chem. A* 4: 7382.
8 Li, L., Wu, Z.P., Sun, H. et al. (2015). *ACS Nano* 9: 11342.
9 Lv, J.L., Liang, T.X., Yang, M. et al. (2017). *Composites, Part B* 123: 28.
10 Lv, J.L., Yang, M., Liang, T.X., and Miura, H. (2017). *Mater. Lett.* 197: 127.
11 Xu, M., Fei, L.F., Lu, W. et al. (2017). *Nano Energy* 35: 271.
12 Zhou, H.P. and Hu, J. (2017). *Mater. Lett.* 195: 26.
13 Cai, P., Zou, K.Y., Deng, X.L. et al. (2020). *Front. Chem.* 8: 11.
14 Brousse, T., Marchand, R., Taberna, P.L., and Simon, P. (2006). *J. Power Sources* 158: 571.
15 Konno, H., Kasashima, T., and Azumi, K. (2009). *J. Power Sources* 191: 623.
16 Li, J.M., Chang, K.H., and Hu, C.C. (1800). *Electrochem. Commun.* 2010: 12.
17 Aravindan, V., Gnanaraj, J., Lee, Y.S., and Madhavi, S. (2014). *Chem. Rev.* 114: 11619.
18 Cao, W.J., Li, Y.X., Fitch, B. et al. (2014). *J. Power Sources* 268: 841.
19 Jain, A., Aravindan, V., Jayaraman, S. et al. (2013). *Sci. Rep.* 3: 6.
20 Yang, Z.W., Guo, H.J., Li, X.H. et al. (2016). *J. Power Sources* 329: 339.
21 Sun, X.Z., Zhang, X., Zhang, H.T. et al. (2014). *J. Power Sources* 270: 318.

22 Gokhale, R., Aravindan, V., Yadav, P. et al. (2014). *Carbon* 80: 462.
23 Zhao, Y.J., Liu, J.Z., Horn, M. et al. (2018). *Sci. China Mater.* 61: 159.
24 Bispo-Fonseca, I., Aggar, J., Sarrazin, C. et al. (1999). *J. Power Sources* 79: 238.
25 Schroeder, M., Menne, S., Segalini, J. et al. (2014). *J. Power Sources* 266: 250.
26 Minakshi, M., Duraisamy, S., Tirupathi, P.R. et al. (2014). *Int. J. Electrochem. Sci.* 9: 5974.
27 Yuan, M.R., Liu, W.Q., Zhu, Y.F., and Xu, Y.J. (2014). *Russ. J. Electrochem.* 50: 1050.
28 Park, M.S., Lim, Y.G., Hwang, S.M. et al. (2014). *ChemSusChem* 7: 3138.
29 Ren, J.J., Su, L.W., Qin, X. et al. (2014). *J. Power Sources* 264: 108.
30 Park, C.M., Jo, Y.N., Park, J.W. et al. (2014). *Bull. Korean Chem. Soc.* 35: 2630.
31 Karthikeyan, K., Amaresh, S., Lee, S.N. et al. (2014). *ChemSusChem* 7: 2310.
32 Kaiyappan, K., Amaresh, S., and Lee, Y.S. (2014). *ACS Appl. Mater. Interfaces* 6: 11357.
33 Amatucci, G.G., Badway, F., Du Pasquier, A., and Zheng, T. (2001). *J. Electrochem. Soc.* 148: A930.
34 Kim, H., Park, K.Y., Cho, M.Y. et al. (2014). *ChemElectroChem* 1: 125.
35 Ren, J., Zhang, Y., Bai, W.Y. et al. (2014). *Angew. Chem. Int. Ed.* 53: 7864.
36 Kim, M., Xu, F., Lee, J.H. et al. (2014). *J. Mater. Chem. A* 2: 10029.
37 Barcellona, S., Ciccarelli, F., Iannuzzi, D., and Piegari, L. (2014). *IEEE Trans. Sustainable Energy* 5: 785.
38 Pan, S.W., Zhang, Z.T., Weng, W. et al. (2014). *Mater. Today* 17: 276.
39 Aravindan, V., Sundaramurthy, J., Jain, A. et al. (1858). *ChemSusChem* 2014: 7.
40 Lotfabad, E.M., Ding, J., Cui, K. et al. (2014). *ACS Nano* 8: 7115.
41 Zhu, J., Xu, Z., and Lu, B.N. (2014). *Nano Energy* 7: 114.
42 Tian, L.L., Wei, X.Y., Zhuang, Q.C. et al. (2014). *Nanoscale* 6: 6075.
43 Ahn, S.K., Yang, J.J., Kim, H.I. et al. (2014). *Chem. Lett.* 43: 898.
44 Yuan, M.R., Liu, W.Q., Zhu, Y.F., and Xu, Y.J. (2014). *Russ. J. Electrochem.* 50: 594.
45 Ohki, Y. (2014). *IEEE Electr. Insul. Mag.* 30: 54.
46 Xu, J., Mi, C.C., Cao, B.G. et al. (2014). *IEEE Trans. Veh. Technol.* 63: 1614.
47 Zhang, J., Shi, Z.Q., and Wang, C.Y. (2014). *Electrochim. Acta* 125: 22.
48 Naskar, P., Kundu, D., Maiti, A. et al. (2021). *ChemElectroChem* 8: 1393.
49 Tanaka, K., Ohzeki, K., Yamabe, T., and Yata, S. (1984). *Synth. Met.* 9: 41.
50 Yata, S., Hato, Y., Sakurai, K. et al. (1987). *Synth. Met.* 18: 645.
51 Li, G., Yang, Z., Yin, Z. et al. (2019). *J. Mater. Chem. A* 7: 15541.
52 Naoi, K., Ishimoto, S., Miyamoto, J.-i., and Naoi, W. (2012). *Energy Environ. Sci.* 5: 9363.
53 Pasquier, A.D., Plitz, I., Gural, J. et al. (2004). *J. Power Sources* 136: 160.
54 Hosaka, T., Kubota, K., Hameed, A.S., and Komaba, S. (2020). *Chem. Rev.* 120: 6358.
55 Peña, O. (2015). *Physica C* 514: 95–112.
56 Levi, E. and Aurbach, D. (2010). *Chem. Mater.* 22: 3678–3692.
57 Wan, L.F., Perdue, B.R., Apblett, C.A., and Prendergast, D. (2015). *Chem. Mater.* 27: 5932–5940.

58 Liang, Y., Feng, R., Yang, S. et al. (2011). *Adv. Mater.* 23: 640–643.

59 Liu, B., Luo, T., Mu, G. et al. (2013). *ACS Nano* 7: 8051–8058.

60 Kaewmaraya, T., Ramzan, M., Osorio-Guillén, J.M., and Ahuja, R. (2014). *Solid State Ionics* 261: 17–20.

61 Patridge, C.J., Wu, T.L., Jaye, C. et al. (2010). *Nano Lett.* 10: 2448–2453.

62 Ling, C., Zhang, R., Arthur, T.S., and Mizuno, F. (2015). *Chem. Mater.* 27: 5799–5807.

63 Spahr, M.E., Novak, P., Haas, O., and Nesper, R. (1995). *J. Power Sources* 54: 346–351.

64 Novak, P. and Desilvestro, J. (1993). *J. Electrochem. Soc.* 140: 140–144.

65 Tao, Z.L., Xu, L.N., Gou, X.L. et al. (2004). *Chem. Commun.* 2080–2081.

66 Su, S., Huang, Z., NuLi, Y. et al. (2015). *Chem. Commun.* 51: 2641–2644.

67 Kim, H.S., Arthur, T.S., Allred, G.D. et al. (2011). *Nat. Commun.* 2: 427.

68 Tian, H., Gao, T., Li, X. et al. (2017). *Nat. Commun.* 8: 14083.

69 NuLi, Y., Zheng, Y., Wang, F. et al. (2011). *Electrochem. Commun.* 13: 1143–1146.

70 Orikasa, Y., Masese, T., Koyama, Y. et al. (2014). *Sci. Rep.* 4: 5622.

71 Kuang, C., Zeng, W., and Li, Y. (2019). *J. Nanosci. Nanotechnol.* 19: 12–25.

72 Park, M.S., Kim, J.G., Kim, Y.J. et al. (2015). *Isr. J. Chem.* 55: 570–585.

73 Ling, C., Banerjee, D., and Matsui, M. (2012). *Electrochim. Acta* 76: 270–274.

74 Arthur, T.S., Singh, N., and Matsui, M. (2012). *Electrochem. Commun.* 16: 103–106.

75 Malyi, O.I., Tan, T.L., and Manzhos, S. (2013). *J. Power Sources* 233: 341–345.

76 Wang, Z., Su, Q., Shi, J. et al. (2014). *ACS Appl. Mater. Interfaces* 6: 6786–6789.

77 Er, D., Detsi, E., Kumar, H., and Shenoy, V.B. (2016). *ACS Energy Lett.* 1: 638–645.

78 Banerjee, S. and Pati, S.K. (2016). *Chem. Commun.* 52: 8381–8384.

79 Wu, N., Lyu, Y.C., Xiao, R.J. et al. (2014). *NPG Asia Mater.* 6: e120.

80 Aurbach, D., Lu, Z., Schechter, A. et al. (2000). *Nature* 407: 724–727.

81 Gregory, T.D., Hoffman, R.J., and Winterton, R.C. (1990). *J. Electrochem. Soc.* 137: 775–780.

82 Muldoon, J., Bucur, C.B., Oliver, A.G. et al. (2013). *Energy Environ. Sci.* 6: 482–487.

83 Zhang, H., Cao, D., and Bai, X. (2019). *J. Power Sources* 444: 227299.

84 Zhang, H., Ye, K., Zhu, K. et al. (2017). *Chem. Eur. J.* 23: 17118–17126.

85 Nam, K.W., Kim, S., Lee, S. et al. (2015). *Nano Lett.* 15: 4071–4079.

86 Mizuno, Y., Okubo, M., Hosono, E. et al. (2013). *J. Phys. Chem. C* 117: 10877–10882.

87 Yoo, H.D., Shterenberg, I., Gofer, Y. et al. (2014). *J. Electrochem. Soc.* 161: A410–A415.

88 Sun, Y., Chen, H., Xing, Y. et al. (2018). *Colloids Surf., A* 553: 539–545.

89 Zhang, H., Ye, K., Zhu, K. et al. (2017). *ACS Sustainable Chem. Eng.* 5: 6727–6735.

90 Cao, X., Wang, L., Chen, J., and Zheng, J. (2018). *ChemElectroChem* 5: 2789–2794.

91 Tian, Z., Tong, X., Sheng, G. et al. (2019). *Nat. Commun.* 10: 4913.
92 Juran, T.R. and Smeu, M. (2017). *Phys. Chem. Chem. Phys.* 19: 20684–20690.
93 Lipson, A.L., Pan, B., Lapidus, S.H. et al. (2015). *Chem. Mater.* 27: 8442–8447.
94 Ponrouch, A., Frontera, C., Bardé, F., and Palacín, M.R. (2016). *Nat. Mater.* 15: 169–172.
95 Ponrouch, A., Tchitchekova, D., Frontera, C. et al. (2016). *Electrochem. Commun.* 66: 75–78.
96 Aurbach, D., Skaletsky, R., and Gofer, Y. (1991). *J. Electrochem. Soc.* 138: 3536–3545.
97 Wang, D., Gao, X., Chen, Y. et al. (2018). *Nat. Mater.* 17: 16–20.
98 Shyamsunder, A., Blanc, L.E., Assoud, A., and Nazar, L.F. (2019). *ACS Energy Lett.* 4: 2271–2276.
99 Li, Z., Fuhr, O., Fichtner, M., and Zhao-Karger, Z. (2019). *Energy Environ. Sci.* 12: 3496–3501.
100 Ta, K., Zhang, R., Shin, M. et al. (2019). *ACS Appl. Mater. Interfaces* 11: 21536–21542.
101 Lee, C. and Jeong, S.K. (2018). *Electrochim. Acta* 265: 430–436.
102 Wu, N., Yao, W., Song, X. et al. (2019). *Adv. Energy Mater.* 9: 1803865.
103 Konarov, A., Voronina, N., Jo, J.H. et al. (2018). *ACS Energy Lett.* 3: 2620–2640.
104 Zhou, Z., Zhou, X., Zhang, M. et al. (2020). *Small* 16: 2003174.
105 Han, S.D., Kim, S., Li, D. et al. (2017). *Chem. Mater.* 29: 4874–4884.
106 Tian, Y., Amal, R., and Wang, D.W. (2016). *Front. Energy Res.* 4: 34.
107 Wang, H., Wang, M., and Tang, Y. (2018). *Energy Storage Mater.* 13: 1.
108 Wang, F., Borodin, O., Gao, T. et al. (2018). *Nat. Mater.* 17: 543.
109 Zhang, H., Liu, Q., Fang, Y. et al. (2019). *Adv. Mater.* 31: 1904948.
110 Chen, S., Ma, L., Zhang, K. et al. (2019). *J. Mater. Chem. A* 7: 7784.
111 Maughan, P.A., Tapia-Ruiz, N., and Bimbo, N. (2020). *Electrochim. Acta* 341: 136061.
112 Huang, Z., Chen, A., Mo, F. et al. (2020). *Adv. Energy Mater.* 10: 2001024.
113 Sun, G., Xiao, Y., Lu, B. et al. (2020). *ACS Appl. Mater. Interfaces* 12: 7239.
114 Yin, J., Zhang, W., Wang, W. et al. (2020). *Adv. Energy Mater.* 10: 2001705.
115 Li, J., Wang, N., Tian, J. et al. (2018). *Adv. Funct. Mater.* 28: 1806153.
116 Yin, J., Zhang, W., Alhebshi, N.A. et al. (2020). *Small Methods* 4: 1900853.
117 Holleck, G.L. (1972). *J. Electrochem. Soc.* 119: 1158–1161.
118 Karpinski, Z.J. and Osteryoung, R.A. (1984). *Inorg. Chem.* 23: 1491–1494.
119 Zhang, X. and Devine, T.M. (2006). *J. Electrochem. Soc.* 153: B375–B383.
120 Sun, H., Wang, W., Yu, Z. et al. (2015). *Chem. Commun.* 51: 11892–11895.
121 Lin, M.C., Gong, M., Lu, B. et al. (2015). *Nature* 520: 324–328.
122 Lei, H., Tu, J., Tian, D., and Jiao, S. (2018). *New J. Chem.* 42: 15684–15691.
123 Wang, F., Liu, Z., Wang, X. et al. (2016). *J. Mater. Chem. A* 4: 5115–5123.
124 Li, K., Shao, Y., Liu, S. et al. (2017). *Small* 13: 1700380.
125 Li, H.S., Zhang, X., Zhao, Z.C. et al. (2020). *Energy Storage Mater.* 26: 83.
126 Zhu, J.H., Roscow, J., Chandrasekaran, S. et al. (2020). *ChemSusChem* 13: 1275.
127 Wang, L., Yang, G.R., Peng, S.J. et al. (2020). *Energy Storage Mater.* 25: 443.
128 Meng, F., Long, T., Xu, B. et al. (2020). *Front. Chem.* 8: 652.

129 Chen, J.T., Yang, B.J., Liu, B. et al. (2019). *Curr. Opin. Electrochem.* 18: 1.
130 Jia, R., Shen, G.Z., and Chen, D. (2020). *Sci. China Mater.* 63: 185.
131 Zhai, H.F., Xia, B.Y., and Park, H.S. (2019). *J. Mater. Chem. A* 7: 22163.
132 Zhao, X.Y., Zhang, Y.B., Wang, Y.F., and Wei, H.G. (2019). *Batteries Supercaps* 2: 899.
133 Han, D.L., Zhang, J., Weng, Z. et al. (2019). *Mater. Today Energy* 11: 30.
134 Kim, J., Choi, M.S., Shin, K.H. et al. (2019). *Adv. Mater.* 31: 20.
135 Zhang, T.Y., Yang, L., Yan, X.B., and Ding, X. (2018). *Small* 14: 20.
136 Glushenkov, A.M. and Ellis, A.V. (2018). *Adv. Sustainable Syst.* 2: 25.
137 Ding, J., Hu, W.B., Paek, E., and Mitlin, D. (2018). *Chem. Rev.* 118: 6457.
138 Li, F. and Zhou, Z. (2018). *Small* 14: 25.
139 He, Y.Z., Han, X.J., Du, Y.C. et al. (2018). *Nano Res.* 11: 2573.
140 Ding, J., Wang, H.L., Li, Z. et al. (2015). *Energy Environ. Sci.* 8: 941.
141 Ruan, J.F., Yuan, T., Pang, Y.P. et al. (2018). *Carbon* 126: 9.
142 Qie, L., Chen, W.M., Xiong, X.Q. et al. (2015). *Adv. Sci.* 2: 6.
143 Wang, X.L., Li, G., Hassan, F.M. et al. (2015). *Nano Energy* 15: 746.
144 Li, D.D., Zhang, L., Chen, H.B. et al. (2016). *J. Mater. Chem. A* 4: 8630.
145 Lu, M.J., Yu, W.H., Shi, J. et al. (2017). *Electrochim. Acta* 251: 396.
146 Zou, L., Lai, Y.Q., Hu, H.X. et al. (2017). *Chem. Eur. J.* 23: 14261.
147 Simon, P., Gogotsi, Y., and Dunn, B. (2014). *Science* 343: 1210.
148 Nishihara, H. and Kyotani, T. (2012). *Adv. Mater.* 24: 4473.
149 Zhao, L.P., Qi, L., and Wang, H.Y. (2013). *J. Power Sources* 242: 597.
150 Gonzalez-Gil, A., Palacin, R., and Batty, P. (2013). *Energy Convers. Manage.* 75: 374.
151 Ding, J., Li, Z., Cui, K. et al. (2016). *Nano Energy* 23: 129.
152 Xu, J.T., Wang, M., Wickramaratne, N.P. et al. (2015). *Adv. Mater.* 27: 2042.
153 Seh, Z.W., Li, W.Y., Cha, J.J. et al. (2013). *Nat. Commun.* 4: 6.
154 Wang, Y.P., Zhao, K.J., Wang, K. et al. (2020). *J. Alloys Compd.* 844: 8.
155 Han, M., Srivastava, S.B., Yildiz, E. et al. (2020). *ACS Appl. Mater. Interfaces* 12: 42997.
156 Yan, H.X., Yang, M., Liu, L. et al. (2020). *J. Alloys Compd.* 843: 9.
157 Li, Y.Z., Wang, H.W., Wang, L.B. et al. (2019). *Small* 15: 13.
158 Dong, S.Y., Shen, L.F., Li, H.S. et al. (2016). *Adv. Funct. Mater.* 26: 3703.
159 Que, L.F., Yu, F.D., He, K.W. et al. (2017). *Chem. Mater.* 29: 9133.
160 Xu, D.M., Chao, D.L., Wang, H.W. et al. (2018). *Adv. Energy Mater.* 8: 11.
161 Li, H., Peng, L., Zhu, Y. et al. (2016). *Nano Lett.* 16: 5938.
162 Moeez, I., Jung, H.G., Lim, H.D., and Chung, K.Y. (2019). *ACS Appl. Mater. Interfaces* 11: 41394.
163 Zou, K.Y., Cai, P., Liu, C. et al. (2019). *J. Mater. Chem. A* 7: 13540.
164 Zou, K.Y., Cai, P., Tian, Y. et al. (2020). *Small Methods* 4: 8.

2

Characteristics of Sodium-Ion Capacitor Devices

Peng Cai, Wentao Deng, Hongshuai Hou, Guoqiang Zou, and Xiaobo Ji

Central South University, College of Chemistry and Chemical Engineering, No. 932, Lushan South Road, Yuelu District, Changsha 410083, China

2.1 Basic Features

The composition and structure of SICs are similar to SIBs, including cathode and anode materials, current collectors, electrolytes, and separators [1–3]. In SICs, the electrode materials serve as the core of energy storage, and the SICs consist of a battery-type electrode and a capacitor-type electrode [4]. The current collectors play the role of carrying active materials and collecting currents [5]. The working voltage range of the electrolytes directly affects the energy density of the devices [6]. The separator of the SICs may be the same as that of SIBs, such as glass fibers. The electrolytes play a role in conducting ions, and the applicable temperature range of the electrolyte affects the operating temperature restricts of the devices in practical application [7]. Aqueous electrolytes and organic electrolytes are commonly utilized. Aqueous-based sodium-ion capacitors have the advantage of better stability, lower cost, and higher safety [8, 9]. However, the working potential region is narrow, leading to a poor energy density in the devices. Fortunately, the working voltage ranges of the organic electrolytes are wider [10]. For example, a classic SICs configurations usually utilize hard carbon as anode materials and porous carbon as cathode materials in an electrolyte of 1 M $NaClO_4$ (EC:DMC:EMC = 1 : 1 : 1 vol%). Nonaqueous-based SIC systems can work in an enlarged voltage window (4 V with insertion-type anodes in classic SIC configurations). For example, tetrasodium ethylenediaminetetraacetate tetrahydrate-derived carbon is served as an anode and is pre-sodiated to reach the low potential (0.01 V vs. Na^+/Na). The assembled SICs devices reach a high potential of 4 V when coupled by the N-doped carbon cathode (4.5 V vs. Na^+/Na) [11]. The obtained high working potential range in an organic electrolytes hybrid devices provide the opportunity to achieve high energy density. However, in SIB devices, the polarization of the Na ion concentration is a rate performance constraint. In the discharge processes, Na ions are de-intercalation from the anodes and are intercalated simultaneously into the cathodes [12–18]. On the contrary, the driving forces for the ClO_4^- and Na^+ to the opposite-side electrodes

Sodium-Ion Capacitors: Mechanisms, Materials, and Technologies, First Edition.
Edited by Guoqiang Zou, Xiaobo Ji, and Hongshuai Hou.

decrease the concentration gradients to create an advantageous effect on the fast discharge capability in SICs devices. The working mechanism different from SIBs in charge and discharge processes also shows the excellent rate capability of SICs. Therefore, the typical advantages of SICs include: (i) higher energy density as compared to electric double-layer capacitors (EDLCs) due to the large battery-type capacity, (ii) higher power density as compared to SIBs due to fast capacitor-type adsorption–desorption processes.

In SICs, electrode materials include battery-type electrodes and capacitor-type electrodes. Capacitive-type materials refer to electrode materials that physically absorb and desorb ions on the electrode surface to achieve energy storage. Common capacitor materials mainly include EDLC-type activated carbon (AC) and pseudocapacitor materials. The capacitance of EDLC is attributed to the separation and accumulation of charge at the electrode/electrolyte interface. Therefore, the capacitance of EDLC mainly depends on the surface/interface characteristics and pore size distribution of the easily active material. AC is used as an EDLC capacitor material because it is easy to obtain, simple to process, rich in variety, nontoxic, and high chemical stability. EDLC has a high power density and excellent cycle stability on the electrode surface during the charge/discharge process, but usually shows a lower energy density. This is due to the limitation of the pore structure, not all the Brunauer–Emmett–Teller (BET) surface area of the porous electrode material is electrochemically accessible when in contact with the electrolyte. In view of the energy storage advantages of pseudocapacitors, materials based on the pseudocapacitive energy storage mechanism have become a hot spot in recent research. The contribution of pseudocapacitance to the electrode capacity overcomes the sodium storage mechanism of sodium battery electrode materials limited by the solid-state diffusion control of Na^+, and finally promotes a substantial improvement in the sodium storage rate performance and cycles at a large current density [19].

Battery-type materials refer to electrode materials that realize energy storage through insertion–deinsertion reactions. Common battery-type materials mainly include: cathode: sodium ion conductors, $Na_3V_2(PO_4)_3$ and other sodium-based materials; anode: alloys, transition metal chalcogenides, transition metal oxides, and so on. Transition metal oxides are potential anode materials. V_2O_5, Fe_3O_4, and Nb_2O_5 have been used in SICs for many applications. However, low conductivity limits their full performance. Among many sodium ion storage materials, titanium-based oxides with low volume expansion and low voltage are very promising anode materials. Similarly, in order to improve the conductivity of titanium-based oxides, researchers have done a lot of work. For example, in 2017, Le et al. used the microwave-assisted solvothermal method to prepare single-crystal TiO_2 on graphene. In the sodium ion half-cell capacity contribution test, the capacitance contribution reached 73% at a sweep rate of 3 $mV\,s^{-1}$ [20]. The SICs assembled with this material and AC cathode exhibit an energy density of 64.2 $Wh\,kg^{-1}$, and the capacity retention rate reaches 90% after 10 000 cycles. In the same year, Zhou et al. used electrospinning technology to prepare TiO_2@CNT@C nanorod anode materials and assembled them with biomass-derived carbon cathodes to form SICs. When the scan rate is 3 $mV\,s^{-1}$, the capacitance contribution of TiO_2@CNT@C is

81%. SICs can obtain high energy of 81.2 Wh kg^{-1} and 12 400 W kg^{-1} in the potential window of 1.0–4.0 V. Power density and the capacity retention rate of SICs reached 85.3% after 5000 cycles [21].

The electrolyte of SICs can generally be divided into two types: solid electrolyte and liquid electrolyte. The safety of solid electrolytes is high, but the cost is also high, which limits their commercial application. According to different solvents, liquid electrolytes can be divided into aqueous electrolytes and organic electrolytes. Aqueous electrolyte with good conductivity is relatively cheap; they have high conductivity, but their low working voltage and strong corrosiveness are their main defects. Organic electrolytes are mostly composed of ClO_4^-, BF_4^-, and other anions and active metal ions (Na^+) cations. The conductivity of the organic electrolyte is lower than that of the aqueous electrolyte, which affects the rate performance of the device to a certain extent, but its working voltage is relatively high, which can reach about 4.0 V. It can be seen from $E = 1/2CV^2$ that by increasing the working voltage, the energy density of the device can be improved. Therefore, the organic electrolyte has higher commercial use, and its corrosiveness is small, which further broadens the practical application range of the organic electrolyte [22].

The separators of the SICs can avoid the short circuit caused by the contact between the positive and negative electrodes during the charging and discharging process. Therefore, the microporous material with the pore diameter smaller than the electrode active material is mostly used. Because of its microporous structure to provide channels for the transmission of electrolyte ions, it is required to have good wettability, insulation, and high mechanical strength. At present, the more commonly used diaphragms are polypropylene, glass fiber diaphragms, etc. [23].

In solid-state SICs, the solid-state electrolyte plays the dual role of electrolyte and diaphragm, simplifying the structure of the device and making it more compact. It is worth noting that many sodium-ion electrode materials exhibit pseudocapacitance characteristics and inclined platforms on the charge and discharge curves [24]. These characteristics make SICs more attractive for the development of novel devices. Since battery-type materials suitable for SICs often have lower potentials, in order to obtain higher energy density, researchers often use battery-type materials as anodes and capacitive materials as SIC cathodes. However, some SICs still choose battery-type materials as cathodes and capacitor-type materials as anodes.

In addition, similar to the charging and discharging mechanism of LICs, the charging and discharging mechanism of SICs can also be divided into electrolyte consumption mechanism and sodium ion exchange mechanism according to whether the salt in the electrolyte is consumed or not, as follows:

(1) *Electrolyte consumption mechanism*: This type of SIC generally uses battery-type anode materials and capacitive/pseudocapacitive cathode materials (AC or MXene). During the charging process, the anions in the electrolyte migrate to the cathode (such as AC) and physically adsorb on the electrode surface, while the sodium ions migrate to the anode (such as TiO_2 or hard carbon) and finally achieve intercalation. During the discharge process, sodium ions are released from the anode into the electrolyte, while the cathode also releases adsorbed

anions to achieve charge balance. SICs with typical electrolyte consumption mechanisms include TiO_2//AC and hard carbon//AC.

(2) *Sodium-ion transport mechanism*: In this type of SIC, the sodium ion source is provided by the cathode (such as $Na_3V_2(PO_4)_3$), and the anode is usually composed of a capacitively active material (such as AC). During the charging process, sodium ions are deintercalated from the cathode material and enter into electrolytes, while the sodium ions in the electrolyte migrate to the anode to generate electric double layer capacitance. During the discharge process, the adsorbed sodium ions are released from the anode into the electrolytes and finally inserted in the cathode material. Typical SICs with this mechanism include AC//$Na_3V_2(PO_4)_3$ and so on. Since researchers mostly use battery-type materials as anode materials for SICs, consistent with what we have observed, the charging and discharging mechanism of most new SICs is an electrolyte consumption mechanism. This means that two points in the energy storage mechanism of SICs must be fully emphasized: (i) Unlike batteries (LIBs or SIBs, etc.), electrolytes should be regarded as important parts in SICs; (ii) electrolyte salts were consumed, and the electrolyte gradually returned to its initial state during the discharge process. Therefore, the design of the dynamic characteristics of the material during the charging process becomes extremely essential for SICs.

2.2 Working Principle

An EDLC is a device that stores electric charge at the interface between a porous carbon electrode and an electrolyte solution. Compared with batteries, the charge storage of EDLCs is non-Faraday, which is achieved by the physical adsorption and analysis of ions in the pores of the carbon electrode when an external voltage is applied. When charge accumulates in the electrode, it is balanced at the interface by equal and opposite ionic charges in the electrolyte [25]. This physical mechanism of charge storage contributes to the characteristics of rapid charge and discharge and long cycle life. The concept of EDLC was first proposed and formed by Helmholtz in the nineteenth century when he studied the opposite charge at the interface of colloidal particles [26–28]. According to the proposed model, opposite charges are layered at the electrode/electrolyte interface and separated by a distance of one atom, similar to a traditional capacitor with a bipolar plate. This simple model was revised by Gouy and Chapman into a diffusion model. This model regards cations and anions as continuously distributed in the electrolyte, forming a diffusion layer driven by thermal motion. Considering that the ions are not firmly attached to the surface, the distribution of opposite ion charges in the electrolyte surrounding the charged solid is equal. The thickness of the diffusion layer is partly determined by the kinetic energy of the ions. However, the disadvantage of this model is that it will get an excessive capacitance value, because the capacitance generated between two different separate charge arrays is inversely proportional to the distance between them, so when the point charge ions are close to the electrode on the surface, a huge capacitance value will be obtained. Later, Stern proposed a combined model between the Helmholtz model and the Gouy–Chapman model, emphasizing the two regions of ion distribution—the Stern layer (dense layer) and the diffusion layer Internal area. In the Stern layer, electrolyte ions and solvated molecules are strongly adsorbed on the electrode surface, which is consistent with the law of the Helmholtz model, while the diffusion layer follows the charge distribution law described in the Gouy–Chapman model. After that, Grahame made detailed corrections and supplements to the Stern layer (dense layer). He divided the dense layer into the inner Helmholtz layer (IHP) and the outer Helmholtz layer (OHP). In IHP, ions are tightly attached to the surface of the electrode material. The ions in this area are all electrolyte ions without solvent molecules. In OHP, there are solvated molecules that combine electrolyte ions and solvent molecules. The Grahame model reveals in detail the actual changes in the electrolyte during the charging and discharging process of an EDLC, and it is a relatively complete electric double layer theoretical model at present. With the research of new electrolytes and the design and development of electrode materials, the electric double layer theory is still being further improved.

For a further comprehension of the association of EDLC-type and battery-type electrodes with this property of devices, some fundamental schematic representations are presented in Figure 2.1. As depicted in Figure 2.1a, the charge–discharge curves of a traditional EDLC-type electrode are linear, but for a classic SIB-type electrode related to a phase change, the voltage keeps unchanged in the charge and discharge processes on the basis of the phase rule and adhering to the Nernst equation, as shown in Figure 2.1b. The energy stored in the capacitor can be expressed by $0.5q_c \times \Delta V_c$ (q_c means charge and ΔV_c means voltage), which is lower

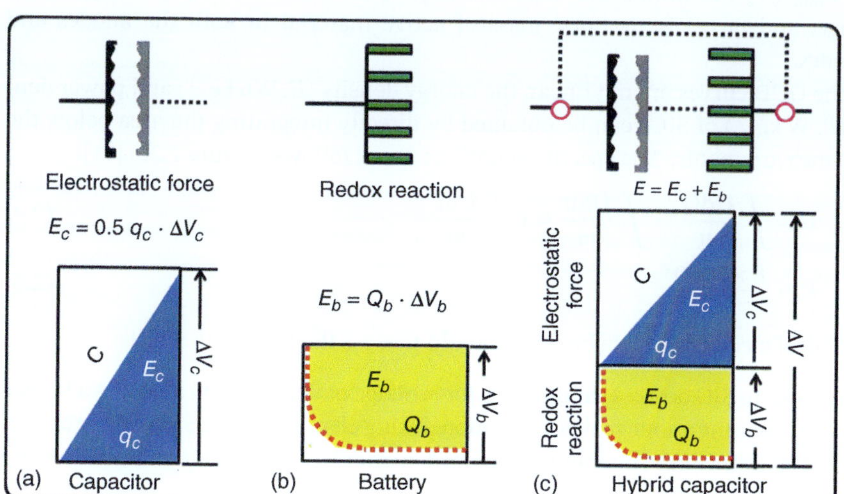

Figure 2.1 Schematic representations of a single electrode system. (a) Capacitor, (b) battery, and (c) typical SIC according to the charge-potential profile with corresponding equations for energy storage. Source: Reproduced with permission of Dubal et al. [29]. Copyright 2015, The Royal Society of Chemistry.

than that of SIBs ($E_b = Q_b \times \Delta V_b$). Hence, the schematic representations illustrate the electrochemical characteristics of capacitors and batteries. If the two systems are combined, the high power performance of the capacitor and the high energy characteristics of the battery will bridge the gap between SIBs and SCs in Figure 2.1c. In these devices, it is important to utilize a capacitor-type electrode possessed of a higher working potential (ΔV) at the start states of the charging process to achieve the working voltage of the battery-type electrode (ΔV_b) leading to an increase in energy storage compared with a single electrode. Moreover, an obvious problem of this hybrid devices is that the battery-side electrode with intrinsically sluggish kinetics is difficult to carry out at a rapid enough rate to meet the capacitor-side electrodes [29].

2.3 Equations

The specific capacitance (C, F g^{-1}), energy density (E, Wh kg^{-1}), and power density (P, W kg^{-1}) of the SICs, based on the galvanostatic charge–discharge curve (GCD) tests, are calculated utilizing the equations (Figure 2.2a,b) [30]:

$$C = It/(\Delta V \times m) \tag{2.1}$$

$$E = C\left(V_{\max}{}^{2} - V_{\min}{}^{2}\right)/(2 \times 3.6) \tag{2.2}$$

$$P = E \times 3600/t \tag{2.3}$$

where I (A) is the discharge current, t (seconds) is the discharge time, $V_{\max}$ (V) and $V_{\min}$ (V) the initial and final discharge potentials, ΔV (V) is the potential change, and m (g) is the total mass of active material in both the anodes and cathodes.

If the GCD curves are not linear, the energy density (E, Wh kg^{-1}) and power density (P, W kg^{-1}) of SICs can be obtained by directly integrating the area below the GCD measurements. The specific equations are as follows (Figure 2.2c) [31]:

$$E = \int \frac{U\mathrm{d}Q}{m} = \int \frac{UI\mathrm{d}t}{m} = I\int \frac{U\mathrm{d}t}{m} \tag{2.4}$$

$$P = \frac{E \times 3600}{t} \tag{2.5}$$

An ideal electrode in a hybrid SIC should possess the following merits:

(1) Anodes in SIC devices should have low voltage plateau, which achieves a higher $V_{\max}$ since the anode realizes sodiation during charging of the SICs. In the case of other parameters remaining unchanged, higher $V_{\max}$ can produce higher energy density and power density.
(2) Improve the kinetics in anodes, and allows large energy-density anodes and high power-density cathodes to be further matched to reach excellent rate capabilities. By improving the rate performance of the anodes, the dynamic imbalance between the electrodes can be effectively solved.

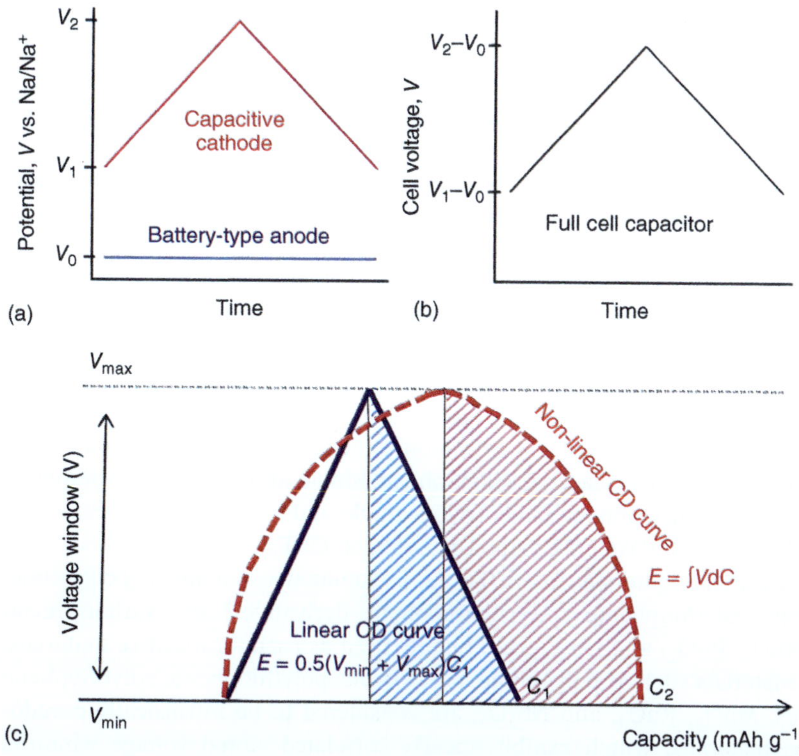

Figure 2.2 An idealized behavior of a sodium-ion capacitor. (a) Potential profiles on the battery-type anode (blue) and the capacitive cathode (red) originating from the insertion and electric double layer mechanisms, respectively. (b) Cell voltage profile of the full cell capacitor. Source: Reproduced with permission of Glushenkov and Ellis [30]. Copyright 2018, WILEY-VCH. (c) Schematic of electrochemical profile of a typical SIC with the linear (solid line) and nonlinear (dashed line) charge/discharge curve. Source: Reproduced with permission of Wang et al. [31]. Copyright 2017, WILEY-VCH.

(3) According to the capacity equation ($1/C_{SIC} = 1/C_{cathode} + 1/C_{anode}$), the capacity of the cathodes will drag down the capacity of the anodes with relatively high capacity. In this case, the traditional carbon-based cathodes need to be appropriately designed or modified to improve the specific capacity awhile maintaining the inherent high power and excellent cycling stability.

(4) The essential charge balances between the cathodes and anodes are the key to achieving high energy density. The relationship between the specific capacity (C) and mass (m) of the electrode is $Q = C \times m$. Consequently, the theoretical mass ratio of the cathode and anode electrodes can be calculated as:

$$m_+/m_- = (C_- \times \Delta E_-)/(C_+ \times \Delta E_+) \tag{2.6}$$

where (+) stands for the cathodes, (−) stands for the anodes, m is the electrode, C is the specific capacitance, and ΔE is the potential range in the charge–discharge process for both electrodes, respectively.

Sections 2.2 and 2.3 discussed the relevant formulas, which is an understanding from the perspective of whole SICs. The following step is to understand SICs from mechanistic parts, and this part is the beginning of Chapter 3. The development of electrochemical capacitors (also known as supercapacitors or ultracapacitors) from electrostatic capacitors has been a game changer in capacitor technology [32]. The usual high frequency (response time 10^{-9} seconds) application areas for electrostatic capacitors are mainly in the field of power electronics [33]. However, supercapacitors have shifted to low frequency (response time 1 seconds) electrochemical energy storage applications [34]. The formation of EDL on the high-surface-area active electrode materials of supercapacitors is the key to energy storage applications. Mechanistically, supercapacitors can be divided into two categories, namely EDLC and pseudocapacitances. The former stores energy by non-Faraday charge separation through EDL formation, while the pseudocapacitance (pseudo-Faraday mechanism) is a surface Faraday energy storage mechanism. Carbonaceous materials such as AC, carbon black (CB), graphene (single and multilayer), graphene oxide (GO), reduced graphene oxide (rGO), single and multi-walled carbon nanotubes (SWCNT and MWCNT), carbon nanofibers (CNFs), carbon nanorods, etc., are used as active materials for EDLC [35]. Conductive polymers (polyaniline, polypyrrole, polythiophene, and polythiophene derivatives) and various metal oxides (MnO_2, RuO_2, and Nb_2O_5) are mainly used as pseudocapacitive materials. Notably, materials such as polyaniline, polypyrrole, polythiophene, polythiophene derivatives, MnO_2, RuO_2, and Nb_2O_5, are considered to be intrinsically pseudocapacitive materials, which exhibit linearly correlated stored voltage windows and exhibit quasi-rectangular CV curves [36]. Such materials that resemble EDL materials in terms of electrochemical curves but have a pseudocapacitive storage mechanism are referred to as intrinsically pseudocapacitive materials. This intrinsic pseudocapacitance exhibits typical capacitive charge storage independent of material properties such as morphology, particle size, and crystallinity. On the other hand, typical battery electrode materials such as various metal oxides (with the exception of MnO_2, RuO_2, Nb_2O_5), hydroxides, carbides, nitrides, phosphides, phosphates, phosphites, and sulphide generics, can be considered as extrinsic pseudocapacitive materials which exhibit capacitor-like properties with nanoscale modifications. Nanostructures favor fast kinetics of ion and electron transport, and therefore capacitor-like properties can be achieved. In the 1990s, Conway identified three pseudocapacitance processes [33]: (i) underpotential deposition occurs when cations in the electrolyte form adsorbed monolayers on the surface of different metal electrodes at a certain potential, which is much higher than their redox potential; (ii) redox pseudocapacitance occurs when ions are specifically adsorbed on the electrode surface by electrochemical means (i.e. the electrode/electrolyte interface) with concomitant Faraday charge transfer; and (iii) insertion pseudocapacitance occurs when ions are inserted into the redox-active host material, accompanied by Faraday charge transfer without crystal phase transition. In this case, the distinction between Faraday and pseudo-Faraday redox processes needs to be clarified. In short, most electron transfer in active materials is attributed to diffusion-controlled Faraday processes (in batter-type materials), whereas surface-constrained electron transfer is called a pseudo-Faraday process. Finally, cyclic voltammetry with various scan rates conforms to the following law: Eq. (2.7) [36]:

$$i = av^b \tag{2.7}$$

i and v are the current and scan rate, respectively, and a and b are the characteristic constants. According to the Randles–Sevcik equation, $b = 0.5$ indicates a diffusion-controlled Faraday mechanism, while $b = 1$ indicates a surface-controlled capacitance/pseudocapacitance mechanism. The ratio of diffusion to surface contribution can be attributed to Eq. (2.8) [37]:

$$i(V) = k_1 v + k_2 v^{1/2} \tag{2.8}$$

k_1 and k_2 are coefficients related to surface control (capacitance/pseudocapacitance) and diffusion control processes, respectively. In terms of construction, supercapacitors fall into three categories, e.g. EDLCs, pseudocapacitors, and hybrid supercapacitors. The coupling of EDLCs or pseudocapacitors is also referred to as an asymmetrical supercapacitor. The coupling of EDLCs and pseudocapacitors is also referred to as an asymmetrical supercapacitor. Devices with matched capacitive and battery-type electrodes can be called hybrid ionic capacitors. Thus, there is sufficient scope for optimizing the energy and power characteristics of hybrid ion capacitors through an intelligent choice of materials for the EDLC-type electrodes, the pseudocapacitor electrodes, and the battery-type electrodes.

By definition, SICs are composed through the electrochemical coupling of batteries and supercapacitors that convert chemical energy into electrical energy. Batteries store energy through chemical transformations (bond breaking and formation) via a diffusion-controlled Faraday electron transfer mechanism [38]. The energy storage mechanism in supercapacitors (especially EDLC) is a surface-controlled non-Faraday physical charge separation with no active electron transfer between the electrode–electrolyte interfaces. In the field of pseudocapacitors, energy storage is attributed to a surface-controlled pseudo-Faraday mechanism. From a kinetic point of view, Faraday processes are slower than pseudo- and non-Faraday processes. Conversely, the energy accumulated in the Faraday process is much greater compared to the pseudo- and non-Faraday mechanisms. Based on this argument, batteries are classified as energy devices, while the entire family of supercapacitors is referred to as power devices. It is worth noting that each battery and supercapacitor system has a unique energy density (normalized energy by mass or volume) and power density (normalized power by mass or volume) based on the principle of operation. As a result, each electrochemical energy system plays a role in the market segment based on energy and power requirements. In addition to energy and power specifications, the market is also concerned with self-discharge rates and lifetime characteristics, in addition to commercial issues such as safety, cost, and environmental hazards. Note that, in general, energy equipment has a lower lifetime than power equipment, while power equipment has a higher self-discharge rate than energy equipment.

The most likely applications for supercapacitors are broadly divided into two parts: (i) high power/low energy demand and (ii) medium power/medium energy demand.

In the first part, supercapacitors, especially high-power EDLCs, are envisaged for the start-up of electric vehicles, which will be driven by high-energy density LIBs or by hydrogen energy via, for example, fuel cells. Regenerative braking, which converts the kinetic energy of a decelerating vehicle into electrical energy to be stored in an energy storage device and then reused for acceleration to minimize the load on the fuel tank or battery, is another potential end use for EDLCs in the electric vehicle industry. These application areas are in the midst of a huge research and development process. In contrast, the second segment is more feasible and can be widely used today. Stationary large-scale energy storage systems are a major concern in the field of renewable energy (solar, wind, etc.). The main drawback of solar or wind energy harvesting systems is their intermittency due to changing weather conditions. Reliable energy storage modules are therefore needed to make efficient use of renewable energy sources. In monsoons or in the absence of strong winds, the batteries are unable to recover their full state of charge in short periods of sunlight. As a result, the service life of these batteries is significantly reduced. To date, rechargeable batteries, particularly lead-acid and nickel-based alkaline batteries, have been used primarily for storing renewable energy. These batteries are not suitable for this purpose because of their sluggish kinetics, i.e. poor charge acceptance. In this context, supercapacitors, especially hybrid ion capacitors, can be used in renewable energy storage devices. With a higher power capacity than batteries, hybrid-ion capacitors can accept charges (from the sun or wind) in a short time, even in rainy seasons or slow winds. Hybrid energy storage systems are therefore the only option in the field of renewable energy storage. Hybrid systems can be built internally, like hybrid supercapacitors, or by external assembly of batteries and supercapacitors. Nevertheless, both application areas aim at a sustainable future through the development of electric vehicles (minimizing greenhouse gas emissions) and the storage of renewable energy (minimizing fossil fuel-based economies). However, the reader should consider that there is no clear boundary between the two applications and that a specific candidate with specific advantages for one market segment may be suitable for the other and vice versa.

In hybrid-ion capacitors, the Faraday reaction of the cell type is related to the EDLC or pseudocapacitor counterpart. Therefore, the kinetic mismatch between cell type and capacitor type electrodes is an important issue, which minimizes the utilization of the active material [35]. Quantitatively, the "*b*-value" reflects the charge storage kinetics of the electrodes. In practice, a *b*-value close to 0.5 indicates slow kinetics for cell-based diffusion control, while a *b*-value close to 1 indicates fast kinetics for capacitor-based surface control. If any material shows slow kinetics ($b \sim 0.5$), the following measures have been taken to enhance the kinetic properties. Thin film design, porosity enhancement, nanostructuring, morphology control, core/shell design, performing composite design etc., are the conventional route to material modification for fast ion and electron transport [36]. As a result, the kinetic mismatch between battery and capacitor-type electrodes will be reduced. However, the diffusion-controlled Faraday reaction makes the hybrid system sluggish. Therefore, the selection of battery components is crucial. The lithium-ion family of batteries contains the highest energy and power densities of any other commercially viable battery. It is therefore reasonable to select Li-ion chemicals for hybrid ion capacitors to achieve appreciable energy and power characteristics. The oscillation of Li^+ between cathodes and anodes (similar to a rocking chair) via an intercalation/de-intercalation mechanism is the operating principle of most LICs [37]. Today, Li-ion technology in nonaqueous media is sufficiently mature for consumer electronics, grid energy storage and, to some extent, electric vehicles. Today, lithium-ion technology cannot fully enter the consumer electronics and electric vehicle sectors. This is because of its high cost. Therefore, SICs, PICs, MICs, ZICs, CICs, and AICs can be considered as low-cost alternatives to LICs for energy storage applications. Scientists have cleverly designed hybrid-ion capacitors that use this battery chemistry in both nonaqueous and aqueous electrolytes to make the devices cheaper, safer, and more environmental friendly.

For the battery system, different characteristic aspects of Li^+, Na^+, K^+, Mg^{2+}, Ca^{2+}, Zn^{2+}, and Al^{3+} are used as charge carriers for the respective energy storage system [37]. It is worth noting that the low mass-to-charge ratio and the high density of charge carriers are responsible for the respective high weight-to-volume and volume-to-capacity of the metal electrodes. First, we would like to discuss the pathway for the development of hybrid-ion capacitors under the intercalation/de-intercalation mechanism; as most of the seminal results in LICs, SICs, PICs, etc., have evolved in this way. This will be followed by a discussion of energy storage via conversion reactions and alloying/de-alloying processes [38].

As the Shannon's ionic radii of the monovalent charge carriers increase from Li^+ to Na^+ to K^+, the ion-solvent interaction (represented by the Stokes radii) in the electrolyte decreases from Li^+ to Na^+ to K^+. This event indicates that the molar conductivity of K^+ is better than that of Na^+ than Li^+ [39]. The solventized charge carriers are desolvated at the electrode–electrolyte interface prior to insertion into the main frame and the activation energy for ion diffusion at the electrode–electrolyte interface is strongly dependent on the ion-solvent interaction [39]. In summary, Na^+ diffuse faster at the interface than Li^+ ions due to the weaker Lewis acidity. Ionic radii and Stokes radii of Li^+ and Na^+ in different solvents are sufficiently close that their charge storage chemistry is very similar. For both Li^+ and Na^+, almost similar classes of frameworks and electrolytes are available to rationalize the ion system. Thus, the higher power characterized Na^+ ion system (higher molar conductivity and lower desolvation energy) will be complemented by the higher energy characterized Li^+ ion system (higher voltage and lower mass of lithiated compound) [39]. In the case of the K^+ ion system, a lower desolvation energy is expected than for the Na^+ ion, because of the lower Lewis activity. However, due to the large ion size, the desolvation energy factor alone may not determine the activation energy of K^+ diffusion in the electrode–electrolyte interface alone. Therefore, the activation energy diffusion of K^+ may depend on the relative relationship between ion size and desolvation energy in different electrolyte systems. On the other hand, the choice of a body framework for K^+ ion systems is challenging because similar frameworks in Li^+ and Na^+ systems may be structurally unstable during the solid-state diffusion of large K^+. The power capability of K^+ systems depends not only on the higher molar conductivity of the solventized ions in the electrolyte and the lower desolvation energy at the interface, but also on the activation energy of the active material at the K^+. The fast solid state diffusion and the rapid kinetics of structural changes ion intercalation/de-intercalation also play an important role. Due to the low density and high quality of the K-based active material, the energy realization of the K^+ system is also poor. However, from the standard electrode potential data available in Table 1.1; it is worth noting that in most organic media, K^+ systems can provide a wider voltage window compared to Li^+ and Na^+ systems [40]. In aqueous media, K^+ exhibits the second lowest (after lithium) standard electrode potential of all charge carrier ions (vs. SHE).

In multivalent charge carriers the main advantages are as follows: low equipment costs due to the association with the most abundant elements and a higher theoretical capacity due to the multiple electron transfer of the individual ions [41]. On the other hand, the higher charge density of the ions is responsible for the strong ion-electrolyte interactions, so that lower molar conductivity and higher desolvation energies are inevitable. Furthermore, the slow kinetics of solid-state diffusion of multivalent ions is practical for strong electrostatic interactions of the anion with the host framework. Therefore, suitable electrolyte solutions with weak ion-solvent interactions and good electrode materials for fast solid-state ion diffusion are needed to explore the good electrical energy quality of multivalent ion-based devices [42]. In the Mg^{2+}, Ca^{2+}, Zn^{2+}, and Al^{3+} systems, mainly metal anode is used in nonaqueous media, but suitable electrolytes should be selected to minimize nonconductive passivation layers on the metal electrodes. Of all candidates, only the Zn^{2+} ionic system is suitable in mild aqueous media with metal negative electrodes because the stripping/plating potential of Zn (−0.76 V vs. SHE) is higher than the hydrogen precipitation reaction [43].

Multivalent hybrid-ion capacitors are not a focal part of this section, so a small number of future perspectives are discussed: (i) the synthesis of high-performance electrode materials for multivalent metal-ion hybrid capacitors. To date, few have been reported on MICs, CICs, and AICs due to the limited choice of cell electrodes for these systems. More types of electrode materials for MIBs, CIBs, and AIBs should be developed and investigated for the corresponding multivalent metal-ion hybrid capacitors. In particular, for practical applications of hybrid capacitors based on metal electrodes (e.g. ZICs based on zinc foil), the energy density of the whole device (and not only the cathode material) has to be considered. This requires metal electrodes with long-term stability and fast kinetics, even if their mass in the hybrid-ion capacitor is relatively low (relative to the mass of the counter electrode). In addition, for these hybrid-ion capacitors, electrode materials other than AC materials should be further investigated. We hope that our book of electrode materials for supercapacitors (EDLC and pseudocapacitors) and multivalent metal-ion batteries in Section 2.2 will provide some insight. (ii) A detailed description of the actual electrochemical processes inside a multivalent metal ion hybrid capacitor. The multivalent ion storage mechanism is a relatively novel concept. Due to the unique properties of multivalent ions (different from monovalent ions in terms of charge number, polarization strength, and ion diameter) they are not yet fully understood in battery systems. In addition to this, the interaction between multivalent ions and supercapacitor materials such as carbon may not be pure adsorption/desorption. Therefore, an integrated approach including electrochemical measurements and physicochemical analyses should be used to infer the energy storage mechanism of multivalent metal-ion hybrid capacitors. This will benefit science by enriching the understanding of multivalent ion storage mechanisms. An important scientific question for multivalent hybrid-ion capacitors is whether the anode can be used without a metal electrode and the design of pre-metallization strategies. (iii) Development of multivalent metal-ion hybrid capacitors in multiple configurations. For ZICs, Zn electrode-based ZICs and metal oxide/carbon-based ZICs have been proposed, while for MICs, CICs, and AICs, other types of multivalent metal-ion hybrid capacitor configurations need to be developed to suit different applications. The construction of various configurations of multivalent metal-ion hybrid capacitors relies on research into high performance electrode materials, electrolytes, and multivalent ion storage mechanisms. (iv) Research on multivalent metal-ion hybrid capacitors for flexible and wearable energy storage. As mentioned above, multivalent metal ion hybrid capacitors offer a number of advantages, such as significantly higher energy density compared to conventional supercapacitors and higher safety than batteries. Therefore, multivalent metal-ion hybrid capacitors with good deformability based on a rational electrode/device structure design have a wide range of applications for flexible/wearable energy storage. These flexible hybrid capacitors can also be combined with flexible solar cells, flexible sensors, and other electronics for multifunctional smart applications in future societies.

Unlike intercalation/de-intercalation mechanisms, the breaking and creation of new bonds are associated with transformation reactions. Two types of conversion reactions are described below in Eqs. (2.9) and (2.10) [44]:

Conversion:

$$\mathrm{TM}_a\mathrm{X}_b + (b \times c)\mathrm{A}^+ + (b \times c)\mathrm{e}^- \leftrightarrow a\mathrm{TM}^0 + b\mathrm{A}_c\mathrm{X} \tag{2.9}$$

Transformation:

$$\mathrm{X} + a\mathrm{A}^+ + a\mathrm{e}^- \leftrightarrow \mathrm{A}_a\mathrm{X} \tag{2.10}$$

TM = transition metals, A = Li, Na, K, etc., X = Anionic substances of type A (F^-, Cl^-, Br^-, I^-, S^{2-}, Se^{2-}, Te^{2-}, O^{2-}, etc.) and neutral substances (S, Se, Te, Br, I, etc.) [44]. The advantages of such conversion materials are high capacity, high energy, low cost, and environmental friendliness; at the same time, the disadvantages are low Coulombic efficiency (CE), high impedance, resulting in large potential hysteresis, large volume changes during charge carrier insertion/extraction, and poor subsequent cycle life [45]. Alloying materials are materials that electrochemically form alloys and composite phases with ionic charge carriers at lower reduction potentials [46]. The alloying/dealloying process is similar to the transformation reaction. Extremely high volume and weight capacities are achieved with alloyed materials, similar to the transformation reaction [47]. These high values were obtained due to the several-fold higher acceptance of charge carrier ions by one-mole conversion/alloying materials compared to insertion materials [48]. For example, in the LiC_6 insertion phase (specific capacity: 372 mAh g^{-1}), 0.17 mol of Li can be inserted into 1 mol of graphite, while 1 mol of Si in $Li_{15}Si$ can accept 3.75 mol of Li_4-alloying compound (specific capacity: 3600 mAh g^{-1}) [49]. Thus, the volume change upon embedding/detachment of Li^+ ions is greater for Si alloy materials (320%) than for graphite-intercalated materials (10%) [50]. Short lifetimes are inevitable as a result of the large strain due to the large volume change, which leads to material comminution. Nanostructuring, designing carbon composites, and fabricating core-shell structures can withstand large volume changes and increase the durability of the device [51].

Based on the above discussion, it is possible to summarize the different classes of materials used for cathode and anode chemistry. Cathode materials are designed for intercalation chemistry where the structural voids and the space for solid-state charge carrier diffusion should be synchronized. For suitable intercalated/de-intercalated cathodes, the following main types of crystal structures were identified: layered, spinel, olivine, and polyanionic materials. In addition, promising conversion anodes are represented by the general formula MX_z, where M = Fe^{3+}, Fe^{2+}, Ni^{2+}, Cu^{2+}, Co^{2+}, Mn^{3+}, etc. X = halogen ions (F^-, Cl^-, Br^-, I^-), sulphide ions (S^{2-}, Se^{2-}, Te^{2-},etc.), and oxide ions. The anode materials are subdivided as follows: intercalated/de-intercalated (graphite, hard carbon, soft carbon, graphene, CNT, MXenes, NASICON structures, etc.), alloyed/de-alloyed (Si, SiO, Sn, SnO, Ge, Sb, P, Pb, etc.), conversion reaction materials (Fe_2O_3, Fe_3O_4, CoO, Co_3O_4, NiO, RuO_2, Mn_xO_y, Cu_2O/CuO, MoO_2/MoO_3, MX_y [M = Fe, Mn, Ni, Cu, Co; X = P, S, N, etc.]) and metal electrodes. It is worth noting the use of charge carrier ions inserted into the cell-type anode active material prior to full cell assembly (pre-sodiation) in order to achieve a closer approximation to a stable voltage metal electrode and to reduce the irreversible capacity of parasitic reactions. At the same time, this step will reduce the consumption of charge ion carriers from the electrolyte as it provides an additional source.

In order to better understand the process of pre-sodiation, a deep understanding of the interfacial reaction mechanism is necessary. The formation of SEIs on the anode is essential for the successful implementation of energy systems. The SEIs are formed by reducing solvent molecules at the start of the first charge and protect the anode from further reduction [52]. "Electronically insulating and ionically conducting" is the pioneering feature of SEIs. A similar passivation layer, known as the cathode–electrolyte interfaces (CEIs), is also important for device performance due to the oxidation of the electrolyte solvent on the surface of the cathode. Although the exact composition of the SEIs and CEIs varies depending on the electrode/electrolyte systems [53]. However, the formation and maintenance of proper SEIs and CEIs are essential throughout the operation of the device, so the synchronization of the electrode material and electrolyte composition is decisive. For example, alloyed or transformed negative electrodes (Si, SiO, etc.) have a large volume change during charging/discharging and the rupture of the SEIs. The reduction products provide stability and flexibility for the SEI to maintain volume changes [54]. Faraday reactions in battery-based electrodes are accompanied by phase changes, which are responsible for the slow reaction kinetics and therefore inevitably lead to poor multiplicative performance as well as limited cycles. To avoid these problems, nanostructures and the preparation of conductive composites of active materials are often used in practical devices [55]. Based on the nature of the electrode materials, aqueous, organic, and ionic liquid (IL) media are often used as suitable electrolytes. More recently, salt-coated aqueous electrolytes have been used in aqueous devices to increase the operating voltage beyond the thermodynamic hydrolysis limit and thus greatly increase the energy density. In order to achieve optimized performance, different modifications must be made to the cathode, anode, and electrolyte materials. To create hybrid ionic capacitors, the cathode or anode chemistry will be combined with EDLC or pseudocapacitive materials in a suitable electrolyte system. However, recent advances in Na^+-based carriers for SICs will be detailed in Chapters 5 and 6 [56].

In batteries, poor coulombic efficiency is typically associated with the irreversible and continuous formation of SEIs and CEIs. SICs typically have smaller Li/Na reservoirs than conventional LICs, have more stringent requirements to minimize IR losses due to high power, and have several orders of magnitude more cycles [57]. All three factors make the formation of SEIs more important for SICs than for LICs. What is not well recognized in the SICs field is that the formation of CEIs due to a wide voltage window leads to pore blockage and loss of active surface area in the AC. 2.7 V acetonitrile supercapacitors with high-end (i.e. pure and therefore minimal parasitic reactions) AC will retain their surface area during cycling or full charge retention [58]. However, at a typical charge cell voltage of 4.0 V, this is not expected to be the case with carbonate electrolyte SICs. On SIC cathodes, it is questionable to attribute all charge storage to pure EDLC. Although even after the carbon has been overgrown by CEIs and EDLC, It still makes some contribution [59]. Furthermore, it is expected that the SEIs products resulting from the reduction of carbonate electrolytes in SICs should be similar to other types of electrode materials used in battery anodes, such as graphite or hard carbon. Therefore, materials with large surface-to-volume ratios will result in higher irreversible capacity losses [60]. These reducing films passivate the anode surface and prevent further decomposition of the electrolyte solution. However, any volume changes experienced during electrochemical cycling continuously weaken and destroy the SEIs, exposing fresh material to the electrolyte in each cycle, which creates new SEIs. SEIs consist mainly of electrolyte reduction products such as Na_2CO_3 [61]. The instabilities of SEIs eventually lead to a loss of overall capacity and the failure of the SICs devices. In addition to solvent reduction products such as Na_2CO_3 and alkyl carbonate, the anode SEI also consists partly of NaF, which is a (chemical) decomposition product of the $NaClO_4$ salt, but can also be formed by reaction with traces of water to form HF and finally NaF. In the classical description of SEI in graphite or hard carbon, it is considered to be a complex of various redox and chemical decomposition product phases [62]. Recent work on the subject has pointed to radial gradients in composition and structure within the SEIs, as well as to dynamic growth-shrinkage properties of the charge and discharge. Although the SEIs structure/chemistry of the sodium ion has been less explored, they appear qualitatively similar to those of lithium. The irreversible reduction of solvent molecules leads to the reduction of solvent molecules to Na_2CO_3, Na alkyl carbonates, Na alcohol salts, etc.

In practice, scientists perform electrochemical characterization by various electroanalytical methods using electrodes, such as CV, linear sliding voltammetry (LSV), EIS, GCD, overpotential estimation by the Tafel method [63]. CV analysis is used for various purposes, such as identification of the active material during charging/discharging processes, understanding the stability of electrolytes under operating conditions, quantifying the surface and diffusion controlled current contribution of hybrid systems, and estimating the specific capacitance of materials. The electrochemical stability of reactive materials, electrolytes, diaphragms, and battery containers under actual operating conditions can be assessed by the LSV method. EIS results and associated equivalent circuit modeling are carried out. Examples include resistance, capacitance, inductance, diffusion-related Warburg components, and response times (time constants). The chrono-current method is commonly used to obtain the charge acceptance of an active material at a fixed applied voltage. The chronopotentiometric method is widely used to estimate the charge storage performance (capacity/capacitance, energy density, and cycle stability) of active materials. The kinetic behaviors of the electrodes can be obtained by estimating the reaction overpotential values by the Tafel method. Based on these characteristics, promising materials are considered for further studies of the performance of two-electrode battery assemblies with various electrolytes.

References

1 Sennu, P., Chua, R., Dintakurti, S.S.H. et al. (2020). *J. Power Sources* 472: 10.

2 Liu, J., Xu, Y.G., and Kong, L.B. (2020). *J. Colloid Interface Sci.* 575: 42.

3 Qiu, R.Y., Fei, R.X., Guo, J.Z. et al. (2020). *J. Power Sources* 466: 8.

4 Liu, S.Y., Hu, F.Y., Shao, W.L. et al. (2020). *Nano-Micro Lett.* 12: 15.

5 Sun, Q.J., Cao, Z., Wang, S.H. et al. (2020). *J. Power Sources* 461: 7.

6 Zhang, Y.D., Jiang, J.M., An, Y.F. et al. (2020). *ChemSusChem* 13: 2522.

7 Liu, L., Zhao, Z.C., Hu, Z.Q. et al. (2020). *Front. Chem.* 8: 9.

8 Minakshi, M., Mitchell, D.R.G., Munnangi, A.R. et al. (2018). *Nanoscale* 10: 13277.

9 Krishnamoorthy, K., Pazhamalai, P., Sahoo, S. et al. (2017). *ChemElectroChem* 4: 3302.

10 Tang, J.Y., Huang, X., Lin, T.G. et al. (2020). *Energy Storage Mater.* 26: 550.

11 Zou, K.Y., Cai, P., Liu, C. et al. (2019). *J. Mater. Chem. A* 7: 13540.

12 Yan, Z.H., Yang, Q.W., Wang, Q.H., and Ma, J.M. (2020). *Chin. Chem. Lett.* 31: 583.

13 Jin, T., Han, Q.Q., and Jiao, L.F. (2020). *Adv. Mater.* 32: 14.

14 Qi, S.H., Xu, B.L., Tiong, V.T. et al. (2020). *Chem. Eng. J.* 379: 32.

15 Zhang, D.M., Jia, J.H., Yang, C.C., and Jiang, Q. (2020). *Energy Storage Mater.* 24: 439.

16 Dong, L., Yang, W., Yang, W. et al. (2019). *Nano-Micro Lett.* 11: 94.

17 Qi, S.H., Wu, D.X., Dong, Y. et al. (2019). *Chem. Eng. J.* 370: 185.

18 Liu, Q.N., Hu, Z., Chen, M.Z. et al. (2019). *Small* 15: 24.

19 Yuan, T., Jiang, Y., Sun, W. et al. (2016). *Adv. Mater.* 26: 2198.

20 Le, Z., Liu, F., Nie, P. et al. (2017). *ACS Nano* 11: 2952.

21 Zhu, Y.E., Yang, L., Sheng, J. et al. (2017). *Adv. Energy Mater.* 7: 1701222.

22 Zhang, D.K., Liu, Y., Xi, Y.K. et al. (2018). *Technology* 18: 324.

23 Hao, J.Y. and Wang, X.W. (2014). *China Pulp. Pap* 33: 62.

24 Ding, J., Hu, W., Paek, E. et al. (2018). *Chem. Rev.* 118: 6457.

25 Saikia, B.K., Benoy, S.M., Bora, M. et al. (2020). *Fuel* 282: 17.

26 Pan, Y., Xu, K., and Wu, C. (2019). *Nano. Rev.* 8: 299.

27 Wang, Y., Zhang, L., Hou, H. et al. (2021). *Mater. Sci.* 56: 173.

28 Siwatch, P., Sharma, K., Arora, A., and Tripathi, S.K. (2019). *J. Energy Storage* 21: 801.

29 Dubal, D.P., Ayyad, O., Ruiz, V., and Gomez-Romero, P. (2015). *Chem. Soc. Rev.* 44: 1777.

30 Glushenkov, A.M. and Ellis, A.V. (2018). *Adv. Sustainable Syst.* 2: 25.

31 Wang, H.W., Zhu, C.R., Chao, D.L. et al. (2017). *Adv. Mater.* 29: 18.

32 Duff, W. (1916). *A Text-Book of Physics*. Philadelphia: P. Blakiston's Son & Co.

33 Conway, B.E. (1999). *Electrochemical Supercapacitors: Scientific Fundamentals and Technological Applications*. New York: Kluwer Academic/Plenum Publishers.

34 Bird, J. (2010). *Electrical and Electronic Principles and Technology*. Routledge.

35 Béguin, F. and Frackowiak, E. (2009). *Carbons for Electrochemical Energy Storage and Conversion Systems*. Boca Raton: CRC Press.

36 Gogotsi, Y., Gogotsi, Y., and Presser, V. (2006). *Carbon Nanomaterials*. Boca Raton: CRC Press.

37 Brezesinski, T., Wang, J., Polleux, J. et al. (2009). *J. Am. Chem. Soc.* 131: 1802–1809.

38 Winter, M. and Brodd, R.J. (2004). *Chem. Rev.* 104: 4245–4270.

39 Hosaka, T., Kubota, K., Hameed, A.S., and Komaba, S. (2020). *Chem. Rev.* 120: 6358–6466.

40 Das, S.K., Mahapatra, S., and Lahan, H. (2017). *J. Mater. Chem. A* 5: 6347–6367.

41 Aurbach, D., Lu, Z., Schechter, A. et al. (2000). *Nature* 407: 724–727.

42 Konarov, A., Voronina, N., Jo, J.H. et al. (2018). *ACS Energy Lett.* 3: 2620–2640.

43 Gummow, R.J., Vamvounis, G., Kannan, M.B., and He, Y. (2018). *Adv. Mater.* 30: 1801702.

44 Yu, S.H., Feng, X., Zhang, N. et al. (2018). *Acc. Chem. Res.* 51: 273–281.

45 Goriparti, S., Miele, E., De Angelis, F. et al. (2014). *J. Power Sources* 257: 421–443.

46 Nitta, N., Wu, F., Lee, J.T., and Yushin, G. (2015). *Mater. Today* 18: 252–264.

47 Kim, J., Kim, H., and Kang, K. (2018). *Adv. Energy Mater.* 8: 1702646.

48 Zuo, X., Zhu, J., Müller-Buschbaum, P., and Cheng, Y.J. (2017). *Nano Energy* 31: 113–143.

49 Teki, R., Datta, M.K., Krishnan, R. et al. (2009). *Small* 5: 2236–2242.

50 Wang, J., Xu, T., Huang, X. et al. (2016). *RSC Adv.* 6: 87778–87790.

51 Kim, S.H., Kim, Y.S., Baek, W.J. et al. (2018). *ACS Appl. Mater. Interfaces* 10: 24549–24553.

52 Edström, K., Gustafsson, T., and Thomas, J.O. (2004). *Electrochim. Acta* 50: 397–403.

53 Liu, S., Liu, D., Wang, S. et al. (2019). *J. Mater. Chem. A* 7: 12993–12996.

54 Heiskanen, S.K., Kim, J., and Lucht, B.L. (2019). *Joule* 3: 2322–2333.

55 Suo, L., Oh, D., Lin, Y. et al. (2017). *J. Am. Chem. Soc.* 139: 18670–18680.

56 Aurbach, D. (2000). *J. Power Sources* 89: 206–218.

57 Broussely, M., Biensan, P., Bonhomme, F. et al. (2005). *J. Power Sources* 146: 90–96.

58 Komaba, S., Murata, W., Ishikawa, T. et al. (2011). *Adv. Funct. Mater.* 21: 3859–3867.

59 Aurbach, D., Markovsky, B., Weissman, I. et al. (1999). *Electrochim. Acta* 45: 67–86.

60 Aurbach, D., Markovsky, B., Levi, M. et al. (1999). *J. Power Sources* 81–82: 95–111.

61 Etacheri, V., Geiger, U., Gofer, Y.G. et al. (2012). *Langmuir* 28: 6175–6184.

62 Xu, K. (2004). *Chem. Rev.* 104: 4303–4417.

63 Naskar, P., Kundu, D., Maiti, A. et al. (2021). *ChemElectroChem* 8: 1393.

3

Fundamental Understanding of Sodium-Ion Capacitors Mechanism

Peng Cai, Wentao Deng, Hongshuai Hou, Guoqiang Zou, and Xiaobo Ji

Central South University, College of Chemistry and Chemical Engineering, No. 932, Lushan South Road, Yuelu District, Changsha 410083, China

3.1 EDLC-Type Mechanism of SCs and Battery-Type Mechanism of SIBs

Electrochemical double layer capacitor (EDLC) mechanism is generated by the confrontation of charges caused by the directional arrangement of electrons or ions at the electrode/electrolyte interfaces. For an electrode/electrolyte system, EDLC is generated at the interfaces of the electronically conductive electrodes and the ionically conductive electrolyte solutions. As electric forces are imposed on the two electrodes, the cations and anions in the solution migrate to the negative and positive electrodes, respectively, forming EDLC on the surface of the electrodes. After the electric forces are removed, the negative and positive charges on the electrodes are combined with the oppositely charged ions in the solution [1]. At this time, for a certain electrode, there will be a certain distance of anionic charge, which is equal to the charge on the electrode, so it remains electrically neutral. In addition, since the electrode is connected to an external circuit, the charge migration causes the cations and anions in the electrolyte to migrate into the solution, making the solution electrically neutral. This is the mechanism of charging and discharging processes of the EDLC mechanism (Figure 3.1).

In the sodium ion battery (SIBs) systems, the energy storage is generated by electrochemical intercalation/conversion/alloying reactions on the main anodes and intercalation of cathodes, which are typically composed of crystals and micron-sized electrochemically active material particles. In conventional SIBs composed of hard carbon anodes and sodium metal oxide cathodes, the diffusion and/or phase changes of sodium ions in the host lattices limit the achievable charge and discharge rates. This leads to the relatively low specific power density of SIBs (approximately 500–2000 $W\,kg^{-1}$), and their inability to completely absorb or release their charge within tens of minutes, which is desirable for many applications. During the ion-insertion processes, the volume changes are related to the phase change of the electrodes, which further causes the stress of the crystal lattice and limits the cycling stability of the SIBs.

Sodium-Ion Capacitors: Mechanisms, Materials, and Technologies, First Edition.
Edited by Guoqiang Zou, Xiaobo Ji, and Hongshuai Hou.

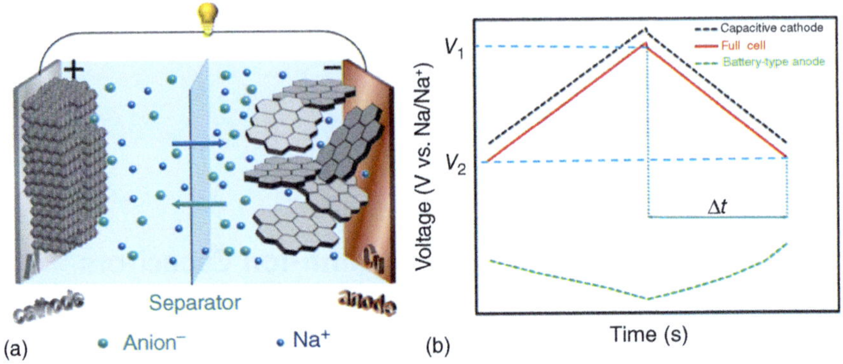

Figure 3.1 (a) Schematic illustration of the typical configuration (cathode: activated carbon, anode: hard carbon) and (b) charge–discharge curves of SICs. Source: Wang et al. [2]. Copyright 2021, Wiley-VCH.

3.2 Pseudocapacitance Mechanism

In order to deliver high energy and high power performance in cathodes and anodes, pseudocapacitive materials have been explored in sodium ion capacitor (SICs) devices. As a result, a thorough understanding of pseudocapacitance is critical to gaining a comprehensive understanding of the fields of SIC devices.

3.2.1 Motivation for the Search for Pseudocapacitance

The motivation for looking for pseudocapacitive electrode materials is to overcome the dynamic limitations of ion intercalation and to achieve higher energy density than charge adsorption. In recent years, in the field of SICs research, more and more work has utilized pseudocapacitive electrode materials to replace EDLC materials (Figure 3.2) [3]. Pseudocapacitive materials could undergo reversible electrochemical reactions and, in some cases, could even be compared with EDLCs, still exhibiting high charge storage capabilities comparable to batteries. The cycling stability of pseudocapacitive materials is also in the middle for SIBs and EDLCs. Generally, anode materials with pseudocapacitive electrochemical characteristics are not restricted by solid-state diffusion on a time scale of a few minutes, and their electrochemical responses are highly reversible, thereby achieving high efficiency at high current densities. A deep understanding of energy storage mechanisms such as pseudocapacitance can further develop an understanding of the kinetics of electrochemical behaviors (Figure 3.3) [4]. The electrochemical characteristics of pseudocapacitive materials are neither purely capacitive processes nor a large number of Faraday processes. Energy storage devices involving pseudocapacitive materials occupy an intermediate position between EDLC-type mechanisms and battery-type mechanisms.

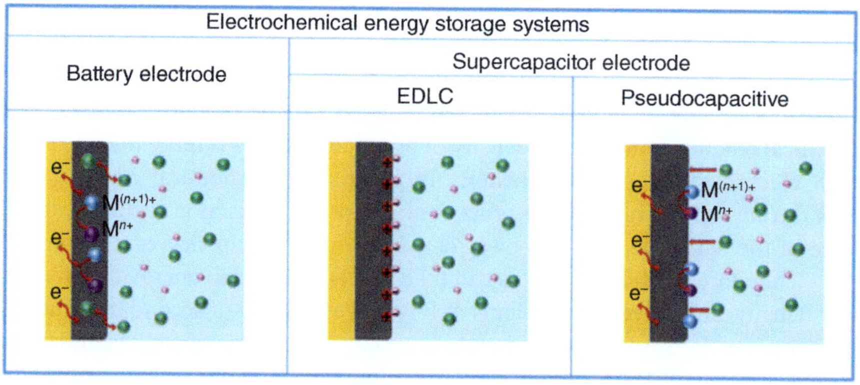

Figure 3.2 Schematic diagram comparing the fundamental charge storage mechanisms in battery-type electrodes and supercapacitor-type electrodes. This schematic diagram explains the idea that pseudocapacitance mechanism materials can be utilized to replace EDLC materials. And further putting forward the focus of this book: A clear understanding of the mechanism in the SIC device. Source: Reproduced with permission Jiang et al. [3]. Copyright 2019, WILEY-VCH.

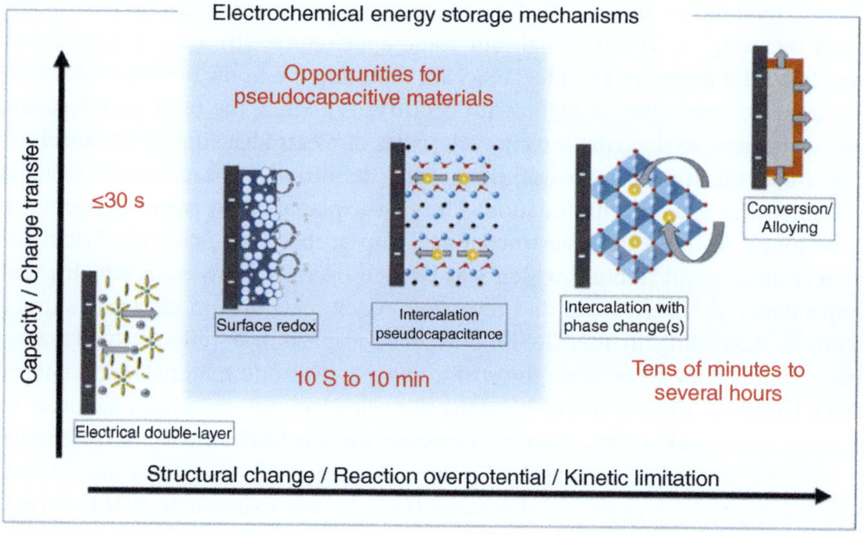

Figure 3.3 Classification of electrochemical energy storage mechanisms as a function of their characteristic capacity and the associated degree of structural changes in the electrodes. Source: Reproduced with permission Fleischmann et al. [4]. Copyright 2020, American Chemical Society.

3.2.2 Definition and Types of Pseudocapacitance

According to the working principle, supercapacitors can be divided into two categories [5, 6]. The first type is to rely on physical adsorption on the electrode surface

EDLCs for desorption energy storage are characterized by extremely high power density, but they also have the problem of low energy density. The second type is called pseudocapacitors, which realize energy storage through fast and reversible redox reactions near the surface, so it is also called Faradic pseudocapacitors. Due to chemical reactions, the energy density of pseudocapacitors is more than 10 times that of electric double-layer capacitors. In addition, the chemical reaction that occurs during the energy storage process of the pseudocapacitor occurs near the surface of the electrode, and its power density is higher than that of a secondary battery. Therefore, pseudocapacitors can take into account both energy density and power density, and solve the problems of low power density of secondary batteries and low energy density of electric double-layer capacitors. It is of great significance to build a multilevel electrochemical energy storage system to meet different needs. Due to their excellent energy storage performance, pseudocapacitors have received extensive attention from scholars at home and abroad in recent years.

The origin of the term "pseudocapacitance" can be found in the relationship between the prefix "pseudo" and "capacitance." In the dictionary, the word "pseudo" (derived from the Greek compound words false) has two meanings. It can be used to describe untrue, false, and deceptive things (for example, pseudoscience, a theory, method, or practice that is considered to have no scientific basis). The second meaning is: (i) not actual, but some appearance; (ii) almost, close to, or trying to be (for example, pseudoclassicism, imitating the usage of classicism in art and literature, prevailing in eighteenth century) [7]. Since the term pseudocapacitance is utilized to describe the characteristics of electrodes similar to capacitors in electrochemical characteristics, the second definition of pseudocapacitance is more correct in the current situation. The first explanation of pseudocapacitance can be found in the book "Electrochemical Supercapacitors: Scientific Principles and Technical Applications," which was written by B.E. Conway, in which pseudocapacitance proposed rapid and reversible redox reactions on electrode surface materials. According to this definition, pseudocapacitance reflects the Faraday process that occurs on or close the surface of the electrode materials, and energy storage based on a redox reaction means that the electrode material's behavior in processes is somewhat like a battery; however, the electrochemical characteristics are reflected in (quasi) rectangular cyclic voltammetry (CV) curves and almost linear galvanostatic charge and discharge (GCD) curves similar to EDLC curves, indicating that the fast Faraday reaction is not limited by solid-state diffusion. However, pseudocapacitance mechanism is different from EDLC mechanism and battery mechanism. EDLC is a non-Faraday mechanism due to charge adsorption, although it is also stored on the electrode surface. Therefore, regardless of whether there is a Faraday reaction, the capacitance and the pseudocapacitance have a clear boundary, which also explains the prefix of the pseudocapacitance. Compared with EDLC materials, pseudocapacitive materials usually have a larger capacitance. Conway listed several Faraday mechanisms that can lead to capacitive electrochemical properties: (i) adsorption pseudocapacitance, (ii) redox pseudocapacitance, and

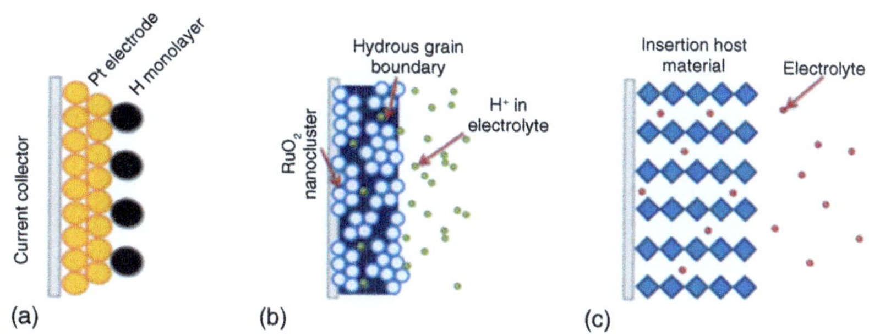

Figure 3.4 Illustration of the three types of mechanisms that give rise to pseudocapacitance, as defined by Conway. (a) Adsorption pseudocapacitance, (b) redox pseudocapacitance, and (c) intercalation pseudocapacitance. Source: Reproduced from Augustyn et al. [7]/Royal Society of Chemistry.

(iii) intercalation pseudocapacitance. These processes are shown in Figure 3.4. It is useful to rediscuss the theory of the first type of pseudocapacitance (adsorption pseudocapacitance) developed by Conway, where the electrosorption of the cation A^+ clarifies the surface of the conductive substrate M [8]:

$$M + A^+ + e^- \leftrightarrow MA_{ads} \tag{3.1}$$

Redox pseudocapacitance occurs when ions are electrochemically adsorbed to or near the surface of the material and accompanied by Faraday charge transfer. When ions are inserted into the tunnel or layer of the redox active material, the insertion pseudocapacitance occurs when the Faraday charge transfer occurs without crystal phase transition (Figure 3.4). The differences between these three mechanisms are due to different physical processes in various types of materials. The similarity in electrochemical characteristics is due to the relationship between the potential and charge ranges generated by the electrode/electrolyte interfaces or the adsorption/desorption processes on the inner surfaces of the materials [7].

$$E \sim E^0 - \frac{RT}{nF}\text{In}\left(\frac{X}{1-X}\right) \tag{3.2}$$

where, E is the potential (V), R is the ideal gas constant (8.314 J mol^{-1} K^{-1}), T is the temperature (K), n is the number of electrons, F is the Faraday constant (96 485 C mol^{-1}), and X is the extent of fractional coverage of the surface or inner structure. Although the above equation describes the thermodynamic feature of pseudocapacitance, the advantage of these types of materials for energy storage lies in their dynamic behavior. This is originated on the basis that the reaction occurs on the surface, is restricted by the surface, or is not affected by solid-state diffusion, and therefore exhibits high rate performance. This is a distinctive difference between materials that exhibit pseudocapacitive behaviors and those that do not exhibit pseudocapacitive behaviors.

3.2.3 Energy Storage Mechanism of Pseudocapacitors

The energy storage mechanism of pseudocapacitors can be divided into underpotential deposition, near-surface redox reactions, and intercalation pseudocapacitances.

3.2.3.1 Adsorption Pseudocapacitance

Adsorption pseudocapacitance (under-potential deposition) means that when the binding force of a certain metal ion (Pt^+, Pd^+, etc.) or hydrogen ion (H^+) in the electrolyte with the heterogeneous noble metal electrode is greater than the binding force of its body, it can be more positive than its own equilibrium potential. Atomic layer is deposited within the range [9]. For example, nanoporous gold (Au) sheets are used to deposit atoms. Taking the cathode as an example, the electrode reaction formula is as follows:

$$E + I^{x+} + xe^- \leftrightarrow EI \tag{3.3}$$

E is the electrode, I^{x+} is the cation in the solution, and EI is the product of I deposited on E. However, the potential range of underpotential deposition is small (0.3–0.6 V) and concentrated on the surface of the electrode material, so the capacity is limited, and it is less used for pseudocapacitor energy storage.

3.2.3.2 Redox Pseudocapacitance

The redox reaction mechanism refers to the electrochemical reaction of ions near the surface of the electrode. Typical pseudocapacitance active materials that follow this reaction mechanism mainly include conductive polymers [10]. Taking conductive polymers as an example, the charge storage and release process in the energy storage process can be expressed by the following formula:

$$\text{conductive polymers(CPs)} + ne^- + nX^+ \leftrightarrow CPsX_n \tag{3.4}$$

Among them, CPs stands for conductive polymers, and X^+ stands for cations (H^+, Na^+, K^+, etc.) in the electrolyte.

3.2.3.3 Intercalation Pseudocapacitance

With the development of electrode materials, researchers have discovered a kind of pseudocapacitance based on a new energy storage mechanism, that is, intercalation pseudocapacitance. It is through the continuous insertion and release of ions in the two-dimensional, layered, or tunnel-shaped electrode material that charge storage is achieved [11]. The corresponding charge storage process can be expressed by the following formula:

$$M_xA_y + aL_i{}^+ + ze \leftrightarrow L_{ia}M_xA_y \tag{3.5}$$

A typical example is a pseudocapacitive energy storage system composed of MXene.

The electrode potential and the number of charge transfers in the pseudocapacitance reaction process of the above three energy storage mechanisms all satisfy the following relationship:

In addition, the capacitance C can also be calculated based on X and E:

$$C = \frac{nF}{m} \times \frac{\mathrm{X}}{\mathrm{E}} \tag{3.6}$$

where m is the molar mass of the active material. Since X and E in formula (3.6) do not belong to a linear relationship, the capacitance value of the pseudocapacitance active material is not a fixed value. Its fast and reversible reaction process makes the current during charging and discharging remain constant, which is very different from secondary batteries. This is also the main reason for being called pseudocapacitance [12]. However, in the actual oxidation–reduction process, the reaction may be delayed and uneven due to the diffusion of products or reactants, and the current will fluctuate to a certain extent.

3.2.4 Pseudocapacitance Electrode Materials

Common pseudocapacitance electrode materials in SICs mainly include conductive polymers and MXene electrode materials, etc. [7, 13–63]. In this section, the basic pseudocapacitance characteristics of these materials will be introduced so that it is easy to understand why these materials exhibit pseudocapacitance in SICs. The specific applications and examples of these materials in the SICs system will be introduced in Chapters 4–6. Subsequently, some materials that exhibit pseudocapacitive characteristics in other systems but have shifted concepts in the SICs system will also be introduced for comparison. Note that these materials (pseudocapacitive materials in other systems for comparison) could be served as good insertion/extraction anode materials. The purpose of separate comparison of these materials is that these materials in the SICs system are different from the previous classification of pseudocapacitive mechanisms, and there is no complete related work and specific examples in SICs. And the concept of pseudocapacitance discussed here is more fully expounded from the perspective of supplementing the AC cathode (derived from supercapacitors), and the concepts should be clearly distinguished.

3.2.4.1 Conductive Polymer

Conductive polymer is another type of typical pseudocapacitance active material, which has the advantages of high theoretical specific capacitance, low cost, simple preparation, light weight, adjustable and controllable polymer skeleton structure, flexibility, and wide operating voltage window. Therefore, pseudocapacitors using them as electrode active materials have the potential for large-scale applications [56, 57]. Normally, the main chain of the conductive polymer contains alternating single and double bonds to form a large conjugated system. Only when the degree of conjugation is high do the p orbitals in the conjugation plane overlap each other, enhancing delocalization and allowing electrons to move freely. In order to improve the conductivity of conductive polymers, the energy level difference between the conduction band (antibonding orbital π^*) and the valence band (bonding orbital π) can be reduced by doping. It mainly includes two doping methods. One is doping with oxidizing substances to reduce the number of electrons in the valence band, that is, p-type doping; the other is doping with reducing substances to increase the number of electrons in the conduction band. That is, n-type doping. The conductivity of the doped conductive polymer can reach 10^2–10^3 S cm^{-1} [57]. In addition to doping, conductive polymers can also be quickly and reversibly dedoped. Therefore, when it is used as the active material of the pseudocapacitance electrode, the rapid and reversible p-type or n-type doping and the corresponding dedoping redox reaction mainly occur in the polymer.

Common conductive polymers used in pseudocapacitance electrodes are polyaniline (PANI), polypyrrole (PPy), polythiophene (PTh), and related derivatives. Gawli et al. [58] used PANI doped with hydrochloric acid (HCl) and weak organic acid phytic acid as an electrode, and achieved a specific capacitance of 350 F g^{-1} (charged in H_2SO_4 electrolyte with a concentration of 1 mol l^{-1}). The discharge current density is 1 A g^{-1}). Zhang and coworkers [59] used cetyltrimethylammonium bromide (CTAB) as a template to prepare PPy nanofibers with a diameter in the range of 30–50 nm. And use this as an electrode to assemble a super capacitor with potassium chloride (KCl) electrolyte with a concentration of 1 mol l^{-1}. The electrochemical performance test shows that the specific capacitance of the electrode is 282 F g^{-1} and the corresponding energy density is 31.8 Wh kg^{-1} at a charge–discharge current density of 1 A g^{-1}. This value is 30% higher than the pseudocapacitance using bulk PPy as the electrode active material, indicating that the nanofiber structure can effectively improve the utilization rate of the pseudocapacitance active material. In addition, Subramania and coworkers used cationic surfactant-assisted dilute polymerization to synthesize PTh nanoparticles and doped them with tartaric acid [60]. The above electrode materials are used as electrodes, and a gel polymer electrolyte based on lithium hexafluorophosphate ($LiPF_6$) and polyvinylidene fluoride-hexafluoropropylene (PVDF-HFP) is used to assemble a super capacitor. The electrochemical energy storage performance test results show that the CV test scan rate is 5 mV s^{-1} with a specific capacitance of 156 F g^{-1}, and the corresponding energy density and power density are 14 Wh kg^{-1} and 522 W kg^{-1}, respectively.

Although a series of studies have been carried out on conductive polymers as active materials for pseudocapacitance electrodes, there are still many problems. First of all, the high conductivity of the conductive polymer can only be achieved when the high degree of doping is completed during the charge and discharge process, and the conductivity of the initial charge and discharge state is poor. Second, when the conductive polymer is directly used as a pseudocapacitor electrode, it is difficult to obtain an effectively dispersed microstructure, resulting in low material utilization and a large gap between the theoretical specific capacitance value. In addition, the repeated insertion and extraction of electrolyte ions will cause the mechanical properties of the conductive polymer to attenuate and seriously damage its cycle life.

3.2.4.2 MXene

In recent years, with the continuous development of electrode materials, new types of pseudocapacitance electrode materials have continuously emerged. MXene is a two-dimensional material with a structure similar to graphene. It realizes charge storage through the intercalation reaction of electrolyte ions. The first MXene developed is $Ti_3C_2T_x$, which has the advantages of good electrical conductivity, high theoretical specific capacitance, and high chemical stability. In addition, the surface of MXene has a large number of functional groups, which is conducive to the electrolyte infiltration into the inside of the electrode, thereby reducing ion transmission resistance. The energy storage performance can be effectively improved by modifying MXene. For example, Gogotsi and coworkers [61] embedded a gel polymer electrolyte based on potassium hydroxide (KOH) and polyvinyl alcohol (PVA) between the $Ti_3C_2T_x$ layers, which effectively improved the flexibility and mechanical properties of the electrode material and also promoted the insertion of cations between the electrode layers. The electrochemical energy storage performance test was performed in a KOH electrolyte with a concentration of $1\,mol\,l^{-1}$, and it was found that the specific capacitance of the electrode was $530\,F\,cm^{-3}$ when the scan rate of the cyclic test was $2\,mV\,s^{-1}$. This value is 51.4% higher than that of $Ti_3C_2T_x$ electrode without KOH/PVA embedded.

3.2.4.3 Pseudocapacitive Materials for Comparison

Transition Metal Oxide In 1971, Trasatti and Buzzanca first discovered the pseudocapacitance energy storage behavior of ruthenium dioxide (RuO_2) [14]. RuO_2 has good conductivity (10^4–$10^5\,S\,cm^{-1}$) [15–17], high theoretical specific capacitance ($\sim$2000 $F\,g^{-1}$) [18], good reversibility, cycle stability, chemical stability, and other advantages. It is an efficient pseudocapacitance electrode material, which has been used in national defense and military fields. The RuO_2 pseudocapacitance reaction is mainly carried out in an acidic environment, that is, H^+ undergoes a fast and reversible redox reaction on the surface of the electrode material. Studies have shown that when ruthenium dioxide hydrate ($RuO_2{\cdot}\sigma H_2O$) is used as an electrode, the corresponding pseudocapacitance energy storage performance is better, because $RuO_2{\cdot}\sigma H_2O$ can provide a fast transmission channel for H^+ [19, 20]. This process can be expressed by the following formula:

$$RuO_x(OH)_y + ne^- + nH^+ \leftrightarrow RuO_{x-n}(OH)_{y+n} \tag{3.7}$$

Rochefort and Pont pyrolyzed ruthenium chloride ($RuCl_3$) in air to directly form a RuO_2 film on a titanium substrate and tested the electrochemical energy storage performance [21]. The results show that when proton-type ionic liquid is used as the electrolyte and the scan rate of the CV test is $20\,mV\,s^{-1}$, the specific capacitance of the electrode is $83\,F\,g^{-1}$. Susanti et al. combined plasma vapor deposition, oxidation, partial pyrolysis, and electrochemical deposition.

The combined method successfully prepared $RuO_2{\cdot}0.46H_2O$ nanorods on the surface of lithium niobate ($LiNbO_3$) [22]. The schematic diagram of the reaction mechanism shows that when the above-mentioned materials are used as electrodes, electrons reach the surface of RuO_2 nanorods from the substrate, and at the same time, H^+ reaches the near surface of the electrode through the gap between the nanorods to complete the redox reaction and achieve energy storage. The electrochemical energy storage performance test showed that when $1\,mol\,l^{-1}$ sulfuric acid (H_2SO_4) was used as the electrolyte and the scan rate of the CV test was $5\,mV\,s^{-1}$, the specific capacitance of the electrode was as high as $520\,F\,g^{-1}$. However, metal ruthenium is expensive (up to ~14 000 RMB kg^{-1}), toxic, and easily causes environmental pollution [23]. Therefore, it is difficult to promote pseudocapacitors with RuO_2 as the active material except in the defense and military fields. For this reason, scholars at home and abroad are committed to researching low-cost pseudocapacitance active materials other than RuO_2, such as manganese dioxide (MnO_2), nickel oxide (NiO), cobalt tetroxide (Co_3O_4), niobium pentoxide (Nb_2O_5), cobalt hydroxide ($Co(OH)_2$), and nickel hydroxide ($Ni(OH)_2$).

Among them, MnO_2 has attracted wide attention from researchers due to its advantages such as abundant content in nature, low cost, nontoxicity, and high theoretical specific capacitance (1370 F g^{-1}) [24–28]. The specific process takes the insertion and extraction of H^+ as an example: when electrons reach the surface of MnO_2, Mn^{4+} is reduced to Mn^{3+}, and at the same time, H^+ in the electrolyte enters the surface of the MnO_2 lattice and combines with O^{2-} in it to form OH^-. MnO_2 is transformed into MnOOH without phase change [29].

Therefore, the energy storage process when MnO_2 is used as a pseudocapacitance active electrode material can be expressed by the following formula [30, 31]:

$$M_nO_2 + e^- + X^+ \leftrightarrow MOOX \tag{3.8}$$

where X^+ represents the cation in the electrolyte, such as H^+, Li^+, Na^+, K^+, and so on. The CV curve of the MnO_2 electrode with 0.1 mol l^{-1} K_2SO_4 as the electrolyte. The red curve in the upper half is charging, and the blue curve in the lower half is discharging. It shows that the reversible conversion between Mn^{4+} and Mn^{3+} needs to undergo successive reversible redox reactions at different potentials [10].

In the actual charging and discharging process, a broad peak may appear on the CV curve due to the hysteresis of the reaction. Hu et al. successfully prepared MnO_2 nanowires, nanoflowers, and nanorod structures using hydrothermal methods, and used them as pseudocapacitance electrodes to carry out energy storage performance tests. They found that sodium sulfate (Na_2SO_4) with a concentration of 0.5 mol l^{-1} was used as the electrolyte, and the MnO_2 nanowires with the smallest geometric size showed the best electrochemical performance [32]. The main reason is that the finer nanostructure is conducive to full contact between the electrode active material and the electrolyte, which effectively improves the electrode activity and material utilization rate. Specifically, the specific capacitance is 345 F g^{-1} under the current density of 1 A g^{-1}, and the capacitance retention rate is as high as 54.7% when the current density is increased to 10 A g^{-1}. Srivastava and coworkers also synthesized MnO_2 nanohair with a spherical structure by hydrothermal method [33]. When Na_2SO_4 with a concentration of 1 mol l^{-1} is used as the electrolyte, the charge and discharge When the test current density is 0.5 A g^{-1}, the electrode achieves a specific capacitance of 272 F g^{-1}. Zhang and Zhang used the redox method to prepare MnO_2 nanoparticles through potassium permanganate ($KMnO_4$) and manganese acetate ($Mn(Ac)_2$), and used 1 mol l^{-1} KOH as electrolyte [34]. The specific capacitance of 314 F g^{-1} is obtained under the conditions of the charge and discharge test current density of 10 mA cm^{-2}. The capacitance retention rate of the electrode after 5000 cycles is 90%. In summary, the specific capacitance obtained by using MnO_2 directly as a pseudocapacitance electrode still has a large gap with the theoretical value. The reason is that MnO_2 itself has poor conductivity and is easy to agglomerate when used directly as an electrode, which leads to the utilization of low-active materials.

Transition Metal Nitride Although transition metal oxides have higher theoretical capacitance values (for example, the theoretical specific capacitance of MnO_2 [23] is ~1370 F g^{-1}), most of them have the disadvantages of poor conductivity and stability (for example, the intrinsic properties of pure MnO_2). The low conductivity is only 10^{-3}–10^{-7} S cm^{-1} [31, 35]. Therefore, in recent years, pseudocapacitors using transition metal nitrides as active materials have received widespread attention due to their high conductivity and high chemical stability [36–40].

Transition metal nitrides have metal bonds, covalent bonds, and ionic bonds at the same time. Among them, the nitrogen atom p-orbital energy level is slightly lower than the Fermi energy level, and the electron movement is relatively free, so the transition metal nitride has good conductivity (10^3–10^4 S cm^{-1}), even better than some metals [36–40]. In addition, transition metal nitrides also have good corrosion resistance and generally do not react with water, HCl, etc. [41]. Common transition metal nitrides include titanium nitride (TiN), chromium nitride (CrN), vanadium nitride (VN), molybdenum nitride (MoN), and tungsten nitride (W_2N), etc. [42–45].

Among them, the electrical conductivity of TiN is as high as 4000–5500 S cm^{-1} [46, 47], so as a pseudocapacitance active material, it can simultaneously achieve higher energy density and power density [48]. The Brousse research group found that the N element in TiN can diffuse into the surface oxide layer (TiO_2) to form oxygen holes, which can provide space for ion insertion and are active sites for redox reactions [50]. The research group used DC magnetron sputtering to directly grow TiN nanofilms on silicon wafers and annealed them at 550 °C for surface activation [51]. The electrochemical energy storage performance test is carried out in a potassium sulfate (K_2SO_4) electrolyte with a concentration of 0.5 mol l^{-1}. When the scan rate of the CV test is 2 mV s^{-1}, the specific capacitance is 8.2 mF cm^{-2}. Hector et al. used titanium (Ti) foil to react with ammonia (NH_3) to prepare TiN nanofilms, and then activated them by electrochemical methods [52]. When KOH with a concentration of 1 mol l^{-1} is used as the electrolyte and the scan rate of the CV test is 1 mV s^{-1}, a specific capacitance of 11.1 mF cm^{-2} is obtained.

In addition to the above pseudocapacitors using TiN as the electrode active material, Bondarchuk et al. studied the pseudocapacitor energy storage performance of VN films. They first used the DC magnetron sputtering method to grow a V film on a silicon wafer, and then achieved the nitridation of the V film in a high-temperature and high-pressure nitrogen (N_2) environment to obtain a VN film [53]. When alkaline KOH or lithium hydroxide (LiOH) with a concentration of 1 mol l^{-1} is used as the electrolyte and the scan rate of the CV test is 3 mV s^{-1}, a specific capacitance of 3 mF cm^{-2} can be obtained. Wang and coworkers also used DC magnetron sputtering to grow CrN films on silicon wafers, and assembled supercapacitors with sulfuric acid (H_2SO_4) with a concentration of 0.5 mol l^{-1} as acid electrolyte [54]. The electrochemical test shows that a specific capacitance of 6.5 mF cm^{-2} is achieved under the condition of a charge–discharge test current density of 1 mA cm^{-2}. The electrode also has good cycle stability, and the capacitance retention rate is as high as 90% after 20 000 cycles of charging and discharging (current density is $2\,mA\,cm^{-2}$). In addition, in order to further improve the specific capacitance of transition metal nitrides, the Brousse research group used synchronous DC magnetic sputtering of Ti and V to form TiVN films on silicon substrates [55]. The electrode and KOH electrolyte with a concentration of $1\,mol\,l^{-1}$ were assembled into a supercapacitor and the electrochemical performance test was carried out. The results showed that the specific capacitance was $15\,mF\,cm^{-2}$ when the scan rate of the CV test was $3\,mV\,s^{-1}$.

Transition Metal Sulfide Common ones are molybdenum disulfide (MoS_2), tungsten disulfide (WS_2), and titanium disulfide (TiS_2). The most representative one is MoS_2, which has a hexagonal crystal layered structure with a layer spacing of about 0.6–0.7 nm. The single-layer structure unit of MoS_2 includes three layers of atoms, with sulfur atoms on both sides and a layer of molybdenum atoms in the middle, forming a sandwich stack. Zhu's research group prepared a metal-phase multilayer hydrated MoS_2 film and tested it in different electrolytes [62]. The results showed that the highest specific capacitance can be obtained when the ion size is close to the interlayer spacing of the MoS_2 film. Assemble the electrode and lithium sulfate (Li_2SO_4) electrolyte with a concentration of $1\,mol\,l^{-1}$ to form a pseudosupercapacitor. The specific capacitance was measured under the condition of a CV scan rate of $50\,mV\,s^{-1}$ is $249\,F\,g^{-1}$.

3.2.5 Evolution of Pseudocapacitance

In this section, the pseudocapacitance of the anode will be fully discussed. Compared with traditional bulk battery materials, nanomaterial electrodes exhibit a larger specific surface area, and the ion diffusion length is significantly shortened. The "bulk" redox reaction may be transformed into a "surface/near surface" level redox reaction. Therefore, some battery materials show characteristics similar to pseudocapacitance in the CV and GCD curves, showing significantly accelerated redox reaction kinetics. In the past, the most representative material was $LiCoO_2$, which is different from traditional pseudocapacitive materials [62]. When their size reaches a critical nanolevel, their battery characteristics will disappear, showing an approximate linear GCD feature. In order to distinguish this type of electrode material from traditional pseudocapacitive materials, Dunn et al. put forward their point that materials like $LiCoO_2$ can be regarded as having "extrinsic pseudo-capacitance." That is, when they are in the bulk phases, they show the behavior of the battery material, but as the size of the electrode decreases, the pseudocapacitance behavior gradually shows up. This characteristic is considered to be the influence of the nanoeffect. In addition to $LiCoO_2$, some intercalation materials have been reported to have a rate performance close to that of traditional pseudocapacitive materials; the redox peak is broad in most cases, and the potential polarization is almost unaffected by the CV scan rates. In the early years, Conway clearly summarized this intercalation process as a transitional behavior between batteries and supercapacitors (SCs). Later, Dunn and Simon accordingly defined it as "intercalation pseudocapacitance." In order to better understand "intercalation pseudocapacitance", Dunn et al. proposed a normalized formula for kinetics analysis based on CV curves [63]:

$$i(V) = k_1 v + k_2 v^{1/2} \tag{3.9}$$

That is: $i(V) = av^b$. After calculation, if the value of $b \approx 1$, it can be quantitatively determined that the electrochemical process in these "intercalation pseudocapacitance" materials are indeed not controlled by diffusion, which is very similar to traditional pseudocapacitance materials. However, this calculation method cannot be utilized alone and directly as a basis for judging pseudocapacitance and materials. And this method is not necessarily suitable for all materials and situations. In recent years, this method of quantitative analysis of kinetics has been widely utilized in SIBs materials and even some battery materials with conversion reaction mechanisms and alloying reaction mechanisms. In these cases, researchers only make this calculation and directly explain this contribution as "pseudocapacitance contribution." It was found that a considerable number of configurations assembled by battery-type cathodes and conversion/alloying-type anodes (small size and fast ions transfer) are mainly achieved through so-called "pseudocapacitive SIBs devices," which is much puzzling. Under this circumstance, the boundaries between "battery materials" and "pseudocapacitance materials" have become blurred, and researchers have become increasingly confused about their concepts. In fact, it is not difficult to find that these reported battery electrodes with significant "pseudocapacitance" contributions are almost all nanomaterials or nanoscale thin-film electrode materials. As early as 30 years ago, Conway proposed that the method of analyzing CV curves can be utilized to understand the kinetic characteristics of battery electrode reactions, and $b \approx 1$ is understood as "surface processes." Therefore, it is easy to understand why the charge storage processes of many current nanomaterial electrodes are hardly controlled by diffusion (small size and surface effects). In other words, these are all rapid redox battery reaction processes that are not controlled by diffusion due to the nanosize effects. In short, for pseudocapacitance materials, $b \approx 1$; when $b \approx 1$, it is not necessarily pseudocapacitance effects, it can be EDLC effects, or it may be battery behaviors under the nanoeffects. Because of this, Brousse and Long et al. strongly suggest that pseudocapacitance materials should be determined by considering the traditional definition of pseudocapacitance and basic electrochemical characteristics [63]. $C = dQ/dV$, this result should be constant in a wide potential window, and the unit is Faraday (F). However, when the electrode materials exhibit battery behaviors, such as the obvious plateau of the GCD curves or the obvious redox peaks of the CV curves, the ratio of dQ/dV is not constant. In this case, the capacity should be coulombic (C) or milliampere hour (mAh) as the unit.

3.2.6 Electrochemical Features of Pseudocapacitance

It is not enough to distinguish between battery materials and pseudocapacitance materials by CV kinetic analysis alone. Therefore, a variety of analysis methods can be considered. The correct use of kinetic analysis also helps to understand the relationship between the "structure and rate" of certain electrode materials. In order to accurately distinguish between pseudocapacitance materials and battery materials, a standard was proposed, that is, electrochemical characteristics (CV, GCD, and EIS curves) and quantitative kinetic analysis (calculation of b value) should be considered at the same time. The CV curve of EDLC material is rectangular, and the GCD curve is straight. For pseudocapacitance materials, the CV curve is approximately rectangular, and the corresponding GCD curve is approximately linear (compared with EDLC, the GCD curve can have an inflection point, but there is no obvious platform) [7]. The CV curve of battery materials shows obvious redox peaks, and the GCD curve provides a clear platform. In order to use this criterion correctly, it needs to be emphasized that all the electrochemical characteristics in the CV and GCD curves should be uniformly displayed over a wide range of scan rate and current density.

For pseudocapacitance materials, these electrochemical characteristics do not vary with the size of the material. When using the Dunn method to calculate the b value, it is recommended: (i) Use the CV curve to calculate the b value at a low scan rate. Especially for many electrodes mixed with a large amount of carbon materials, the proportion of carbon materials will be very significant at a high scan rate. At this time, $b \approx 1$ may not reflect the main material at all, so you need to be cautious; (ii) Correctly understand the meaning of $b \approx 1$: $b \approx 1$ means "surface" process, which can be a capacitor process (EDLC or pseudocapacitance) or it is due to nanosize effect/surface technology and special crystal structure [8]. And you cannot simply equate the capacitive process with $b \approx 1$ (Figure 3.5). This may mean that the redox

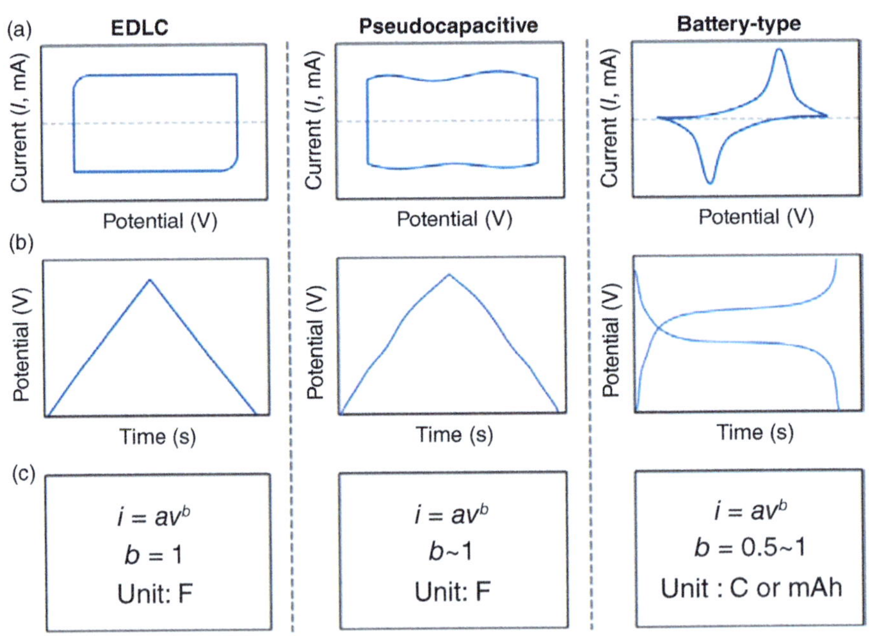

Figure 3.5 Criterion for distinguishing (a) CV, (b) GCD curves, and (c) features of EDLC, pseudocapacitive, and battery materials. Source: Reproduced with permission Jiang et al. [3]. Copyright 2019, WILEY-VCH.

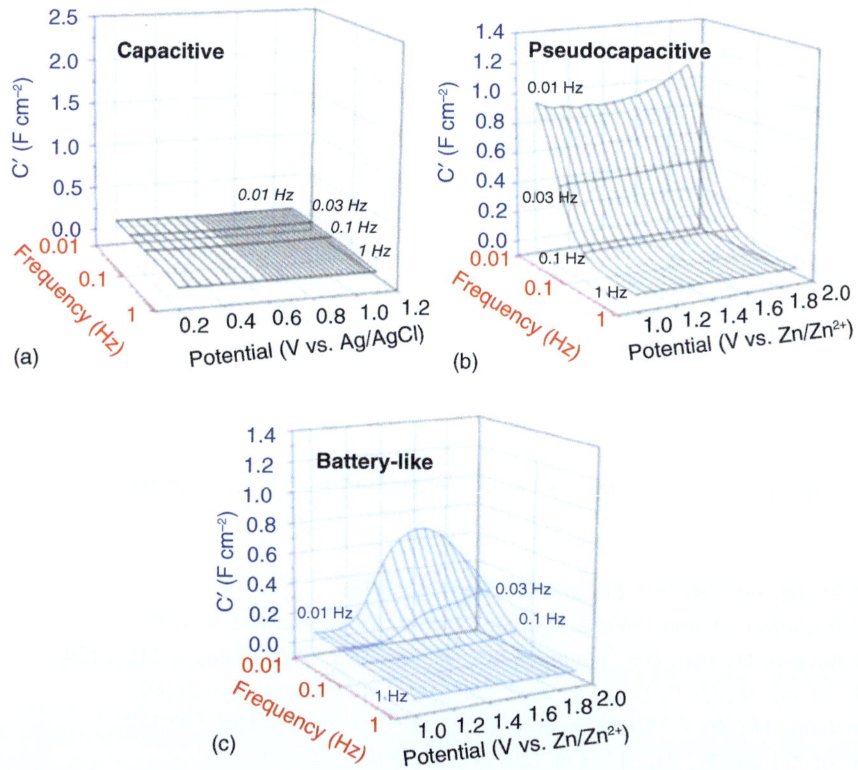

Figure 3.6 Criterion of three dimensional Bode plots for three types of electrodes: C' vs. frequency vs. potential for (a) capacitive materials. Source: Reproduced with permission [4a]. Copyright 2018, ELSEVIER. (b) Pseudocapacitive materials and (c) battery-type materials. Source: Reproduced with permission [4b]. Copyright 2018, RSC.

reaction kinetics of the battery are as fast as the pseudocapacitive material. As shown in Figure 3.6, the actual capacitance, frequency, capacitance, pseudocapacitance, and potential of battery electrodes are normalized by a three-dimensional Bode diagram. Compared with the graph, the *C* value of capacitors and pseudocapacitors is almost constant over a wide potential range at low frequencies. The comparison of capacitance and pseudocapacitance diagrams shows that the latter is more frequency-dependent and has a longer electrical inductance.

References

1 Abruna, H.D., Kiya, Y., and Henderson, J.C. (2008). *Phys. Today* 61: 43.
2 Wang, B., Gao, X., Xu, L. et al. (2021). *Batteries & Supercaps* 4: 538–553.
3 Jiang, Y.Q. and Liu, J. (2019). *Energy Environ. Mater.* 2: 30.
4 (a) Ko, J.S., Sassin, M.B., Rolison, D.R., and Long, J.W. (2018). *Electrochim. Acta* 275: 225–235. (b) Ko, J.S., Sassin, M.B., Parker, J.F. et al. (2018). *Sustainable Energy Fuels* 2: 626–636.

5 Frackowiak, E. and Beguin, F. (2001). *Carbon* 39: 937.
6 Yang, Z.F., Tian, J.R., Yin, Z.F. et al. (2019). *Carbon* 141: 467.
7 Augustyn, V., Simon, P., and Dunn, B. (2014). *Energy Environ. Sci.* 7: 1597.
8 Fleischmann, S., Mitchell, J.B., Wang, R.C. et al. (2020). *Chem. Rev.* 120: 6738.
9 Herrero, E., Buller, L.J., and Abruña, H.D. (1897). *Chem. Rev.* 2001: 101.
10 Simon, P. and Gogotsi, Y. (2008). *Nat. Mater.* 7: 845.
11 Mu, X.P., Wang, D.S., Du, F. et al. (2019). *Adv. Funct. Mater.* 29: 9.
12 Shao, Y., El-Kady, M.F., Sun, J. et al. (2018). *Chem. Rev.* 118: 9233.
13 Jiang, Y. and Liu, J. (2019). *Adv. Funct. Mater.* 2: 30.
14 Trasatti, S. and Buzzanca, G. (1971). *J. Electroanal. Chem.* 29: A1.
15 Chen, L.Y., Hou, Y., Kang, J.L. et al. (2013). *Chem. Mater.* 3: 851.
16 Raming, T.P., Van Zyl, W.E., and Verweij, H. (2001). *Chem. Mater.* 13: 284.
17 Yoshida, N., Yamada, Y., Nishimura, S.I. et al. (2013). *J. Phys. Chem. C* 117: 12003.
18 Soin, N., Roy, S.S., Mitra, S.K. et al. (2012). *J. Mater. Chem* 22: 14944.
19 Dmowski, W., Egami, T., Swider-Lyons, K.E. et al. (2002). *J. Phys. Chem B* 106: 12677.
20 Zheng, J.P. (1995). *J. Electrochem. Soc.* 142: 2699.
21 Rochefort, D. and Pont, A.L. (2006). *Electrochem. Commun.* 8: 1539.
22 Susanti, D., Tsai, D.S., Huang, Y.S. et al. (2007). *J. Phys. Chem. C* 111: 9530.
23 Huang, M., Li, F., Dong, F. et al. (2015). *J. Mater. Chem. A* 3: 21380.
24 Wang, H., Xu, C., Chen, Y. et al. (2017). *Energy Storage Mater.* 8: 127.
25 Yu, G., Xie, X., Pan, L. et al. (2013). *Nano Energy* 2: 213.
26 Parida, K.M., Kanungo, S.B., and Sant, B.R. (1981). *Electrochim. Acta* 26: 435.
27 Toupin, M., Brousse, T., and Bélanger, D. (2004). *Chem. Mater.* 16: 3184.
28 Xu, M., Kong, L., Zhou, W. et al. (2007). *J. Phys. Chem. C* 111: 19141.
29 Arnott, J. and Donne, S. (2007). *J. Electrochem. Soc.* 154: A776.
30 Ragupathy, P., Vasan, H.N., and Munichandraiah, N. (2008). *J. Electrochem. Soc.* 155: A34.
31 Ghodbane, O., Pascal, J.L., and Favier, F. (2009). *ACS Appl. Mater. Interfaces* 1: 1130.
32 Li, W., Liu, Q., Sun, Y. et al. (2012). *J. Mater. Chem.* 22: 14864.
33 Maqbool, Q., Singh, C., Jash, P. et al. (2017). *Chem. Eur. J.* 23: 4216.
34 Zhang, B.H. and Zhang, N. (2003). *Acta Phys. Chim. Sin.* 19: 286.
35 Yu, G., Hu, L., Liu, N. et al. (2011). *Nano Lett.* 11: 4438.
36 Yick, S., Murdock, A.T., Martin, P.J. et al. (2018). *Nanoscale* 10: 7566.
37 Lu, P., Ohlckers, P., Müller, L. et al. (2016). *Electrochem. Commun.* 70: 51.
38 Pan, Z., Xiao, Y., Fu, Z. et al. (2014). *J. Mater. Chem. A* 2: 13966.
39 Yue, Y.H., Han, P.X., Dong, S.M. et al. (2012). *Chin. Sci. Bull.* 57: 4111.
40 Dong, Y.Z. and Liu, Y. (2014). *Nat. Sci. Hei-Univ.* 31: 490.
41 Balogun, M.S., Huang, Y.C., Qiu, W.T. et al. (2017). *Mater. Today* 20: 425.
42 Paldey, S. and Deevi, S.C. (2003). *Mater. Sci.* 342: 58.
43 Zhong, Y., Xia, X.H., Shi, F. et al. (2016). *Adv. Sci.* 3: 1500286.
44 Dong, S.M., Chen, X., Zhang, X.Y. et al. (2013). *Coord. Chem. Rev.* 257: 1946.

45 Theerthagiri, J., Durai, G., Karuppasamy, K. et al. (2018). *J. Ind. Eng. Chem.* 67: 12.

46 Raimondi, M.T. and Pietrabissa, R. (2000). *Biomaterials* 21: 907.

47 Yang, P., Chao, D., Zhu, C. et al. (2016). *Adv. Sci.* 3: 1500299.

48 Zhu, C., Yang, P., Chao, D. et al. (2015). *Adv. Mater.* 27: 4566.

49 Achour, A., Porto, R.L., Soussou, M.A. et al. (2015). *J. Power Sources* 300: 525.

50 Achour, A., Ducros, J.B., Porto, R.L. et al. (2014). *Nano Energy* 7: 104.

51 Achour, A., Chaker, M., Achour, H. et al. (2017). *J. Power Sources* 359: 349.

52 Gray, B.M., Hector, A.L., Jura, M. et al. (2017). *J. Mater. Chem. A* 5: 4550.

53 Bondarchuk, O., Morel, A., Bélanger, D. et al. (2016). *J. Power Sources* 324: 439.

54 Wei, B., Liang, H., Zhang, D. et al. (2017). *J. Mater. Chem. A* 5: 2844.

55 Achour, A., Porto, R.L., Chaker, M. et al. (2017). *Electrochem. Commun.* 77: 40.

56 Shown, I., Ganguly, A., Chen, L.C. et al. (2015). *Nano Energy* 3: 2.

57 Bharti, M., Singh, A., Samanta, S. et al. (2018). *Prog. Mater. Sci.* 93: 270.

58 Gawli, Y., Banerjee, A., Dhakras, D. et al. (2016). *Sci. Rep.* 6: 21002.

59 Wu, Q.F., He, K.X., Mi, H.Y. et al. (2007). *Mater. Chem. Phys.* 101: 367.

60 Gnanakan, S.R.P., Murugananthem, N., and Subramania, A. (2011). *Polym. Adv. Technol.* 22: 788.

61 Ling, Z., Ren, C.E., Zhao, M.Q. et al. (2014). *Proc. Natl. Acad. Sci. U.S.A.* 111: 16676.

62 Geng, X., Zhang, Y., Han, Y. et al. (2017). *Nano Lett.* 17: 1825.

63 Brousse, T., Bélanger, D., and Long, J.W. (2015). To be or not to be pseudocapacitive? *J. Electrochem. Soc.* 162 (5): A5185–A5189.

4

Classification of Sodium-Ion Capacitors Cell Configurations

Peng Cai, Wentao Deng, Guoqiang Zou, Hongshuai Hou, and Xiaobo Ji

Central South University, College of Chemistry and Chemical Engineering, No. 932, Lushan South Road, Yuelu District, Changsha 410083, China

When materials with different energy storage mechanisms are assembled into the same devices, the concepts of asymmetric capacitors or hybrid capacitors are often misused to describe them. Asymmetric SCs cover a wider range of electrode combinations because they can be assembled for SCs that utilize electrodes with the same properties but different mass loads or utilize two electrodes with different materials. Due to the different materials of the electrodes, when pairing two electrodes with different charge characteristics (a capacitor electrode and a battery electrode), the term "hybrid capacitors" should be utilized, and the electrochemical performance of the resulting devices is between SCs and SIBs. Researchers should pay attention to the differences between the two to avoid confusion (Figure 4.1) [1].

4.1 Battery-Type Anode and EDLC Cathode SICs Cell Configurations

In most cases, the configurations of hybrid capacitors generally include a wider range of battery-type anodes and capacitive cathode materials. For example, battery-type anode materials generally include some intercalation mechanism, conversion reaction mechanism, and alloying mechanism materials. For the purpose of comprehending the battery-type anode materials of SIC configurations, the differences and relationships of battery-type anode materials need to be clarified. In this configuration, the specific characteristics of some intercalation mechanisms, conversion reaction mechanisms, and alloying mechanism materials, as well as the electrochemical performance of the half-cell and the assembled SICs devices, are fully discussed (Figure 4.2a) [2]. Dual-carbon configurations are typical battery-type anode and EDLC cathode SICs cell configurations of hybrid capacitors. Dual-carbon configurations are also called classical configurations in some literatures. As is known to all, graphite shows a very low capacity in the sodium storage system, so it is necessary to choose some better battery-type anodes that can replace graphite.

Sodium-Ion Capacitors: Mechanisms, Materials, and Technologies, First Edition.
Edited by Guoqiang Zou, Xiaobo Ji, and Hongshuai Hou.

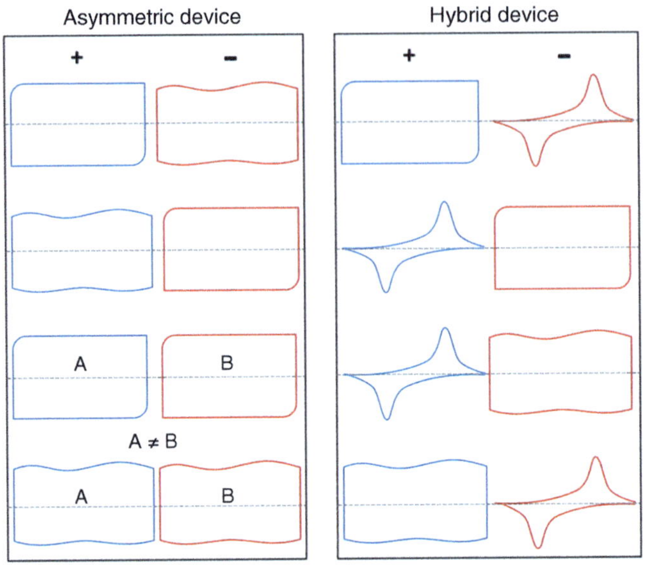

Figure 4.1 Electrochemical characteristics of different device configurations of asymmetric supercapacitors and hybrid capacitors for a better understanding of the mechanism in these devices (A and B are different materials). Source: Reproduced with permission of Jiang and Liu [1]. Copyright 2019, WILEY-VCH.

The first classic configuration of sodium-ion capacitors was successfully assembled and reported in 2012 [3]. In this classic configuration, hard carbon (HC) materials are selected as the anode materials. Due to the low potential plateau (vs. Na^+/Na) in charge–discharge profiles and excellent capacity, HC is a suitable anode material choice in SICs. Therefore, since the HC anodes were utilized in the pioneering work of Kuratani et al., the HC anodes are still regarded as the main choice for battery-type anodes in the classic battery design of SICs. The SICs devices evaluated in this work also contain capacitive carbon as the cathodes. Based on the classic SICs of carbonaceous cathodes and anodes, some recent advanced carbonaceous cathodes and anodes have been summarized. Generally, the relationship between structures and performance of carbonaceous materials plays an important role in the electrochemical performance of the cathodes and anodes in SICs. In these configurations, different dimensions of carbonaceous cathodes and anodes materials (carbon nanospheres, carbon nanofibers, carbon nanosheets, and porous carbon) will be mainly discussed (Figure 4.2b).

4.2 Battery-Type Anode and Pseudocapacitive Cathode SICs Cell Configurations

For a better understanding of the principles in these emerging architectures, the characteristics of pseudocapacitance need to be elaborated. For many readers, this section is a bit confusing because the term " pseudocapacitance" may have different meanings, and some authors utilize this concept differently. The explanation of

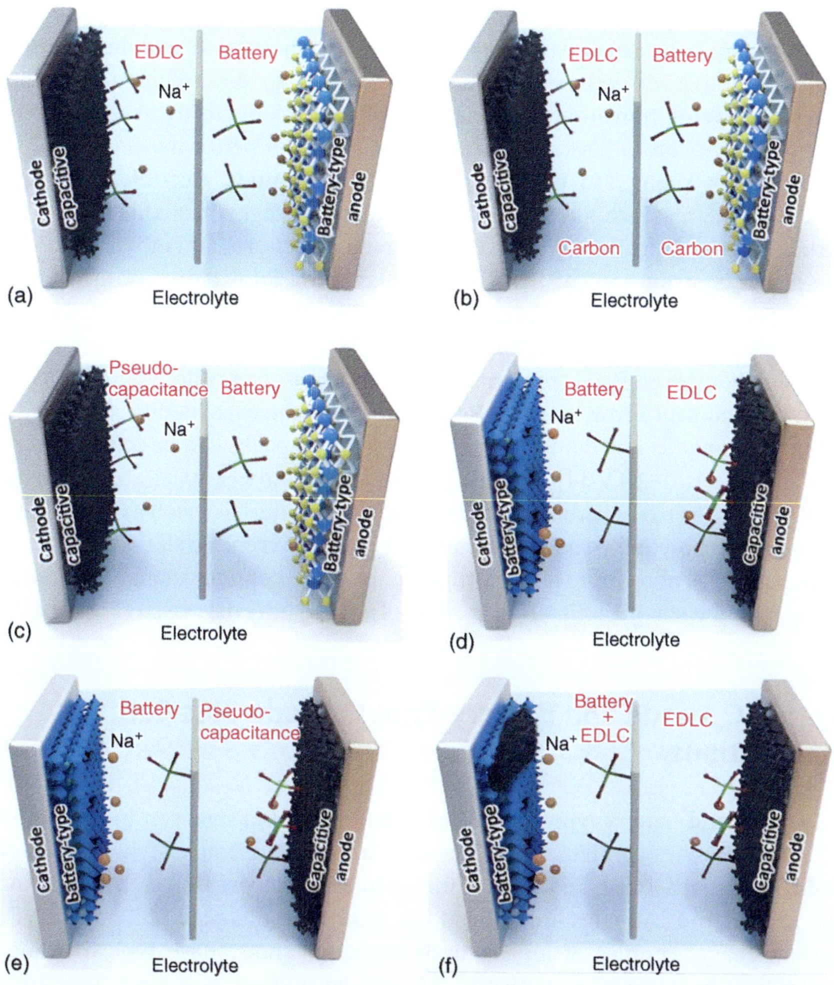

Figure 4.2 Cell structures and charge storage mechanisms of the nine electrochemical energy storage devices: (a) Battery-type anode and EDLC cathode SICs cell configurations, (b) dual carbon configurations, (c) battery-type anode and pseudocapacitive cathode SICs cell configurations, (d) EDLC anode and battery-type cathode SIC cell configurations, (e) pseudocapacitive anode and battery-type cathode SICs cell configurations, (f) capacitive anode and hybrid cathode SICs cell configurations. Source: Adapted from Zhang et al. [2].

the word pseudocapacitive is related to the behavior of electrode materials that behave similar to EDLC-based electrode materials (that is, showing analogous rectangular CV shapes and triangular charge–discharge potentials) [4]. Most of the examples considered in this book are the utilization of capacitive carbon (such as activated carbon or other large surface area porous carbon materials) as the cathode of SICs devices. However, their capacitances are limited, so it is meaningful to explore other materials with approximate charge and discharge characteristics (triangular or roughly triangular curve elements). For example, initial research in this field focuses on utilizing MXenes as an alternative electrode for capacitive carbon. The origin of this idea is that MXene and AC exhibit similar GCD curves in SCs. As an ideal pseudo-capacitance material, MXenes reported the capacitance behavior of aqueous supercapacitors, resulting in an ideal triangular charging and discharging front end and the CV curves close to the rectangular shapes. MXenes is a layered transition metal carbide with excellent electrical conductivity, and has been proven to have promising electrochemical performance not only in aqueous solutions but also in nonaqueous organic battery electrolytes. Despite their electrochemical characteristics, the charge storage mechanism in these materials is mainly attributed to the Faraday redox reactions, and only a small part is due to the contributions of EDLC processes. A special example of this type of pseudocapacitive material is layered transition metal carbide ($Ti_3C_2T_x$). The CV curves of ideal rectangular capacitance are obvious, and their shape is very similar to that of carbon-based EDLC materials. The CV curves of a similar material Ti_2CT_x utilized in organic electrolytes for SIBs are similar to the capacitance curves when the potential range is less than 2 V, and a pair of redox peaks can be seen when the potential range is 2–2.4 V [4]. Due to the similar electrochemical characteristics with high surface area of carbon, pseudocapacitance materials (such as the MXenes and conducting polymers) can be used in SICs devices to replace standard activated carbon electrodes and improve performance. In these configurations, SIC devices utilizing pseudocapacitive materials instead of AC are fully discussed (Figure 4.2c).

4.3 EDLC Anode and Battery-Type Cathode SICs Cell Configurations

EDLC anode and battery-type cathode SICs cell configurations can be also called the "Upside-Down" configurations. In such a device, a battery-type cathode is utilized, and a capacitor-type anode is included as an anode. In fact, the functions of the anodes and cathodes are opposite to those of the classic SIC devices. The charge–discharge profiles of an idealized capacitor anode have typical triangular potential characteristics. Typically, AC can be utilized as a capacitive anode (AC). A battery-type material (with a higher capacity and flat platform in the charge–discharge curves) works at a higher potential as the battery-type cathode. Similar to the classic SICs devices, there is a potential difference between anodes and cathodes. Therefore, integrated devices with capacitive curves can be obtained. The overall cell potential ranges are equal to the potential differences between the cathodes and anodes in the devices. In this configuration, since there are fewer types of cathodes that can be utilized, cathode materials are the main focus of this discussion (Figure 4.2d).

4.4 Pseudocapacitive Anode and Battery-Type Cathode SICs Cell Configurations

The classic "Upside-Down" configurations include a battery-type cathode and a capacitor-type anode. Interestingly, the emerging "Upside-Down" configurations with alternative pseudocapacitive electrodes to replace capacitive anode (AC) can also be assembled. In such a device, an alternative pseudocapacitive anode is utilized, and a battery-type electrode is included as a cathode (Figure 4.2e).

4.5 Capacitive Anode and Hybrid Cathode SICs Cell Configurations

Due to the proper hybridization of the battery component and the capacitor component on a single electrode as cathode, the above-mentioned hybrid electrode is expected to display good electrochemical performance. The origin of this configuration is a report of parallel configurations in lithium storage systems. Du Pasqueir et al. discussed the effects of adding AC to the cathodes of a LIB. Select $LiCoO_2$ as the active material on the cathodes, combined with 5–20% AC. Good rate capability and the energy density of 40 Wh kg^{-1} have been achieved. Amatucci applied for a patent for this device based on the parallel configurations [5]. Here, we will theoretically consider this novel hybrid electrode to comprehend the working processes. In addition, the appropriate structures and ideal conditions of the hybrid electrodes utilized to improve the electrochemical performance of the energy storage system are a good guide for further milestones. Compared with asymmetric CSs, the hybrid electrode system contains classic capacitive electrodes and hybrid nanocomposite electrodes, as depicted in Figure 4.2f. The latter electrodes are composed of battery-type components and capacitive components in the single electrodes, which can store energy by the Faraday and capacitive mechanisms. In such hybrid electrodes, the capacitor components are firstly charged by electrostatic forces in the initial states (q_1) until the electrode potential reaches the oxidation–reduction reaction voltage (ΔV_b) of the battery components. Then, while maintaining the hybrid electrode potential, the battery assembly is charged through an oxidation–reduction reaction until the faradaic assembly reaches the fully charged state (Q_b) shown in Figure 4.3 [6]. After the battery components (Q_b) are fully charged, the capacitor components

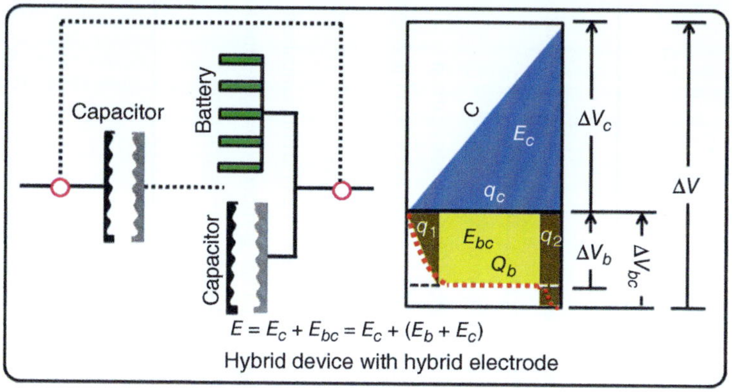

Figure 4.3 Schematic representations of hybrid combination of a hybrid electrode (hybridized battery and capacitor components) and a capacitor electrode according to the charge-potential profile with corresponding equations for energy storage. Source: Reproduced from Dubal et al. [6]/with permission of Royal Society of Chemistry.

Table 4.1 A summary of dual-carbon configurations based on carbon-based anode and cathode materials.

Anode//cathode	Potential range (V)	Max energy density (Wh kg^{-1})	Max power density (W kg^{-1})	Cycling life
Hard carbon//AC	1.5–4.2	>100	7000	70% after 5000 cycles; 2 A g^{-1}
Nanoporous disordered carbon//N-hollow microspheres	0–4.4	157	2356	70% after 1000 cycles; 0.3 A g^{-1}
N,O-carbon//nanosheet carbon	0–4.0	111	14 550	86% after 5000 cycles; 6.4 A g^{-1}
Ordered carbon//nanosheet carbon	1.5–3.5	201	16 500	72% after 10 000 cycles; 6.4 A g^{-1}
3D framework carbon//nanoporous carbon	0–4.0	111	20 000	75.6% after 15 000 cycles; 2 A g^{-1}
Carbon microspheres//activated carbon microspheres	2.0–4.0	52.2	3000	85.7% 2000 cycles; 1.0 A g^{-1}
Nanotube-like hard carbon//AC	1.5–3.5	133.0	14 250	82.5% after 12 000 cycles; 2 A g^{-1}
N-hollow carbon nanowires//AC	0.5–4.0	108	9000	70% after 2000 cycles; 2 A g^{-1}
Carbon nanosheets//activated carbon nanosheet	0–4.0	112	12 000	85% after 3000 cycles; 5 A g^{-1}
Oxygen-functionalized graphene//oxygen-functionalized graphene	0.5–3.5	121.3	8000	86.7% after 2500 cycles; 0.5 A g^{-1}
Hard carbon//3D macroporous graphene	0–4.2	168	2432	85% after 1200 cycles; 1 A g^{-1}
3D carbon framework//AC	0–4.0	133.2	20 000	86% after 4000 cycles; 2 A g^{-1}

are charged again (q_2) until the mixed electrode reaches the maximum potential (ΔV_{bc}). Therefore, the total charge stored in the hybrid electrodes comes from two components (battery [Q_b] and capacitor component [$q_c = q_1 + q_2$]). In addition, the maximum operating voltage window of the entire system can be the sum of the potentials on the capacitive electrodes (ΔVc) and the hybrid electrodes (ΔV_{bc}) (this sum also includes the potential on the battery and capacitive elements, as shown in Figure 4.3). Therefore, the total energy storage (E) of the hybrid system is the sum of the total energy storage of the capacitive electrodes (Ec) and the total energy storage of the hybrid electrodes ($E_{bc} = E_b + E_c$). Due to the structural characteristics of the hybrid electrodes, the power density is greatly improved compared with typical battery electrodes. This is because the capacitive element can also store electrochemical energy through electrostatic processes, which improves the transfer of electrons in the hybrid electrode system to the battery element, thereby generating a better charge transfer reaction at higher rate. The electrochemical performance of hybrid electrode materials depends on the hybrid characteristics of the two materials.

Therefore, reasonable composition and morphological structures are the key factors for realizing the full potential of hybrid electrode materials relative to traditional electrode materials: (i) In the hybrid assembly of batteries and capacitors, the electrodes should be operated as one electrode in a unified manner. Otherwise, the working potential may directly increase without rechargeable batteries part in the charged state due to the fast dynamics of the capacitors part. (ii) The hybrid electrode materials should have a higher surface area and an open porous structure to obtain higher energy density and good ion conductivity. (iii) The network structures are also conducive to enhancing the electron transport capabilities of these hybrid materials while maintaining ionic conductivity. For a more intuitive presentation of the above device classifications, Tables 4.1–4.3 provide some typical characteristics of commonly classified SICs [7].

4.6 Summary

Asymmetric SCs cover a wider range of electrode combinations because they can be assembled for SCs that utilize electrodes with the same properties but different mass loads or utilize two electrodes with different materials. Due to the different materials of the electrodes, when pairing two electrodes with different charge characteristics (a capacitor electrode and a battery electrode), the term "hybrid capacitors" should be utilized, and the electrochemical performance of the resulting devices is between SCs and SIBs. The features of battery-type anode and EDLC cathode SICs cell configurations could own a special configuration (dual carbon configurations). In addition, the cell configuration could deliver large energy density at wide voltage range due to the low intercalation potential. Although EDLC anode and battery-type cathode SICs cell configurations also have dual carbon configurations, the limitation is that the battery-type category is restricted by carbonaceous graphite. Furthermore, the voltage range of anodes in EDLC anode and battery-type cathode

Table 4.2 A summary of SICs devices based on intercalation mechanism anode and EDLC cathode configurations.

Anode//cathode	Potential range (V)	Max energy density (Wh kg^{-1})	Max power density (W kg^{-1})	Cycling life
$Na_2Ti_2O_4(OH)_2$//porous carbon	1.0–4.0	65	5000	93% after 3000 cycles; 1 A g^{-1}
$Na_2Ti_9O_{19}$//porous carbon	1.0–4.0	54	5000	75% after 2000 cycles; 2 A g^{-1}
$Na_2Ti_3O_7$/graphene foam//graphene foam	1.0–3.0	70.6	4000	73.2% after 5000 cycles; 0.5 A g^{-1}
$Na_2Ti_3O_7$//biomass carbon	0.5–3.5	111.2	11 200	86% after 3000 cycles; 3.2 A g^{-1}
Brown TiO_2 nanotubes//AC	1.0–4	68	75 000	80% after 10 000 cycles; 1 A g^{-1}
$Mo_{0.1}Ti_{0.9}O_2$//AC	1.0–4.0	60	10 650	75% after 3000 cycles; 1 A g^{-1}
TiO_2–graphene//AC	1.0–3.8	64.2	1367	90% after 10 000 cycles; 3.35 A g^{-1}
TiO_2/C//nanoporous carbon	1.0–4.0	142.7	25 000	90% after cycles; 1 A g^{-1}
N-TiO_2//AC	1.0–4.0	80.3	12 500	85% after 6500 cycles; 1 A g^{-1}
TiO_2/CFC//CFs	1.2–4.3	73.8	13 750	90% after 4000 cycles; 1 A g^{-1}
TiO_{2-x}/CNT//AC/CNT	1.0–4.0	109	5000	73% after 2000 cycles; 1 A g^{-1}
TiO_2–rGO//porous carbon	1.0–4.2	94.7	4093	64% after 2000 cycles; 2 A g^{-1}
TiO_2@CNT@C//AC	1.0–4.0	81.2	12 400	85.3% after 5000 cycles; 1 A g^{-1}
V_2O_5 nanorods@rGO//AC	0.01–3.0	65	–	74% after 1000 cycles; 0.06 A g^{-1}

V_2O_5@CNT//AC	0–2.8	38	5000	80% after 900 cycles; 6 A g^{-1}
V_2O_3/C//AC	1.0–4.0	63	5805	74.4% after 3000 cycles; 1 A g^{-1}
Nb_2O_5 nanosheets//PSC	1.0–3.0	43.2	5760	80% after 3000 cycles; 1.28 A g^{-1}
Nb_2O_5@C/rGO//(MSP-20)AC	1.0–4.3	76	20 800	63% after 3000 cycles; 1 A g^{-1}
G@mNb_2O_5//AC	0–2.4	56.1	7200	99% after 4800 cycles; 1 A g^{-1}
Gr-Nb_2O_5//AC	0–3.0	112.9	5330	97.1% after 3000 cycles; 2 A g^{-1}
m-Nb_2O_5/CNF//GF/mCNF	0–4.0	124	60 000	80% after 3000 cycles; 2 A g^{-1}
m-WO_{3-x}@nm-rGO//MSP-20	1.0–4.3	67	21 000	stable performance after 1000 cycles at 1 A g^{-1}
$NaTi_2(PO_4)_3$/rGO//AC	0–2.7	53	6680	93% after 5000 cycles; 1.5 A g^{-1}
$NaTi_2(PO_4)_3$ on graphene//graphene	0–3.0	80	8000	90% after 75 000 cycles; 4 A g^{-1}
$NaTi_2(PO_4)_3$//AC	0.01–2.5	56	4096	100% after 20 000 cycles; 5 A g^{-1}
$Na_3V_2(PO_4)_3$//CMK-50 carbon	0–3.0	54	2200	52% after 5000 cycles; 0.7 A g^{-1}
$Na_3V_2(PO_4)_3$@C//CDC	0–3.0	118	850	95% after 10 000 cycles; 0.025 A g^{-1}
$Na_3V_2O_2(PO_4)_2F$@PEDOT//AC	1.0–4.2	158	7000	85.1% after 1000 cycles; 1 A g^{-1}
VP-CNOs//AC	0.5–3.8	120	7400	86% after 20 000 cycles; 0.1 A g^{-1}
MoS_2@porous carbon//porous carbon	0–4.0	112.2	8333	88% after 1000 cycles; 2 A g^{-1}

Table 4.3 A summary of SICs devices based on conversion and alloying reaction anode and EDLC cathode configurations.

Anode//cathode	Potential range (V)	Max energy density (Wh kg^{-1})	Max power density (W kg^{-1})	Cycling life
$NiCo_2O_4$//AC	0–4.5	23.5	308	61.2% after 2000 cycles; 0.15 A g^{-1}
$CoWO_4$/RGO//AC	1.0–4.2	87.2	1120.7	100% after 1000 cycles; 0.1 A g^{-1}
MoO_2@rGO//AC	0.01–3.0	79	3300	78% after 1000 cycles; 0.07 A g^{-1}
3D-IO FeS-QDs@NC//AC	0.5–3.5	151.8	9280	91% after 5000 cycles; 1 A g^{-1}
$FeS_{2-x}Se_x$//AC	0–3.0	67	2543	60% after 1000 cycles; 2 A g^{-1}
Ni_3S_2//AC	0–4.0	65.6	3704.3	98% after 2000 cycles; 0.1 A g^{-1}
NiS_x@PCM//AC	1.0–3.8	99.3	4480	58.6% after 800 cycles; 1 A g^{-1}
Sb_2S_3/carbon fibers//carbon fibers	1.5–4.3	124	5800	90% after 1000 cycles; 1 A g^{-1}
SnS_2/GCA//AC	1.0–4.3	108.3	6053	68.4% after 1500 cycles; 1 A g^{-1}
SnS/rGO//AC	1.0–4.0	115	7500	76% after 2000 cycles; 1 A g^{-1}
SnS/aCMT//aCMT	0.5–4.0	115	35 000	83% after 5000 cycles; 1 A g^{-1}
CNF@VS_2//CNF@GS	0–4.0	116	40 000	82% after 4500 cycles; 5 A g^{-1}
WS_2//AC	0–3.5	63	7840	83% after 6000 cycles; 0.5 A g^{-1}
WS_2@NCNs//NCHS	0.6–4.2	65.8	4700	90% after 2000 cycles; 0.2 A g^{-1}
$MoSe_2$/graphene//AC	0.5–3.0	82	6688	81% after 5000 cycles; 5 A g^{-1}
MoO_2/$MoSe_2$–graphene//AC	1.0–3.4	70.3	14316.1	92% after 7000 cycles; 6 A g^{-1}
$MoSe_2$/SnO_2//AC	0.5–3.0	70	2304	94% after 6000 cycles; 5 A g^{-1}
$NbSe_2$//AC	0.0-3.0	41.7	599.6	93.2% after 3000 cycles; 0.3 A g^{-1}
aSnSe/rGO//AC	1.0-4.0	88	2251	70% after 5000 cycles; 1 A g^{-1}
Co_2P/Sn@NC//AC	0–4.0	112.3	10 000	84% after 10000 cycles; 2 A g^{-1}
C@FeP//AC	0–4.0	133.6	40 000	84.6% after 1000 cycles; 1 A g^{-1}
NCN@Mo_2N//ANCN	0.5–4.2	100.6	2350	87.4% after 1000 cycles; 0.5 A g^{-1}
Mo_2C//AC	0–4.5	76.1	2241	83% after 4000 cycles; 1 A g^{-1}
CoF_2//AC	0.01–4.3	60	8600	66.5% after 200 cycles; 3 A g^{-1}
Na_2FeSiO_4//AC	0–3	70	5400	94% after 3000 cycles; 0.6 A g^{-1}
$TiSb_2$//AC	1.0–3.6	72	20 625	63% after 1000 cycles; 5 A g^{-1}
Cu-CAT//AC	0.01–3.5	132.85	2980	84% after 100 cycles; 0.1 A g^{-1}
PI-2.5//AC(PI-5)	0–4.2	66	1200	72.4% after 1000 cycles; 0.4 A g^{-1}

SICs cell configurations should be carefully selected due to the fact that low potential could change EDLC mechanism into intercalation mechanism. Moreover, the interests of some emerging architectures of SICs cell configurations (such as battery-type anode matching with pseudocapacitive cathode SICs cell configurations and pseudocapacitive anode combining with battery-type cathode SICs cell configurations) are improving cycling stability.

References

1 Jiang, Y.Q. and Liu, J. (2019). *Energy Environ. Mater.* 2: 30.
2 Zhang, H.W., Hu, M.X., Lv, Q. et al. (2020). *Small* 16: 22.
3 Kuratani, K., Yao, M., Senoh, H. et al. (2012). *Electrochim. Acta* 76: 320.
4 Glushenkov, A.M. and Ellis, A.V. (2018). *Adv. Sustainable Syst.* 2: 25.
5 Cericola, D. and Kotz, R. (2012). *Electrochim. Acta* 72: 1.
6 Dubal, D.P., Ayyad, O., Ruiz, V., and Gomez-Romero, P. (2015). *Chem. Soc. Rev.* 44: 1777.
7 Cai, P., Zou, K., Deng, X. et al. (2021). *Adv. Energy Mater.* 11: 2003804.

5

Cathode Materials for Sodium-Ion Capacitors

Xiong Zhang[1,2], Wenjie Liu[1,2], Lei Wang[1,2], Chen Li[1], and Yanwei Ma[1,2]

[1] *Institute of Electrical Engineering, Chinese Academy of Sciences, No. 6 Beiertiao, Zhongguangcun, Beijing 100190, China*
[2] *University of Chinese Academy of Sciences, No. 1, Yanqihu East Road, Huairou District, Beijing 100049, China*

5.1 Introduction

Sodium-ion capacitors (SICs) integrate high-capacity batteries as well as high-rate supercapacitor materials, which can improve energy density without significantly sacrificing their high power or cycling performance [1–3]. According to the different internal configurations, the cathode materials of SICs can be divided into electrical double-layer capacitor (EDLC)-type, pseudocapacitive-type, and battery-type materials, as exhibited in Figure 5.1 [4, 5]. Different types of cathode materials have different effects on the performance of the assembled SIC due to the discrepancy in the charge storage mechanism. Among them, for the configuration of traditional SICs with EDLC-type cathodes (such as activated carbon [AC] or other porous carbon materials), the EDLC materials are the crux to support high current charge and discharge capability, high rate, and high power density of SICs [6–9]. This characteristic is closely related to the capacitive storage mechanism of EDLC materials – the formation of an electric double layer, which is composed of opposite charges between the solid/liquid interface of cathode materials and electrolytes [10]. It is worth noting that the open circuit potential of the EDLC-type material (activated carbon) in the organic sodium salt is 2.7 V vs. Na^+/Na, and the operating voltage window is generally about 2–4.6 V. Above the open circuit potential, the adsorption/desorption behavior of anions occurs, and the adsorption/desorption behavior of sodium ions happens below the open circuit potential. As a result, depending on the electrode's sodium potential window, the EDLC-type carbon material can sometimes be used as the cathode or anode. When pseudocapacitive materials (MXene and conducting polymers) are used in SIC cathodes, they can provide higher capacity and have similar charge–discharge characteristics compared to EDLC materials, aiming to replace commercial activated carbon electrodes for better performance [11]. For the SIC constructed with the battery-type cathode, EDLC material (mainly activated carbon) is generally chosen as the anode. At this time, the configuration of the SIC

Sodium-Ion Capacitors: Mechanisms, Materials, and Technologies, First Edition.
Edited by Guoqiang Zou, Xiaobo Ji, and Hongshuai Hou.

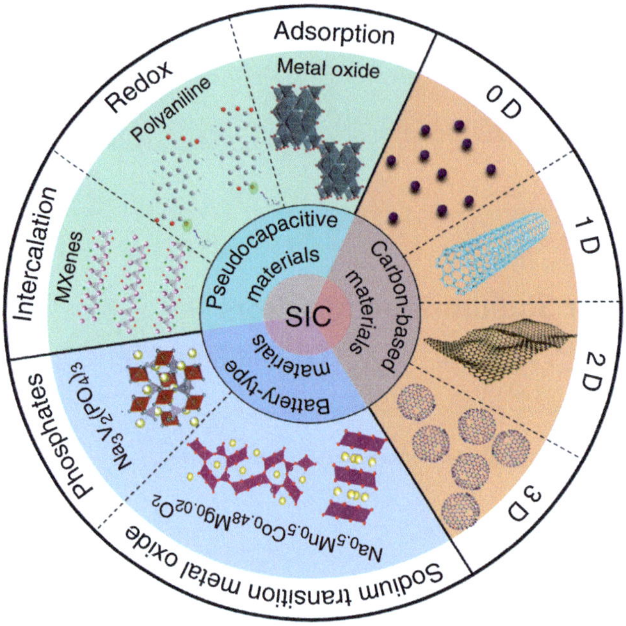

Figure 5.1 Various materials for SIC cathode.

can be described by the "Upside-Down" model, which is just the opposite of the classic SIC model. When the battery-type material is used alone as the cathode and matched with the EDLC-type anode, the energy density of the assembled SIC device is unsatisfactory due to the low specific capacities of both the cathode and anode [12]. Therefore, little literature has reported SICs with this configuration, and only some battery materials with typical voltage platforms are reported to be applied to SIC cathodes. Pseudocapacitor and battery-type cathode materials will be described in detail in Sections 5.3 and 5.4, respectively.

SIC is similar to the lithium-ion capacitor (LIC) in terms of working principles, both of which are the intelligent combination of the EDLC mechanism and the battery-type Faraday reaction storage mechanism [13]. However, because sodium ions have a significantly higher relative atomic mass and effective ionic radius (10.2 Å) than lithium ions (7.6 Å), the initial specific capacity and rate performance of the same capacitive material cathode in organic sodium salt electrolyte are both lower than its performance in organic lithium salt electrolytes [14, 15]. In addition, the large particle size of sodium ions will cause significant structural variations in most electrode materials during the electrochemical reaction process, which further reduces the charge and discharge rates [16]. That is to say, the large particle size of sodium ions leads to a much larger kinetic discrepancy between the cathode and anode of SICs than that of LICs. Hence, the electrode material is the most critical factor affecting the performance of SIC [17]. The preparation of high-performance SIC puts forward higher demands for the research and development of electrode materials.

According to the short-plate effect, the energy characteristics of SICs depend on capacitive materials, and the power output relies on battery-type Na-storage materials. The optimal energy density and power density can be achieved by adjusting the capacity and kinetic matching of the anode and cathode materials. In recent years, researchers have been working on developing capacitive materials with high specific capacities and battery-type materials with quick kinetic properties to compensate for the differences in capacity and kinetics between the two electrodes. Therefore, the research progress and main achievements of SIC cathode materials (EDLC-type, pseudocapacitive-type, and battery-type materials) are systematically introduced, and the challenges and future prospects of cathode materials for SIC research are also proposed.

5.2 EDLC Cathode Materials

For EDLC materials, electrons continuously flow from the anode to the cathode through the external circuit during the discharge process, thus creating an instantaneous surplus of electrons at the cathode and causing the potential of the cathode to decrease continuously. Meanwhile, the anions adsorbed on the surface of the cathode are transferred to the bulk phase of the electrolyte [18]. Upon the charging process, electrons pour into the anode from the external circuit to generate an instantaneous electron surplus, which reduces the negative electrode potential. At the same time, the potential of the cathode gradually increases due to the outflow of electrons from the interface, while the anions are continuously adsorbed on the cathode surface to maintain the charge neutrality of the electrical double layer [2, 19]. In order to store or release energy during charging and discharging, the cathode interface can quickly establish an electric double layer, which results in fast charge rate characteristics. The EDLC-type cathodes are the place where charges are quickly stored, which ensures the high power density of SICs [20]. Therefore, the ideal EDLC material should have the following characteristics: (i) high conductivity, which is conducive to the rapid transfer of electrons in the bulk of electrode and material particles, reducing the internal resistance and polarization, and also promoting the fast transport of charges in the electrical double-layer interface [17, 21]; (ii) high specific surface area, which provides sufficient charge separation site for the formation of the electric double layer and improves the capacity of the cathode [22]; (iii) the abundant pore structure is beneficial for the rapid diffusion of electrolyte ions into the electrode cavity to form an electric double layer and improve the high-rate charge–discharge performance of the electrode; (iv) high volume density, which is essential to improving the volume, energy, and power density of the device [23]; (v) electrochemical inertness, avoiding irreversible side reactions with the electrolyte, which can support the long-term cycle life demand [24].

Carbon-based materials have the advantages of high electrical conductivity, large specific surface area (>1000 $m^2 g^{-1}$), high pore volume (>0.5 $cm^3 g^{-1}$), excellent chemical stability, a wide range of sources, low-cost, and easy processing. They are widely regarded as ideal EDLC cathode materials and have also been widely used in SICs [8, 19, 25]. From a structural point of view, the EDLC cathode material of SICs can be divided into four types: 0-dimensional (0D), 1D, 2D, and 3D carbon materials.

5.2.1 0D Carbonaceous Cathodes

0D nanomaterials are substances with a nanometer scale (<10 nm) in the 3D direction of space, such as nanoparticles, atomic clusters, and quantum dots, and are generally made up of a small number of atoms or molecules. 0D materials can provide electron–hole pairs to improve electronic conductivity. Meanwhile, its high specific surface area is conducive to heteroatom doping and surface functional group modification, which can improve electrochemical activity and stability [26]. Hence, 0D materials offer a high prospect in the field of electrochemical energy storage materials. However, typical 0D materials, such as carbon quantum dots and carbon black, have not yet been reported to be designed and constructed as the cathodes of SICs. In the follow-up, the larger-sized 3D carbon microspheres or hollow carbon microspheres are introduced in the hope of further inspiring the application of carbon dots in SICs.

5.2.2 1D Carbonaceous Cathodes

5.2.2.1 Carbon Nanotubes

Carbon nanotubes (CNTs) are coaxial, seamless, and circular tubes with open or closed ends rolled from a single layer of graphene. According to the number of graphene layers, CNTs can be divided into single-walled carbon nanotubes (SWCNTs) and multi-walled carbon nanotubes (MWCNTs), as shown in Figure 5.2a,b. The diameter of SWCNT is about 0.8–2 nm, and for MWCNT, it is about 5–100 nm, with an interlayer spacing of 0.34 nm. The length of CNTs can be less than 100 nm or as long as tens of centimeters, thus covering the microscopic scale of molecular level and the macroscopic scale of real life [27].

The carbon atoms in CNTs are dominated by sp^2 hybridization to form a hexagonal graphene honeycomb structure, but there is a certain proportion of sp^3 hybridization at the same time, so the graphene honeycomb will bend to a certain extent, forming a spatial topology [28]. Pentagons, heptagons, and other defects are often introduced into the sidewalls of the CNT during actual production, which often degrades the

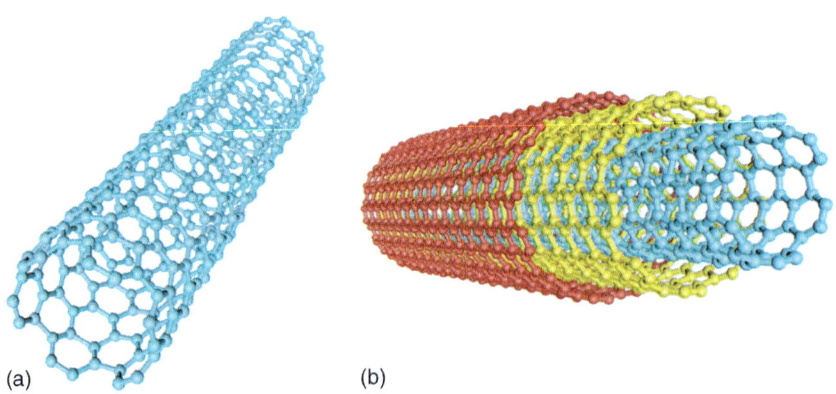

Figure 5.2 Structure diagram of (a) SWCNT and (b) MWCNT.

desired properties of the CNTs. Generally, SWCNTs have higher chemical inertness with pure surfaces, while MWCNTs are much more active and combine a large number of surface groups, such as carboxyl groups. Since Iijima first discovered CNTs in 1991 [29], CNTs with high charge mobility (10 000 $cm^2 V^{-1} s^{-1}$) and good heat transfer properties (3500 $W m^{-1} K^{-1}$) have been widely used in optics, energy storage, and other fields [30, 31]. At present, the main preparation methods for CNTs include the chemical vapor deposition (CVD) method, arc discharge method, laser evaporation method, and catalytic pyrolysis method.

So far, there is no report of CNT alone as the SIC cathode. The main reasons are: (i) The preparation of CNTs is mainly by CVD and a large number of metal catalysts will remain in the product. The residual metal catalysts make it easy to cause irreversible decomposition of the electrolyte. (ii) CNTs are difficult to disperse uniformly in the electrode, which requires sufficient shearing and stirring during the preparation of electrode slurry, consuming a lot of time. (iii) Compared with cheap activated carbon, the production cost of high-quality CNTs is still relatively high. Therefore, the future application of CNTs in SICs needs to be further studied.

5.2.2.2 Carbon Nanofibers

Carbon nanofibers (CNFs) are also 1D carbon materials, which are formed by the spatial accumulation of nano-sized graphite crystallites at different angles along the fiber axis, and the fiber diameter can reach 50–200 nm [32]. Most of the CNFs reported so far are composed of polymer precursors or carbon allotropes with a carbon content of at least 92%. CNFs can be prepared by electrostatic spinning, spray method, melt spinning, plasma-enhanced CVD, wet spinning techniques, etc., with abundant precursor sources including polyacrylonitrile, pitch, natural cellulose, lignin, polymer ethylene oxide, polyethylene glycol-polypropylene glycol-polyethylene glycol (F127), and polystyrene–polybutadiene–polymethyl methacrylate triblock copolymer [33, 34]. Tens of nanometers to several micrometers can be regulated for the prepared CNFs' diameter, and surface area and porosity can also be changed.

As a new type of energy storage material, CNFs, due to their large aspect ratio, short diffusion path of ions in the axial direction (generally no more than 50 nm), and high electronic conductivity, can promote fast electron/ion transport and have excellent structural stability and high-rate capability. Zhang and coworkers developed a CNF-based electrode by tailoring the synthetic process using bacterial cellulose (BC) hydrogel as the precursor, as shown in Figure 5.3a [35]. Firstly, the CNFs were obtained by freeze-drying BC hydrogel and heating it to 1000 °C in a tube furnace. Then, the activated CNFs cathode materials with abundant nanopores and a large specific area were fabricated by KOH activation (Figure 5.3b,c). Two broad diffraction peaks of 1D-CNFs at 23° (002) and 43° (101), respectively, represent the typical amorphous structures. As for electrochemical performances, the cyclic voltammetry (CV) curves of activated-CNFs exhibit quasi-rectangular shapes, suggesting the typical EDLC behaviors, as shown in Figure 5.3d. The specific capacities are 82.5, 75.48, 70.89, 68.36, 66.81, and 66.05 $mAh g^{-1}$ at the current densities of 0.05–0.1, 0.2, 0.5, 1, and 2 $A g^{-1}$ in the voltage range of 2.0–4.2 V, respectively, (Figure 5.3e). The capacity

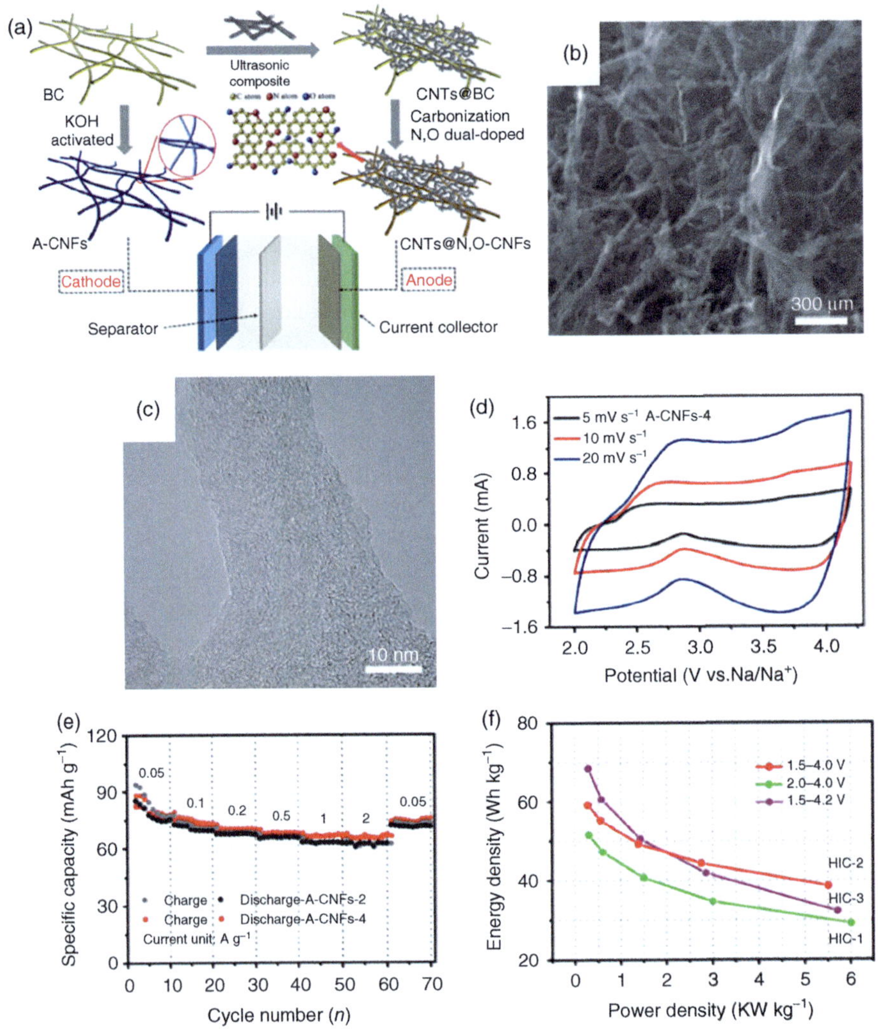

Figure 5.3 (a) Schematic diagram of the synthesis of 1D CNFs and assembly of SIC. (b) Scanning electron microscopy (SEM) and (c) transmission electron microscopy (TEM) images of CNFs. (d) CV curves of the activated-CNFs cathode at various scan rates. (e) Rate performance. (f) Ragone plots of SICs at various voltage ranges. Source: Xu et al. [35]/Royal Society of Chemistry/CC BY 3.0.

retention of 1D-activated CNFs is 78% after 1000 cycles at 0.05 A g^{-1}. To investigate the electrochemical performances of SICs, activated-CNFs were used as the cathode and CNFs/CNTs composite as the anode for SICs assembling. Based on the total mass of the active materials, the assembled SICs can offer a high energy density of 59.2 Wh kg^{-1} and a power density of 38.7 Wh kg^{-1}, as exhibited in Figure 5.3f. The capacity retention of the device is 48.6% at a current density of 0.5 A g^{-1} in the voltage range of 1.5–4.0 V after 5000 cycles. Hence, the BC-derived CNFs electrodes reveal promising electrochemical performance as the cathode for SIC devices.

5.2.3 2D Carbonaceous Cathodes

2D carbon materials are effective in enhancing the electrochemical performance of SICs due to the fast charge and mass transport kinetics in the planar direction.

5.2.3.1 Reduced Graphene Oxide

As a typical 2D material, graphene is composed of sp^2-hybridized carbon atoms, presenting a hexagonal lattice honeycomb structure, usually with a thickness of one carbon atom [36], as shown in Figure 5.4. Perfect monolayer graphene has an ultra-high specific surface area (2630 $m^2\ g^{-1}$) and high intrinsic electrical conductivity (15 000 $cm^2\ V^{-1}\ S^{-1}$) [37, 38]. Conductivity and specific surface area are key parameters that determine the performance of EDLC cathodes [39]. Therefore, the electrochemical performance of graphene cathodes is highly dependent on the material structure and the preparation process of the electrode. Usually, there are two main methods for synthesizing graphene: one is a top-down method, such as preparing graphene by exfoliating graphite by liquid phase or high-energy mechanical shearing; the other is CVD, epitaxial growth, or advanced organic chemistry methods to synthesize graphene from the bottom-up synthesis [40]. At present, the widely used method is to exfoliate graphite through a liquid-phase redox reaction to obtain graphene oxide (GO) and then conduct a reduction treatment to obtain reduced graphene oxide (rGO) [41]. The process is simple and low-cost, and large-scale commercial preparation has been realized at present.

Zhao et al. investigated rGO as a cathode for fabricating SICs, as exhibited in Figure 5.5a [42]. The CV curves of the interconnected 3D macroporous rGO cathode display slightly biased rectangular shapes, and the bent galvanostatic charge and discharge (GCD) curves indicated the combination of EDLC response and a surface redox reaction, which ensures fast charge storage dynamics. The rGO cathode can achieve a high capacity of 70 $mAh\ g^{-1}$ with nearly 100% coulombic efficiency after the 200th cycle, which is superior to the capacity of commercial AC (44 $mAh\ g^{-1}$). The high capacity of rGO comes from the reversible chemisorption of sodium ions by residual functional groups on the surface. Moreover, the flexible SICs fabricated by rGO cathode and polyimide/rGO–rGO (PI/rGO–rGO) anode with the mass ratio of 3 : 1 can deliver gravimetric energy densities of 55.5 $Wh\ kg^{-1}$ (power density of 395 $W\ kg^{-1}$) and power densities of 3.4 $kW\ kg^{-1}$ (energy density

Figure 5.4 Structure diagram of graphene.

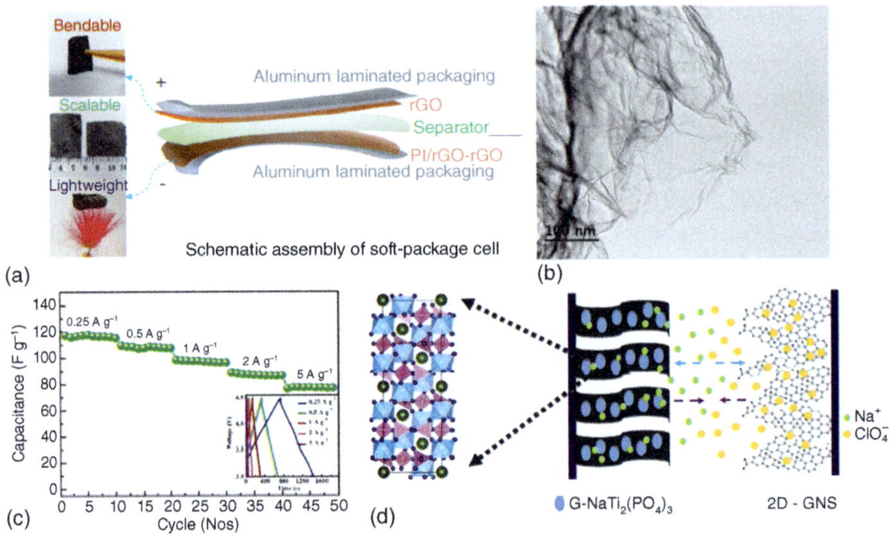

Figure 5.5 (a) The schematic assembly of a soft-package hybrid SIC by rGO cathode and PI/rGO–rGO anode. Source: Zhao et al. [42]/Reproduced with permission from American Chemical Society. (b) TEM image and (c) rate capability of GNS cathode, (d) Working principle of the GNS//GNTP SIC. Source: Thangavel et al. [43]/Reproduced with permission from John Wiley & Sons.

of 21.5 Wh kg^{-1}). For the volumetric performance, volumetric energy/power densities of 58.8 Wh l^{-1}/3546 W l^{-1} for the assembled SIC can be acquired. Besides, a microporous graphene cathode-based quasi-solid-state SIC was designed by Wang et al. [44] A sodium-ion-conducting gel polymer was used as the electrolyte. The SIC can achieve an impressive energy density of 168 kW kg^{-1}, a high operating cell voltage of 4.2 V, as well as a stable cycle life (specific capacitance retention of 85% after 1200 cycles).

Surface modification and elaborate structure design also provide good support for enhancing the electron/ion transport of graphene. Dong et al. revealed that the surface oxygen functional groups and wrinkled structures in rGO are beneficial for high-capacity cathodes [45]. Oxygen-functionalized crumpled graphene (OCG) cathodes with dense and porous structures provide abundant ion-accessible active sites and shorten the ion diffusion route, thus achieving a high capacity of 105 mAh g^{-1} at 0.1 A g^{-1}. After 3000 cycles, the capacity retention maintains about 91% even at 1 A g^{-1}. A quasi-solid-state SIC based on OCG as both the cathode and anode could deliver an impressive energy density of 121.3 Wh kg^{-1}, a power density of 8000 W kg^{-1}, and 86.7% capacitance retention after 2500 cycles. This work has important inspiration for the development of high-capacity cathodes. Besides, Thangavel et al. reported a SIC device fabricated by few-layered graphene nanosheets (GNSs) with a highly folded and crumpled morphology (Figure 5.5b) as the cathode and $NaTi_2(PO_4)_3$ grown directly on GNSs (GNTP) as the anode [43]. Benefiting from the high electronic conductivity and abundant mesopores that ensure the quick transfer of ions and electrons at high current, the GNS cathode can deliver a high capacitance of 118 F g^{-1} at 0.25 A g^{-1} and 78 F g^{-1}at 5 A g^{-1}, much higher than that of activated carbon-type electrodes, as shown in Figure 5.5c. The operating mechanism of the GNS//GNTP SIC is provided in Figure 5.5d; based on the fast sodium kinetics of the two electrodes, the assembled SIC exhibits an energy density of 80 Wh kg^{-1}, high specific power of 8 kW kg^{-1}, and a prolonged cycle life with 90% retention after 75 000 cycles. This result demonstrates the highly attractive and promising GNS as the cathode for SICs.

5.2.3.2 Carbon Nanosheets

2D carbon nanosheets (CNSs) with unique sheet-like morphology can be used as ideal cathode materials for SICs. Chen et al. created hierarchically porous carbon sheets by using various biomass as SICs cathode material precursors [46]. The carbon derived from the wood-like cotton stalk (BPC-5) exhibited an interconnected macroscopically sheet-like structure and randomly distributed short-range graphitic domains, as shown in Figure 5.6a,b. The high surface area and hierarchically porous structure (mainly mesopores and micropores) of BPC-5 facilitated charge

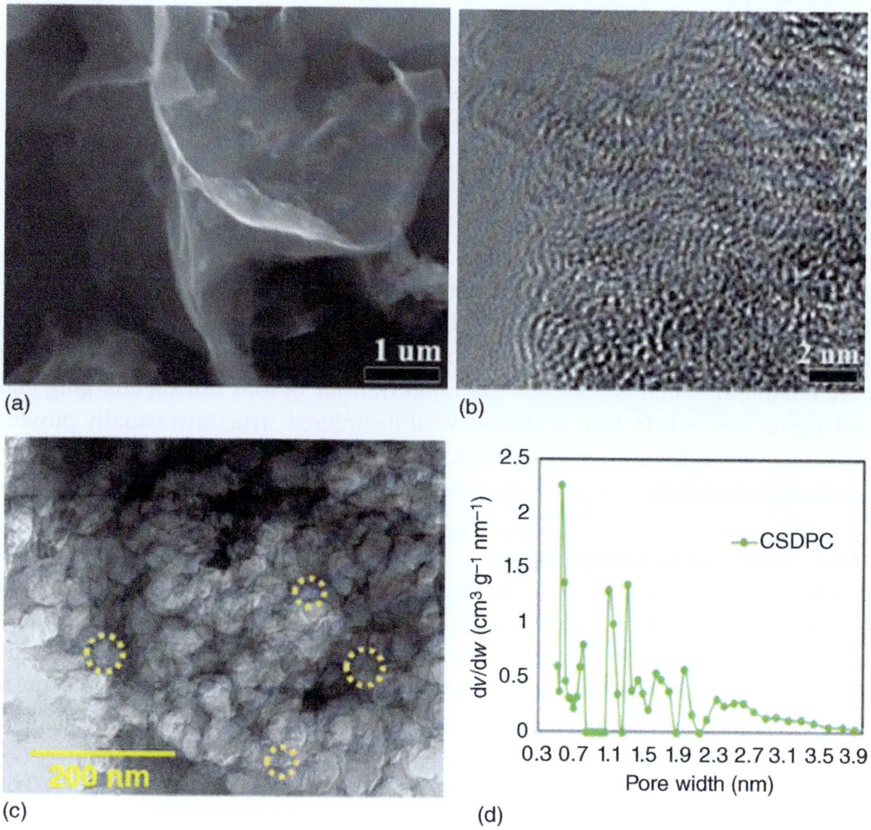

Figure 5.6 (a) SEM and (b) TEM images of BPC-5. Source: Chen et al. [46]/Reproduced with permission from Elsevier. (c) TEM image and (d) pore size distribution analysis of CSDPC cathode. Source: Surendran et al. [47]/Reproduced with permission from Elsevier.

transmission and provided sufficient active sites for energy storage. Thus, the BPC-5 cathode achieves the highest reversible capacitance of 160.5 F g^{-1} at 0.2 A g^{-1} (2–4.2 V), good rate and stability, superior to that of commercial AC and other biomass-derived porous carbon electrodes. Furthermore, a hybrid SIC was assembled by combining sheet-like BPC-5 cathode with $Na_2Ti_{2.97}Nb_{0.03}O_7$ anode, which delivers high specific energy of 169.4 Wh kg^{-1} at 120.5 W kg^{-1}, demonstrating excellent electrochemical performances. Similarly, Surendran et al. utilized biomaterial coconut sprout (CS) as a single precursor by KOH to prepare porous carbon sheets (CSDPC) with a high specific surface area (about 2000 m^2 g^{-1}) as a SIC cathode [47]. As presented in Figure 5.6c,d, the sheet-like morphology of CSDPC has constructed a favorable interconnected carbon network with an abundant microporous/mesoporous structure (0.5–50 nm), which is conducive to the construction of interconnected network structures, thereby promoting the penetration of electrolyte ions and enhancing its electrochemical properties as a promising cathode for SIC. Therefore, the specific capacity of CSDPC can be achieved at 90 mAh g^{-1} at 0.1 A g^{-1} with good rate capability. Moreover, the CSDPC cathode-based SIC can realize specific energy of 88 Wh kg^{-1} at a power density of 273 W kg^{-1}. Even at a power density of 1 kW kg^{-1}, the SIC delivers an energy density of 72 Wh kg^{-1}. The device also shows remarkable capacity retention of 75% after 8500 cycles. Facile preparation of high-performance biomass-derived carbon materials as the cathode is very promising for fabricating eco-friendly and high-efficiency SIC devices.

5.2.4 3D Carbonaceous Cathodes

Compared with 1D and 2D materials, 3D architectures are more conducive to enhancing the interfacial reaction kinetics by providing abundant reaction sites, thus ensuring high electrochemical activity [48]. 3D carbon materials generally have amorphous and disordered structures with a large number of interconnected space channels, which is good for the rapid transfer of ions during charging and discharging [49, 50]. In fact, materials with disordered structures usually possess disordered channels and well-developed pore structures, which are favorable for the adsorption of ions.

5.2.4.1 Hollow Carbon Microspheres

As a typical 3D material, CSs have regular and sharp-edge-free geometry, good fluidity, and high stability, and have applications in gas adsorption, catalysis, and electrochemical energy storage [51, 52]. Especially in the field of energy storage, non-graphitized CSs can promote Na^+ intercalation and delamination by virtue of their enlarged interlayer spacing (0.37 nm) when used as the anode [53]. Furthermore, porous CSs prepared by activating CSs possess improved surface utilization, shortened ion transport path, and can enhance the energy storage of the electric double layer.

Zhang and coworkers reported a scalable and sustainable purification-free route to synthesize CSs from various fresh juices by hydrothermal method [54]. The average diameters of the orange-derived CSs are about 3.3–4.7 μm, and the interlayer space is expanded to 0.375 nm (higher than the theoretical minimum interlayer separation of 0.370 nm) after pyrolysis at 800 °C, which is favorable for the adsorption and desorption of sodium ions. After a facile chemical activation process, the highly porous activated CSs were acquired, which exhibit a high capacity of 67.5 $mAh\,g^{-1}$ at 0.1 $A\,g^{-1}$ and good rate capability (40.7 $mAh\,g^{-1}$ at 5 $A\,g^{-1}$), superior to that of commercial activated carbon. The fabricated full-cell SIC (CS anodes and activated CS cathodes) shows an energy density of 52.2 $Wh\,kg^{-1}$ and a power density of 3 $kW\,kg^{-1}$, with 85.7% capacity retention after 2000 cycles. Porous carbonaceous materials with high porosity and abundant surface reactive sites are beneficial to realizing the excellent capacitive behavior of the cathode. Furthermore, Thangavel et al. successfully designed a high-performance SIC using self-assembly N, S co-doped 3D graphene hollow nanospheres (NS-GHNSs) as both anode and cathode [55]. The NS-GHNS, with a highly connected 3D hollow architecture, was synthesized by a facile sacrificial template method. The presence of N and S atoms can provide more active sites, which may originate from the high electronegativity and larger atomic size of impurity atoms, and further change the structure, charge, and spin density of carbon atoms. In addition, the impurity atoms can also improve the electrolyte wettability of the electrode surface. Benefiting from the hollow 3D architecture, high surface area, and high degree of dual-heteroatom doping that provides extra reactive sites, the synthesized NS-GHNS cathode can deliver capacities of 52 and 19 $mAh\,g^{-1}$ at current densities of 0.2 and 20 $A\,g^{-1}$. The NS-GHNS-based SIC displays a high energy density of 121 $Wh\,kg^{-1}$ with a high capacity retention of 85% after 10 000 cycles, superior to GHNS-based SIC and other devices reported previously.

5.2.4.2 Activated Hard Carbon

Similar to CSs, hard carbon spheres with a uniform structure are not only ideal anode materials for sodium storage but also can be activated to obtain porous carbon materials with high specific surface area and rich heteroatom functional groups, which can provide an efficient pseudocapacitance and thus increase the capacity and rate performance of the cathode [22]. Yin et al. synthesized an activated hard carbon (AHC) with abundant homogeneous nanopores and abundant oxygen-containing functional groups through the KOH activation of controllable carbon sphere precursors (Figure 5.7a) [56]. Due to the uniform morphology and partial carbonization of the carbon sphere precursor, the synthesized AHC has an amorphous and disordered structure with an expanded interlayer spacing of about 0.38 nm after the activation process (Figure 5.7b), which is beneficial for the fast transport of sodium ions. The O1s X-ray photoelectron spectra (XPS) reveal the adequate surface oxygen-containing functional groups of AHC, thus providing additional pseudocapacitance for sodium storage sites. Benefiting from the extra pseudocapacitive contribution combined with traditional EDLC, the AHC cathode delivers capacitances of 141 $F\,g^{-1}$ at 1 $A\,g^{-1}$ and 71 $F\,g^{-1}$ at 20 $A\,g^{-1}$, which are superior to those of the most reported carbon-based cathodes in sodium storage, as described in Figure 5.7c. Therefore, the SIC devices based on AHC cathode and sodiated-AHC anode demonstrate good electrochemical properties. The energy densities of the SICs can achieve 61.2, 51.9, 50.6, 54.1, 48.3, 44.1, and 39.7 $Wh\,kg^{-1}$ at

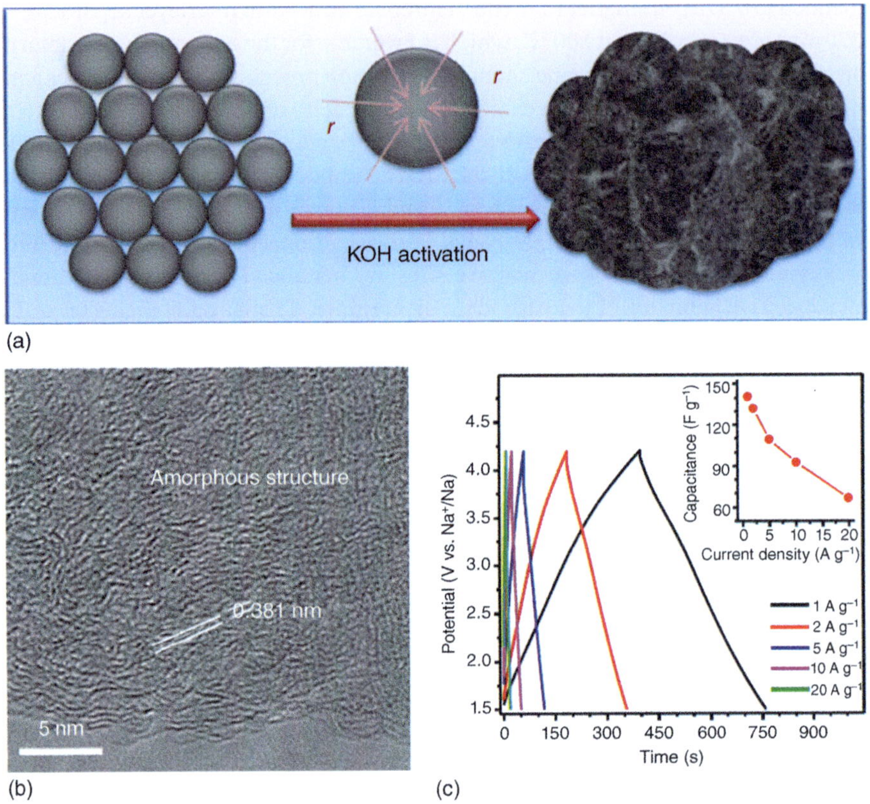

Figure 5.7 (a) KOH activation of carbon spheres, (b) TEM image of AHC, (c) GCD curves and rate capabilities of AHC cathode at various current densities. Source: Yin et al. [56]/ Reproduced from Springer Nature.

current densities of 0.2, 0.5, 1, 2, 3, 4, and 5 A g^{-1}, respectively. Even after 5000 cycles at 2 A g^{-1}, the energy density of SIC remains approximately 43 Wh kg^{-1} (1.44 times the initial state). This result confirms the promising application of AHC in SICs.

5.2.4.3 Disorder Carbon

As a typical disordered carbon material, activated carbon is essentially an amorphous material with a complex structure. The precursors of activated carbon, such as wood, coal, and polymers, are carbonized and activated to form graphite crystallites between the carbon atoms, but there is no regular stacking structure like graphite, and the rest of the carbon atoms are randomly arranged [10, 57–59]. Therefore, different from the microstructure of carbon atoms in crystalline graphite, diamond, and other materials arranged according to certain rules, activated carbon is composed of small graphite-like crystalline regions and amorphous carbon regions connected in disorder. Pores are formed in the interstitial spaces where the crystallites are bonded, giving the activated carbon its characteristic high specific surface area and porous structure [8, 60]. Therefore, activated carbon with high electrical conductivity, large specific surface area (>1000 m^2 g^{-1}), and low cost is the first choice for SICs cathode material, and it is also the only EDLC material that has been commercialized [61].

The first SIC energy storage system was developed by Lu and coworkers in 2012 utilizing commercial activated carbon as the cathode and V_2O_5/CNT nanowire composite as the anode [62]. The energy density of the SICs can reach 40 Wh kg^{-1} under the voltage window of 2.8 V. In the same year, Kuratani et al. used commercial activated carbon and pre-sodomized hard carbon to fabricate SICs and demonstrated good electrochemical performance comparable to the same type of LICs [18]. Afterward, research on SICs developed rapidly. However, the pore structure of the current commercial activated carbon materials is mainly micropores, and the complex and tortuous pore structure is not conducive to the quick transport of electrolyte ions during the charge/discharge process. Therefore, hierarchical pore size distributions with micropores and mesoporous can be constructed in activated carbon to improve the electrochemical performance of cathode materials. In addition, several researchers have proposed that pyridine- and pyrrolic-type nitrogen located at the edge of the carbon layer can provide abundant active sites, enhance surface charge storage through pseudocapacitive behavior, and graphitic nitrogen can enhance the electronic conductivity of carbon materials [24, 63–65]. Hence, much research focuses on the controllable preparation of various types of doped activated carbon materials and their practical application in SIC cathodes. For example, Zou et al. developed a 3D nitrogen-doped hierarchical porous activated carbon (NHPAC) amorphous structure cathode coupled with 3D nitrogen-doped hierarchical porous carbon (NHPC) anode to assemble a full-carbon SIC [66]. The larger interlayer distance of 0.42 nm, high specific surface area (1478 m^2 g^{-1}), and abundant meso/macroporous structure of the NHPAC enable rapid adsorption/desorption of anions on the surface (Figure 5.8a). Besides, the doping of N can supply pseudocapacitive storage behavior with faster kinetics (Figure 5.8b). For the above reasons, the NHPAC cathode reaches a high voltage of 110 mAh g^{-1} at the current density of 0.1 A g^{-1} and has superior rate performance. Moreover, the fabricated SIC with the similar material system of anode/cathode (Figure 5.8c) delivers a high energy density of 115 Wh kg^{-1} at 200 W kg^{-1} and good long-term ability after 10 000 cycles in the voltage window of 0–4.0 V, indicating the prospect of SICs to replace LICs.

Because of the tunable specific surface area and pore structure of low-cost biomass-derived activated carbon that can effectively enhance the electrochemical performances of the cathode, it has aroused extensive concern in the application of SICs [60, 68]. For example, a unique SIC construction with the peanut shells derived activated carbon as both the cathode and anode was firstly reported by Ding et al (Figure 5.8d) [67]. In the process of precursor treatment, the peanut shell is divided into two parts in the process of precursor treatment according to different needs. The peanut shell nanosheet carbon (PSNC) cathode with a macroscopically open sponge structure was synthesized by hydrothermal treatment followed by KOH chemical activation of the outer rough peanut shell (Figure 5.8e). The residual inner portion was firstly carbonized at 1200 °C in argon, followed by activation at low temperature (300 °C) in the air to acquire a peanut shell ordered carbon (PSOC)

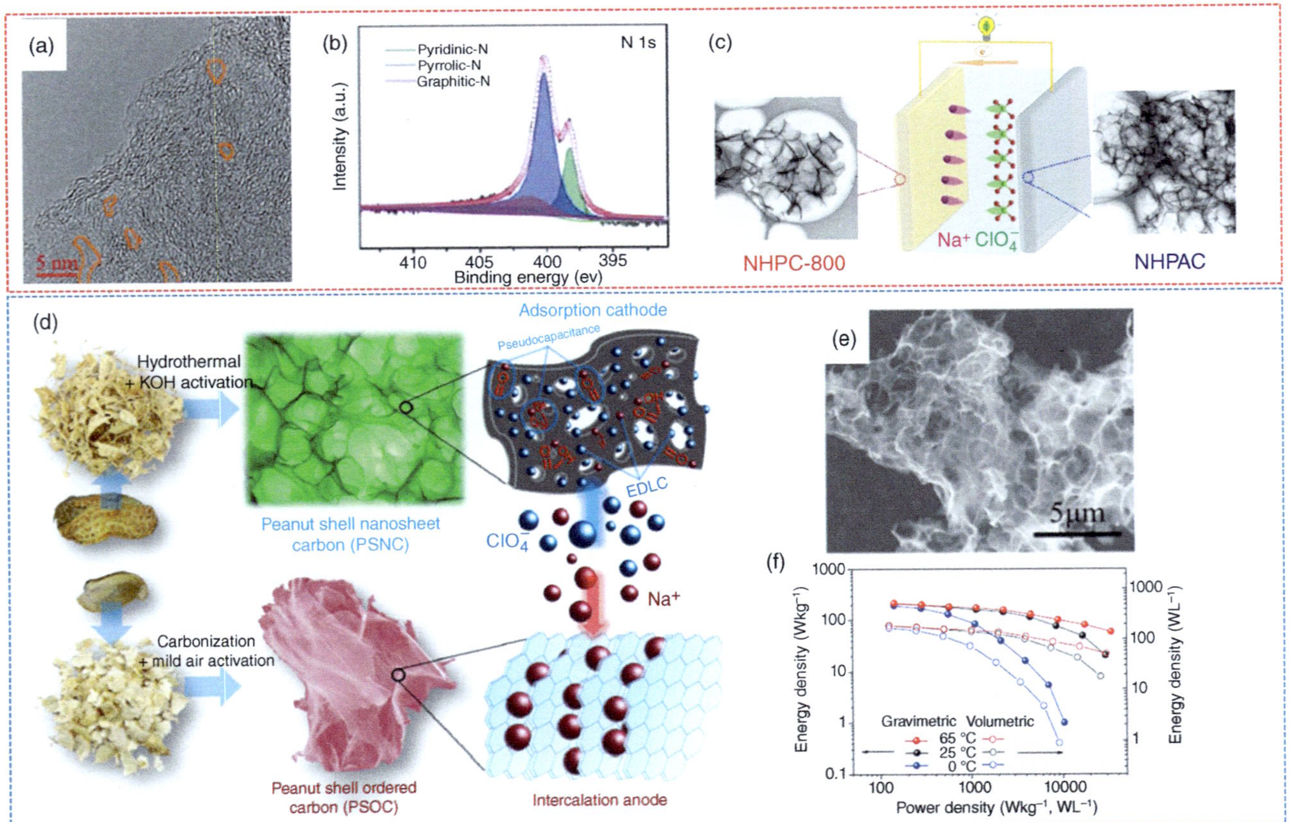

Figure 5.8 (a) HRTEM image and (b) N 1s XPS of NHPAC, (c) schematic of the NHPAC//NHPC-800 SIC device. Source: Zou et al. [66]/Reproduced with permission from Royal Society of Chemistry. (d) the synthesis route of PSNC and PSOC and the corresponding charge storage mechanisms in SIC, (e) SEM image of PSNC cathode, and (f) Ragone plots of the SIC at various temperatures. Source: Ding et al. [67]/Reproduced with permission from Royal Society of Chemistry.

anode. Benefiting from the open hierarchical structure of PSNC and the enlarged interlayer spacing (3.79 Å) of PSOC, the ion transfer kinetics, and capacity of both electrodes are enhanced. The anion adsorption and pseudocapacitive behavior provided by oxygen functionalities both contribute to the high capacity of the PSNC. Thus, the fabricated SIC achieves energy densities of 201–50 Wh kg^{-1} (at power densities of 285–16 500 W kg^{-1}) under temperature ranges of 0–65 °C, and 72% capacity retention after 10 000 cycles at a voltage range of 1.5–3.5 V (Figure 5.8f). Later, a Jasmine rice-derived hierarchical porous carbon (j-HPC) was successfully designed as both the cathode and anode of the SIC by Phattharasupakun et al. [69]. The amorphous j-HPC with high specific surface area (2300 m^2 g^{-1}) and residual nitrogen/oxygen functionalities was synthesized by a solvothermal method and then activated by KOH, which can achieve a capacity of 59.3 mAh g^{-1} at 0.1 A g^{-1} with fast charge transfer kinetics. Based on the j-HPC electrodes, the assembled SIC device provides a high energy density of 116.70 Wh kg^{-1} and power density of 11 121.54 W kg^{-1} as well as maintaining about 90.5% of capacity after 5000 cycles. In conclusion, low-cost biomass-derived porous carbons with high specific surface area, high electrical conductivity, and heteroatom doping are promising for SIC cathode materials.

5.2.4.4 Folded Carbon

Generally, the high specific surface area and small particle size of porous carbon will bring about the problem of low tap density, resulting in a decrease in the volume performance of the prepared electrode, which seriously hinders its practical application [23]. Therefore, it is necessary to achieve a balance between porosity and density. Zhang et al. prepared a high-density porous graphene macroform (HPGM) with folded texture by capillary drying method, as shown in Figure 5.9a. Further, by annealing the original densified graphene electrode in an inert atmosphere at 300–800 °C, the oxygen functional groups and folded texture can be manipulated quantitatively, thus affecting the electrochemical performance of the cathode (Figure 5.9b,c) [70]. On the one hand, for the folded texture, especially after 3D densification, the density of the material can be increased without sacrificing the reversible capacity, while the interconnected structure is beneficial to the rapid charge transfer; on the other hand, for the oxygen functional group, especially carbonyl and carboxyl groups, they can provide extra pseudocapacitance and increase the overall capacity of the cathode. The as-prepared HPGM has a high density of 1.05 g cm^{-3}, a specific capacity of 141 mAh cm^{-3} at 0.05 A g^{-1}, and a slight capacity fade when the current density increases from 1 to 5 A g^{-1} (Figure 5.9d). Moreover, even after 2000 cycles at 1 A g^{-1}, the HPGM cathode still has high cycle stability. This result is beneficial for inspiring the rational design and efficient fabrication of advanced structures and novel materials, thereby better boosting the progress of compact SICs.

In summary, to make SICs more competitive, it is necessary to further enhance the capacity and stability of EDLC cathode materials. Surface functionalization by chemical modification of porous carbon materials is an effective method to improve their capacitive performance. The presence of heteroatoms (such as nitrogen, sulfur,

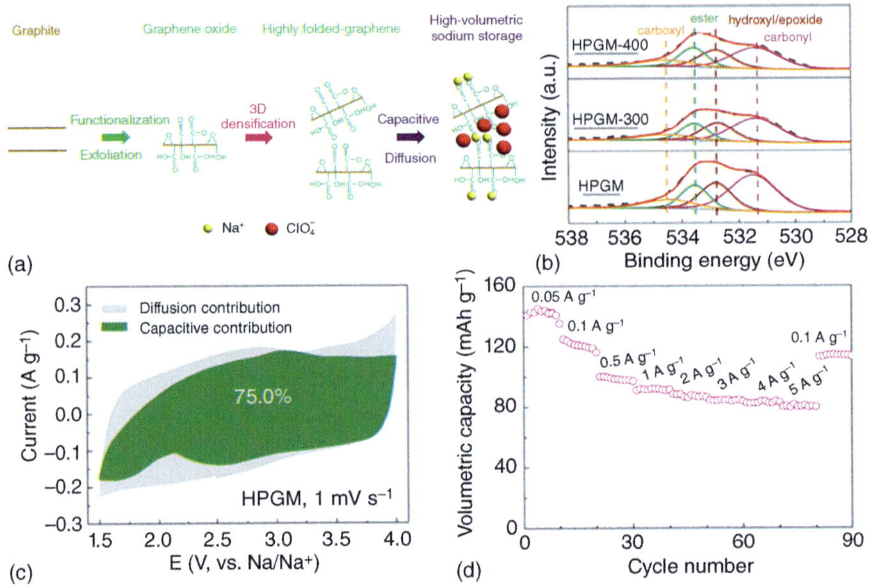

Figure 5.9 (a) Schematic diagram of the process of realizing high-volumetric Na storage using graphite as a starting point. (b) XPS O 1s spectra of HPGM, HPGM-300 and HPGM-400. (c) Capacitive contribution (green region) and diffusion contribution (gray region) in HPGM at a scanning rate of 1.0 mV s^{-1} (d) rate capability of HPGM cathode. Source: Reproduced with permission Zhang et al. [70]. Copyright 2018, WILEY-VCH.

phosphorus, and boron) can increase the wettability between the electrode and the electrolyte, which is conducive to the diffusion of electrolyte ions into the nanopores inside porous carbon, thereby improving the electrochemical performance of the porous carbon. So that the advantages of the high specific surface area of porous carbon can be fully exerted. In addition, an appropriate number of heteroatoms can also improve the pseudocapacitance of the cathode by providing extra reaction active sites. Therefore, it is promising to develop new methods to fabricate carbonaceous cathodes with controllable pore size, high specific surface area, and good electrolyte accessibility. By developing capacitive porous carbon materials with large layer spacing, hierarchical porosity, surface doping, and 3D structure or multicomponent composite, the capacity of porous carbon materials can be improved to compensate for the energy discrepancy between the cathode and anode, so as to achieve high energy density, power density, and long cycle stability of SIC devices.

5.3 Pseudocapacitive Cathode Materials

In the classical system of SICs, capacitive cathodes and battery anodes are often used. The cathode materials are mostly carbon materials with large specific surface area to ensure fast ion adsorption and release [71–73]. However, there is often capacity mismatch between the cathode of carbon materials and the anode of high theoretical capacity, which limits the energy density of the device. In order to provide high performance in the cathode, many pseudocapacitive materials have been explored. The motivation to search for pseudocapacitance electrode materials is to overcome the kinetics limitation of ion intercalation and to achieve higher energy density than charge adsorption [21, 74, 75]. Although the energy storage mechanism of pseudocapacitive materials is a reversible electrochemical reaction, the reaction kinetics are even comparable to EDLCs in some cases, while exhibiting high charge storage capacity comparable to batteries. Overall, pseudocapacitance can be considered a complementary mechanism to EDLC as it is not an electrostatic adsorption mechanism, but it has a similarly shaped CV curve and comparable GCD compared to EDLC curve trend. The electrochemical properties of pseudocapacitive materials are neither purely capacitive processes nor massive Faraday processes. These materials deliver battery-like redox reactions but are confined to the surface or near the surface (surface controlled). Thus, the kinetics of redox reactions in pseudocapacitives are dramatically faster and are not limited by semi-infinite diffusion [76]. Energy storage devices involving pseudocapacitive materials occupy an intermediate position between EDLC-type mechanisms and battery-type mechanisms. Therefore, more research has focused on replacing conventional EDLC cathodes or anodes with pseudocapacitive materials and further assembling hybrid ion capacitors.

The traditional pseudocapacitive materials in the cathode of SICs devices are shown in Figure 5.10, which can be divided into redox pseudocapacitance, intercalation pseudocapacitance, and adsorption pseudocapacitance [77–80]. In 1962, Conwsay and Gileadi first used the concept of pseudocapacitance to describe the reversible capacitance associated with electrochemical adsorption of substances on the surface of electrode, such as O or OH on nickel oxide and H on Pt [81]. The proposed kinetic theory assumes that the heat of adsorption is linearly related to the surface coverage of the electrodeposited species and leads to a capacitance defined in terms of the surface coverage. Conway extended the concept of pseudocapacitance to monolayer adsorption (adsorption pseudocapacitance, Figure 5.10a,d), the Faradaic reaction of hydrated RuO_2 in H_2SO_4 (redox pseudocapacitance, Figure 5.10b,e), and solid-solution electrochemical intercalation (intercalation pseudocapacitance, Figure 5.10c,f). The differences between these three mechanisms are due to different redox reaction processes of various materials. With surface redox pseudocapacitance, the electrolyte ions are electrochemically adsorbed on the electrode surface, creating short diffusion distances and short diffusion times. This mechanism occurs in intrinsically pseudocapacitive materials or can be realized by nanoengineering to elicit a pseudocapacitive response. In 1999, Lee and Goodenough [82] reported that a nanoscale amorphous, hydrated $MnO_2{\cdot}xH_2O$ exhibits capacitor-like behavior in aqueous electrolytes. Since then, many battery-type materials (such as Polyaniline [PANI], $LiCoO_2$, and V_2O_5) have been nanoengineered to induce pseudocapacitance and assembled high-power supercapacitors. The intercalation pseudocapacitance involves the rapid diffusion of ions into and out of the electrode materials, which rely on ion transport channels or interlayer spaces, without the material undergoing a phase change. Conwsay [83] proposed in 1990 that the intercalation of Li^+ in the interlayer of 2D transition metal oxides (TMOs) and chalcogenides (TMCs) is an adsorption process and described this process by

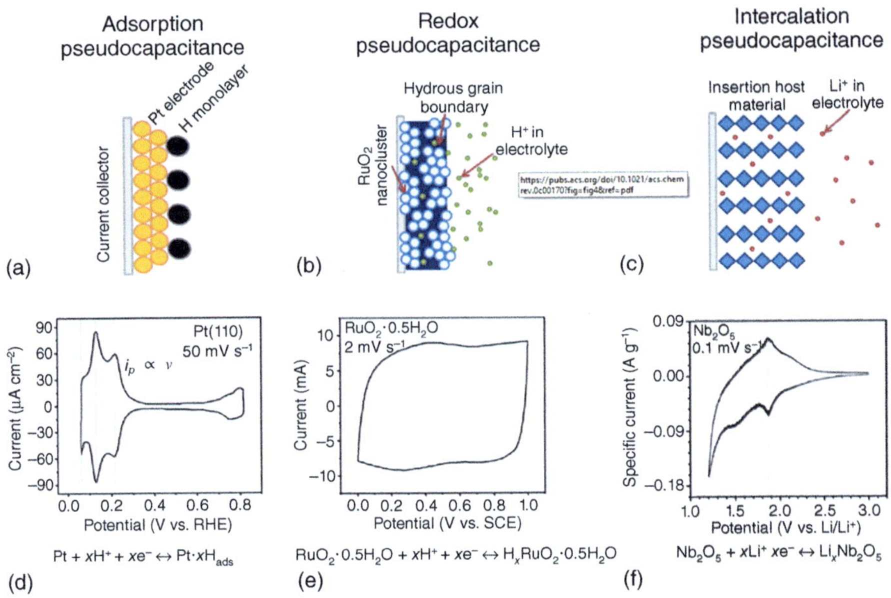

Figure 5.10 (a–c) Illustration of the three types of mechanisms that give rise to pseudocapacitance as defined by Conway and (d–f) their corresponding CV. (a) Adsorption pseudocapacitance, (b) redox pseudocapacitance, and (c) intercalation pseudocapacitance. (d) Cyclic voltammogram of a Pt(110) surface in aqueous 0.1 M $HClO_4$ solution cycled at $50\,mV\,s^{-1}$. RHE stands for the reversible hydrogen electrode, i_p for peak current, and v for sweep rate. The dotted lines indicate the high reversibility of the Faradaic adsorption processes. (e) CV of a $RuO_2{\cdot}0.5H_2O$ electrode in aqueous 0.5 M H_2SO_4 solution cycled at $2\,mV\,s^{-1}$. SCE stands for the standard calomel electrode. (f) CV of T-Nb_2O_5 nanocrystalline film in a nonaqueous Li^+ electrolyte cycled at $0.1\,mV\,s^{-1}$. The dashed line indicates the high reversibility of the Faradaic intercalation process [76]. Source: Copyright 2020, American Chemical Society.

intercalation pseudocapacitance [84]. In this class of TMOs and TMCs, there are crystallographic directions along which Li^+ can diffuse very rapidly, leading to no solid-state diffusion limitations over a wide range of time scales and no first-order phase transformations. The similarity in electrochemical properties is determined by the relationship between the potential and the charge range generated by the interface between the electrode and the electrolyte or the adsorption/desorption process on the inner surface of the material [76]:

$$E \approx E^0 - \frac{RT}{nF} Ln\left(\frac{X}{1-X}\right) \tag{5.1}$$

where E is the potential (V), R is the ideal gas constant ($8.314\,J\,mol^{-1}\,K^{-1}$), T is the temperature (K), n is the number of electrons, F is the Faraday constant ($96\,485\,C\,mol^{-1}$), and X is the extent of fractional coverage of the surface or inner structure. The above equations describe the thermodynamic properties of pseudocapacitive materials, but the strength of this type of energy storage material lies in the kinetic behavior. This is because the reaction only occurs at the surface or shallow layers of the material and is not affected by solid-state diffusion, thus exhibiting high-rate capability.

5.3.1 Adsorption Pseudocapacitive Materials

In fact, adsorption pseudocapacitive cathodes are rarely used in SIC devices. Traditional pseudocapacitance in cathodes of SICs devices includes redox pseudocapacitance and intercalation pseudocapacitance. But adsorption pseudocapacitive materials are fully discussed in the LIBs and SIBs as representative examples. And since the reaction occurs on the cathode surface, it has faster reaction kinetics, which can serve as a supplementary capacity contribution to the electrode capacity. The reaction mechanism of the adsorption pseudocapacitance is that during the charging and discharging process, the alkali metal ions of the electrode material are embedded and desorbed from the electrode material, and the surface adsorbs and desorbs PF_6^- or ClO_4^- in the electrolyte to increase the specific capacitance. As the cathode materials of hybrid ion capacitors, including but not limited to Mn_3O_4, MnO_2, Mn_5O_8, and RuO_2 can exhibit adsorption pseudocapacitance.

For example, Xia and coworkers [74] combined various in situ/ex situ spectroscopic techniques to investigate the electrochemical reactions of electrode surfaces in different electrolyte systems, demonstrating that the adsorption pseudocapacitance of Mn_3O_4 exists in the operating voltage window of 3–4.8 V. The reaction mechanism in the $LiPF_6$ electrolyte system is shown in Figure 5.11. During the charge–discharge process, the adsorption and release of PF_6^- appeared within the voltage window of 3–4.8 V. When the working voltage is lower than 3 V, it is the intercalation and de-intercalation process of Li^+.

Furthermore, the charge compensation mechanism of cathode materials originates from the partial charge transfer of Mn-F during adsorption and desorption,

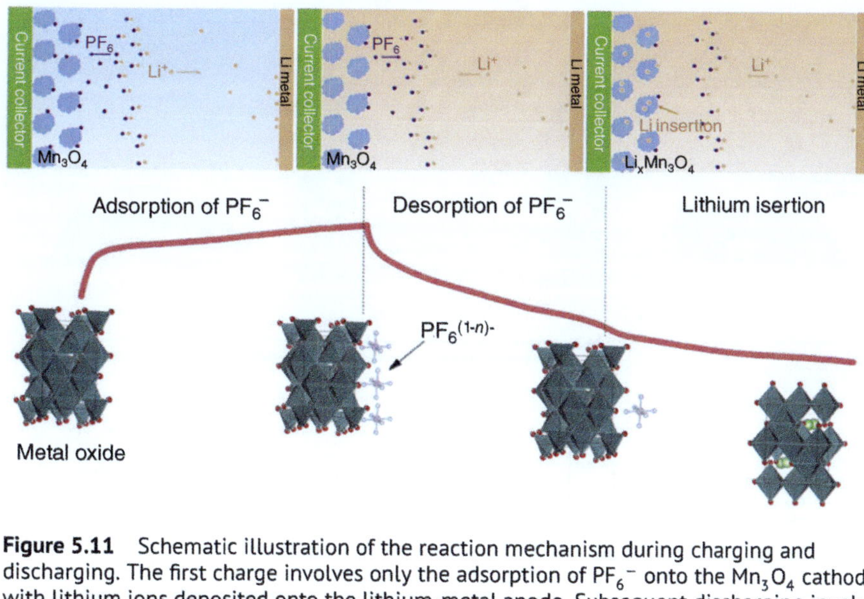

Figure 5.11 Schematic illustration of the reaction mechanism during charging and discharging. The first charge involves only the adsorption of PF_6^- onto the Mn_3O_4 cathode, with lithium ions deposited onto the lithium-metal anode. Subsequent discharging involves PF_6^- desorption and the lithium insertion reaction. Source: Reproduced with permission Li et al. [74]. Copyright 2020, Elsevier.

where F acts as a redox center. PF_6^- was adsorbed on the Mn_3O_4 surface, forming a Mn—F bond, which enabled the charge transfer from F to Mn. This mechanism can further increase the energy density of the material, but it can only be adsorbed on the surface of specific electrode materials. In addition to PF_6^-, $NaClO_4$ can also generate adsorption-type pseudocapacitance on specific electrode surfaces. This additional capacity compensation mechanism allows for a wide variety of cathode materials in SIC designs. Hu et al. [85] implanted Mn single-atom nanohybrids in N, F co-doped ultrathin porous carbon nanosheets (MnSAs/NF-CNs) as anodes and cathodes for SICs. N-coordinated Mn atoms (Mn-N_4) cannot only act as reversible Na^+ storage sites, lowering the energy barrier, but also improve the pseudocapacitance, thus making a great contribution to the accelerated reaction kinetics. Due to the large surface area and pore volume as well as highly hierarchical porosity, MnSAs/NF-CNs are employed as cathode materials. The capacity comes mainly from the capacitive capacity contribution. Thus, the ClO_4^- anions storage mechanism of MnSAs/NF-CNs depend on the adsorption/desorption onto electrode surface and reversible reactions between electrolyte ions and heteroatoms, and a fraction of intercalation process (Figure 5.12a). The MnSAs/NF-CN exhibits linear shape and highly symmetric GCD profiles, demonstrating a typical capacitive characteristic. The MnSAs/NF-CNs as the cathode exhibits a large average reversible specific capacity of 87 mAh g^{-1} at 0.1 A g^{-1}. It still has a specific capacity of 63 mAh g^{-1} when the current density is increased to 10 A g^{-1}. The SICs full cell (denoted as MnSAs/NF-CNs//MnSAs/NF-CNs) device can operate stably within the 0.01–4 V window, the GCD profiles display a slight deviation from the ideal rectangular and linear slope,

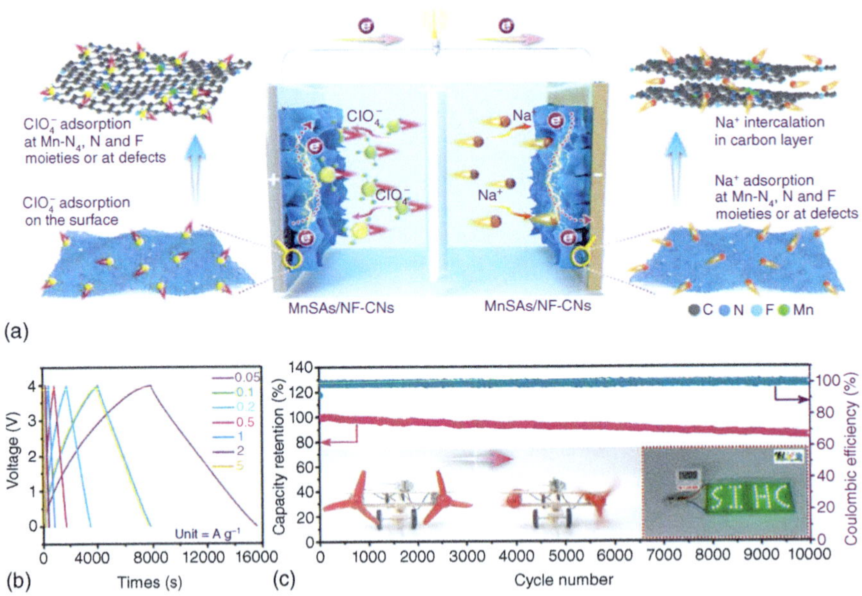

Figure 5.12 (a) Schematic illustration of the MnSAs/NF-CNs//MnSAs/NF-CNs SICs device. (b) GCD curves. (c) Long-term cycle performance of the SICs at 1 A g^{-1}. Source: Hu et al. [85]/ Reproduced with permission from Royal Society of Chemistry.

which is mainly due to the combination of multiple energy storage mechanisms in both the anode and the cathode (Figure 5.12b). The SIC-based MnSAs/NF-CNs delivers a high energy and power density of 197 Wh kg^{-1} and 9350 W kg^{-1}. In addition, these SICs also show an ultralong cycling stability, with 85.2% capacity retention after 10 000 cycles at 1 A g^{-1}, corresponding to only 0.0015% capacity loss per cycle (Figure 5.12c).

5.3.2 Redox Pseudocapacitive Materials

For redox pseudocapacitive materials, the charge storage mechanism is attributed to charge transfer that occurs at or near the material surface. The ions in the electrolyte are electrochemically adsorbed on the electrode surface, with a short diffusion distance and a short diffusion time. At present, the pseudocapacitive materials used as cathodes in SICs are mostly redox pseudocapacitive materials.

5.3.2.1 Conductive Polymers

Conductive polymers with coupled backbones can provide pseudocapacitive active sites with fast and highly reversible redox reactions. Conductive polymers generally have the advantages of nontoxicity, low-cost, flexible feature, structural adaptability, and lower sensitivity to the large sodium ions. The storage of sodium ions in organic molecules is mainly based on C=O reaction, doping reaction, and C=N reaction. Carbonyl reactions (C=O) are ubiquitous in carboxylate, quinone, imide, and anhydride compounds. Organic radicals, organometallics, and microporous polymers undergo doping reactions, while Schiff bases and pteridine derivatives undergo C=N reactions [86–88]. Various conducting polymers, including polyacetylene (PA), polypyrrole (PPy), PANI, and poly (3,4-ethylenedioxythiophene) (PEDOT), have been developed due to their high specific capacitance for SICs. Thangavel et al. [89] used electronically conductive PANI as cathode material in 2017 and assembled SIC composed entirely of organic matter organic hybrid capacitor (OHC). Unlike carbon materials that store charge through adsorption, conducting polymers store ions through a pseudocapacitive reaction. As shown in Figure 5.13a, the protonated state of PANI efficiently stores anions through fast surface pseudocapacitance and thus has similar rate capability and higher energy density compared to bilayer carbon electrodes. Additionally, the electronic conductivity and microscopic nanostructure of PANI can be easily controlled by changing the reaction conditions. And PANI is inexpensive and lighter and has potential as an electrode material for SICs.

From Figure 5.13b, the GCD curves of PANI cathode are not linear in the range of 3–4.5 V vs. Na^+/Na, because the main storage mechanism is the pseudocapacitive reaction of ClO_4^- anion through a fast reversible reaction. The fast pseudocapacitive reaction of conductive polymers is advantageous for accommodating many anions in a short period of time, thereby simultaneously increasing the energy density at higher powers. The PANI cathode delivers a specific capacity of about 68 mAh g^{-1} at 0.25 A g^{-1}, is much higher than that obtained using porous carbon-adsorbed cathode materials. Moreover, PANI shows a superior rate performance of about 51 mAh g^{-1}

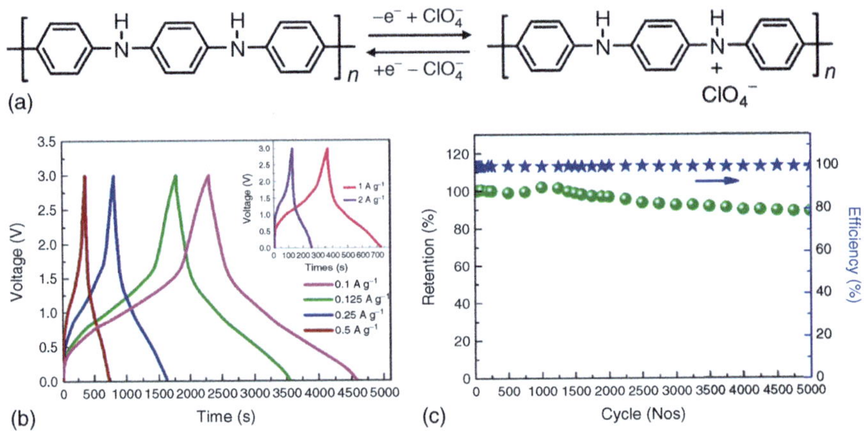

Figure 5.13 (a) Anion storage mechanism in PANI. (b) Charge–discharge profile of OHC at various current densities. The inset shows the CD profile at 1 and 2 $A\,g^{-1}$. (c) Cycling stability of the OHC. Source: Reproduced with permission Thangavel et al. [89]. Copyright 2017, American Chemical Society.

at 1 $A\,g^{-1}$ with an extremely high stability, where about 100% coulombic efficiency and 90% of the initial capacity is maintained beyond 200 cycles. And an all-OHC is assembled by PANI and perylene-3,4,9,10-tetracarboxylic acid dianhydride (PTCD). The PANI//PTCD OHC does not conform to the typical rectangular shape due to the synergetic effect between the two charge storage mechanisms. The OHC provides a high energy density of about 95 $Wh\,kg^{-1}$ and a power density of 7 $kW\,kg^{-1}$. In addition, the OHC exhibits excellent stability over 5000 cycles, with about 89% performance retention (Figure 5.13c).

5.3.2.2 Vanadium-Based Materials

Layered vanadium oxides exhibit pseudocapacitive behaviors and are the most promising cathode materials for SICs due to its high theoretical capacity. The high charge storage capability of vanadium-based materials is attributed to the multi-electron faradaic redox process from vanadium valance. The interlayer space of layered $V_2O_5{\cdot}nH_2O$ is about 1.336 nm, and the larger interlayer space can provide a fast channel for the transport of Na^+, and there is no phase transition during the charge–discharge process, which can provide a high sodium storage capacity of about 200 $mAh\,g^{-1}$. However, many studies have shown that origins of capacity loss for vanadium oxides are the collapse of layer structure, self-trapping of polarons, and decrease of conductivity during Na^+ (de)intercalations [90–92]. Due to the layered structure of vanadate-based cathodes, which is easy to collapse, the cycle life of $V_2O_5{\cdot}nH_2O$ cathode is very poor, which hinders the application of vanadate-based cathode for high energy and power SICs.

In order to solve this problem, as shown in Figure 5.14a,b, Dong et al. [93] used protonated pyridine (PyH^+) as the pillar to embed between $V_2O_5{\cdot}nH_2O$ layers to stabilize the structure. Pyridine-$V_2O_5{\cdot}nH_2O$ exhibits a specific capacity of 112 $mAh\,g^{-1}$ at 4 $A\,g^{-1}$ in a sodium half-cell with good rate capability (50 $mAh\,g^{-1}$ at 30 $A\,g^{-1}$) and

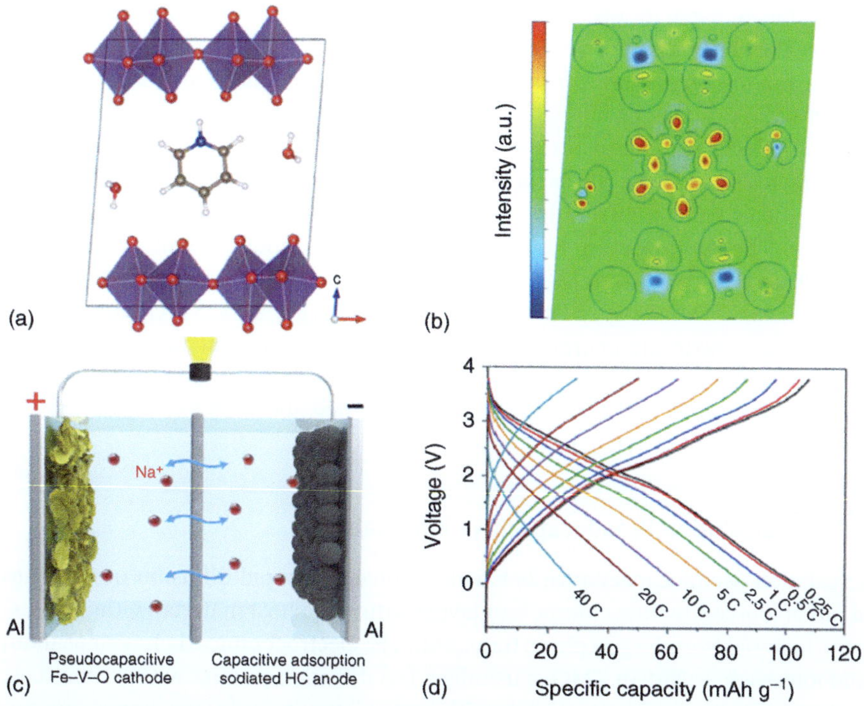

Figure 5.14 (a) Structural characterizations of pyridine-$V_2O_5 \cdot nH_2O$ nanowires optimized atomic configurations. (b) 2D charge-density difference. Source: Reproduced with permission Dong et al. [93]. Copyright 2019, WILEY-VCH. (c) Schematic of the SIC, consisting of pseudocapacitive Fe–V–O NSs cathode and capacitive adsorption sodiated HC anode. (d) The charge–discharge curves of SIC at various rates from 0.25 to 40 °C in a voltage window of 0–3.8 V. Source: Reproduced with permission Wei et al. [94]. Copyright 2018, Elsevier.

excellent cycle life (capacity retention of 67% after 3000 cycles). The SIC assembled with pyridine-$V_2O_5 \cdot nH_2O$ exhibits a maximum energy density of about 96 Wh kg^{-1} at a power density of 59 W kg^{-1} and a high power density of 14 kW kg^{-1} at an energy density of 37.5 Wh kg^{-1}.

The vanadates are derivatives of vanadium-based materials with increased electronic conductivity and enlarged/stabilized layer structure, which can also facilitate the transport of ions, showing promising electrochemical performance. Wei et al. [94] employed layered ferric vanadate (Fe–V–O) instead of conventional carbon material (AC) as cathode to construct SIC. During charging and discharging, the average valence state of vanadium varies from +4.92 to +3.90, indicating a redox transfer of vanadium with a total charge of 15.32 mol e^-/unit formula. The redox of iron reduces the iron valence from +3.00 to +2.66, according to 1.02 mol e^- per unit formula. Charge storage comes mainly from the redox state of V rather than that of Fe. The total redox reaction was very close to the delivered capacity (17 mol Na^+ per unit formula), confirming the pseudocapacitive behavior of Fe–V–O cathode. When the current density is 50 mA g^{-1}, the Fe–V–O delivers a high reversible sodium storage capacity up to 229 mAh g^{-1}, excellent rate capability, and cycling stability. Furthermore, the interlayer space of Fe–V–O is stabilized at about 0.87 nm, and the very small crystal structure change is beneficial to obtain excellent cyclability.

The pseudocapacitance contribution of Fe–V–O cathode is calculated as 83%. According to the analysis, the capacitive contribution occurs at a rate of 0.2 mV s^{-1}, whereas the capacity contribution from the diffusion process occurs at relatively slow rate. The fast reaction kinetics favor Fe–V–O should be on SIC cathode materials. And the SIC assembled by Fe–V–O cathode and HC anode Fe–V–O//HC (Figure 5.14c) delivers a specific capacity of 104 mAh g^{-1} at 0.25 C (Figure 5.14d). The SIC has a maximum energy density of about 194 Wh kg^{-1} at an average power density of 45 W kg^{-1}. The SIC shows stable cyclability over 800 cycles with a capacity retention of 67.2% and coulombic efficiency above 99.5%, indicating the good reversibility of the HC//Fe–V–O SIC.

5.3.3 Intercalation Pseudocapacitive Materials

Intercalation pseudocapacitance behavior involves the rapid diffusion of ions into and out of ion-conducting channels or layers within the host material without causing the material to undergo a phase transition. The electrochemical behavior of alkali metal ions such as Na^+ in TMO is usually a fast diffusion process with a time scale of only a few minutes. The materials with intercalation pseudocapacitance are crucial for improving the electrochemical performance of SICs. And thus, by converting the diffusion process into surface control, the kinetics of the electrode material will also be greatly improved. Furthermore, new mechanisms such as anion intercalation are currently being explored in the search for high-power energy storage materials.

There are many kinds of intercalated pseudocapacitive materials. Among these, MXene materials have been widely studied for their high electrical conductivity and unique surface chemistry [95, 96]. MXene was developed and prepared by Gogotsi et al. in 2011 [97]. The general formula of MXene can be expressed as $M_{n+1}X_nT_x$, in which M represents an early transition metal, X is C and/or N, and $n = 1$, 2, and 3, T is the surface functional groups (–OH, –O, –F, –Cl, or –Br, etc.) (Figure 5.15a) [99, 100]. MXene is mainly obtained by selective etching of MAX, where A is the element in the third or fourth main group in the periodic table, such as Al or Ga. Taking hydrofluoric acid (HF) etching as an example, the etching process can be expressed by the following formula [101]:

$$M_3AX_2(s) + 3HF(aq) \rightarrow AF_3(s) + 3/2H_2(aq) + M_3X_2\ (g) \tag{5.2}$$

$$M_3X_2(s) + 2H_2O(aq) \rightarrow M_3X_2(OH)_2(s) + H_2(g) \tag{5.3}$$

$$M_3X_2(s) + 2HF(aq) \rightarrow M_3X_2F_2(s) + H_2(g) \tag{5.4}$$

There are more than 30 kinds of MXenes that have been synthesized in the laboratory (Figure 5.15b). The diversity of MXene species also exhibits several excellent properties, such as high electrical conductivity, high bulk capacitance,

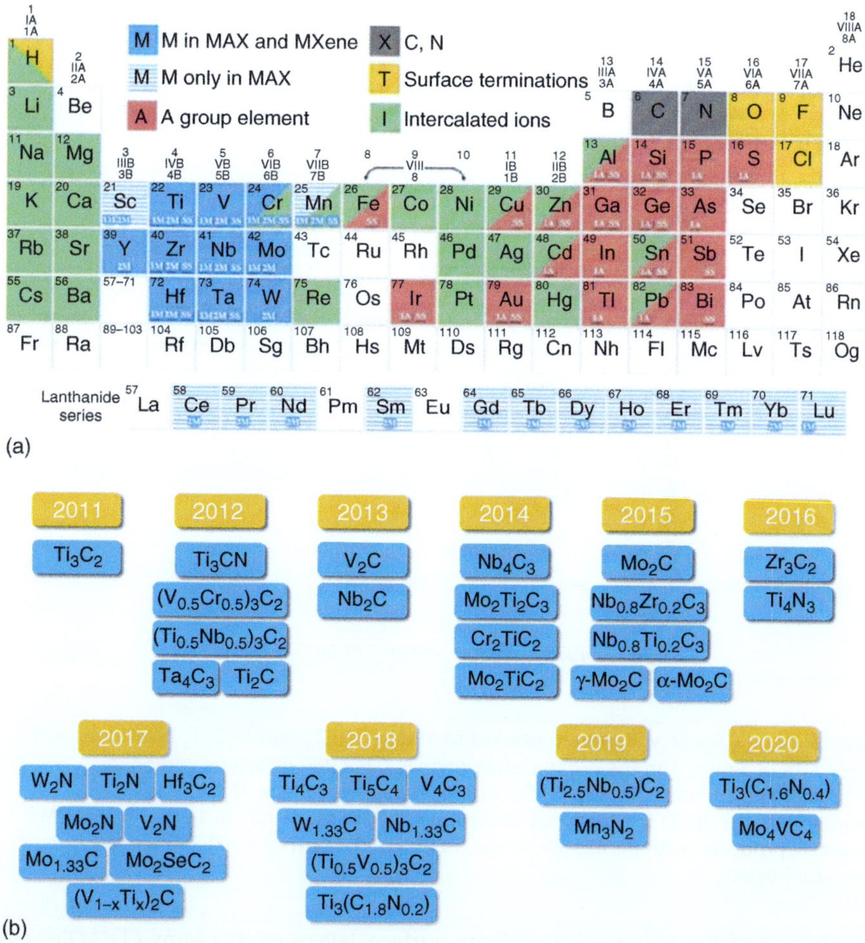

Figure 5.15 (a) Periodic tables presenting compositions of MAX phases and MXenes. Source: Reproduced with permission Gao et al. [98]. Copyright 2021, WILEY-VCH. (b) Experimentally obtained MXenes reported to date.

and electromagnetic interference shielding. Based on the excellent properties of MXene materials, it has broad application prospects in electrochemical energy storage, transparent conductive electrodes, electromagnetic interference shielding, photodetectors, sensing, etc. [98].

In the field of energy storage, MXene-based materials have also attracted great interest. It has excellent electrochemical properties, as well as good mechanical stability, hydrophilicity, and 2D layered structure. The high electrical conductivity and fast ion transport channels between layers can effectively alleviate the kinetic mismatch between cathode and anode. $Ti_3C_2T_x$ MXene has been intensively studied for energy storage applications [102, 103]. CV of this material in sulfuric acid electrolyte shows broadened peaks with little separation in peak position on charge/discharge, on top of a capacitive rectangular box. In this material,

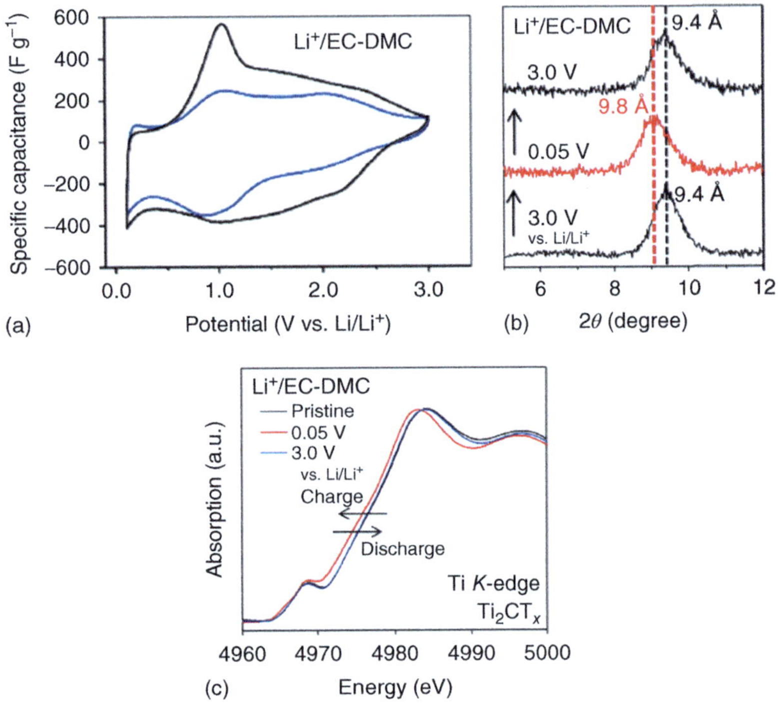

Figure 5.16 (a) Cyclic voltammetry curves of MXene Ti_2CT_x and $Ti_3C_2T_x$ electrodes at a sweep rate of 1 $mV\,s^{-1}$ in the organic electrolyte (1.0 M $LiPF_6$/ethylene carbonate [EC]-dimethyl carbonate [DMC]). (b) Structural changes of MXene Ti_2CT_x upon charge/discharge. (c) Ti K-edge X-ray absorption spectra for Ti_2CT_x in 1 M $LiPF_6$/EC-DMC. Source: Reproduced with permission Okubo et al. [104]. Copyright 2017, American Chemical Society.

the change of the valence state of the surface layers of Ti atoms (Ti^{2+}/Ti^{3+}) is balanced by the reversible intercalation reaction of protons [103]. Okubo et al. [104] verified that Li^+ in organic electrolytes embeds between layers of MXenes and produces charge transfer. As shown in Figure 5.16a, Li^+ shows an oxidation peak at 1.1 V and a reduction peak at 0.9 V, corresponding to the reversible reaction in Eq. (5.5):

$$Ti_3C_2 + 2Li = Ti_3C_2Li_2 \tag{5.5}$$

With the occurrence of lithium intercalation, Li^+ enters the interlayer of MXene and expands the interlayer space. Therefore, on the XRD curve, the (002) peak representing the interlayer distance shifts to the left, and the interlayer space increases from 9.4 to 9.8 Å. When the delithiation reaction is complete, it returns to 9.4 Å (Figure 5.16b). The K-edge X-ray absorption spectrum of Ti also shows that Ti is accompanied by a reversible redox reaction during the charging and discharging process (Figure 5.16c). Theoretically, the specific capacitance (C) of an electrode can be calculated by Eq. (5.6), where $\Delta(\varphi cc - \varphi b)$ is the change in the inner potential difference between a current collector (φcc) and a bulk electrolyte (φb) and ΔQ is the stored charge per the weight of the electrode.

$$C = \frac{\Delta Q}{\Delta(\varphi cc - \varphi b)} \quad (5.6)$$

$$\Delta(\varphi cc - \varphi b) = \Delta\left(\varphi_e^E - \varphi_i^E\right) - \frac{\Delta\left(\mu_e^E + \mu_i^E\right)}{F} \quad (5.7)$$

Using the equilibrium conditions of the electron and ion electrochemical potentials, $\Delta(\varphi cc - \varphi b)$ can also be calculated by Eq. (5.7), in which μ_e^E and μ_i^E are the chemical potentials of an electron and an ion in the electrode, respectively, and φ_e^E and φ_i^E are the inner potentials of the electron and ion in the electrode, respectively. The first term gives an EDL capacitance, whereas the second term gives a non-EDL capacitance (pseudocapacitance, quantum capacitance, or chemical capacitance). For nonaqueous electrolytes, the solvation energy is much weaker than that for hydration. Partial desolvation thus occurs more easily at the MXene–electrolyte interface, and after intercalation, the atomic orbitals of cations hybridize with the orbitals of the surface termination groups of the MXene. The existence of a donor band of cations was supported by DFT calculations. This orbital hybridization forms a donor band to reduce MXene. Importantly, charge transfer between MXene and intercalated cations occurs through orbital hybridization, which should significantly screen the electric field between the electron and cation to reduce $\Delta\left(\varphi_e^E - \varphi_i^E\right)$. In addition, the large potential window of organic electrolytes allows the storage of large amounts of electrons and cations, leading to considerable changes in μ_e^E and μ_i^E due to the band-filling effect and intercation interactions. Therefore, the capacitance of MXene electrodes in nonaqueous electrolytes arises mainly from intercalation pseudocapacitance dominated by the chemical potentials, which are larger than those in aqueous electrolytes.

Dall'Agnese et al. [105] first used V_2CT_x (MXene) as cathode material and assembled a SIC with HC anode in 2015. The etching process and sodium storage mechanism of V_2CT_x MXene are shown in Figure 5.17, the socialization and desocialization

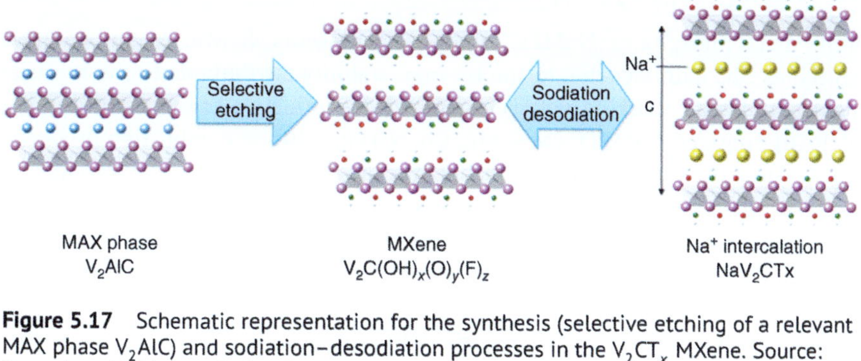

Figure 5.17 Schematic representation for the synthesis (selective etching of a relevant MAX phase V_2AlC) and sodiation–desodiation processes in the V_2CT_x MXene. Source: Reproduced with permission Dall'Agnese et al. [105]. Copyright 2015, American Chemical Society.

mechanism of V_2CT_x is a typical intercalation pseudocapacitance. The profiles of the V_2CT_x in 1–3.5 V are similar to the capacitive curve, and V_2CT_x has a specific capacity of about 100 mAh g^{-1}. The SIC that was assembled by HC anode and V_2CT_x cathode (V_2CT_x//HC) exhibited a maximum cell voltage of 3.5 V and an energy density of 103 Wh kg^{-1}. At the current density of 2 A g^{-1}, the capacity retention was 70% after 300 cycles. $Ti_3C_2T_x$ can also be used as the cathode of SIC, Zhang and coworkers [106] assembled a SIC with MnO_2 anode and $Ti_3C_2T_x$ cathode in $NaSO_4$ aqueous electrolyte. The SIC can safely work with voltage of 0–2.4 V and shows outstanding rate performance; even when the current density increased to 250 times the initial value, the capacity retention at still retained 38%. Moreover, the SIC delivered superior cycling stability, with 95.9% of the capacitance retention and over 90% of the Coulombic efficiency still retained after 450 cycles.

5.4 Battery-Type Cathode Materials

Generally, SICs can be classified into two configurations: one is electrolyte-consuming SIC, in which anions and cations from electrolyte simultaneously act as charge carriers during the charge/discharge process [2, 107]. The typical anode involves a faradaic Na^+ insertion process, while the cathode is an EDLC-type carbon material [108, 109]. Another configuration is the rocking-chair SIC, in which the anode is an EDLC-type material, and the anode is a battery-type Na^+ intercalation material. So far, most studies have focused on the first SIC configuration. But according to Eq. (5.8), the EDLC cathodes are often limited by low capacity, resulting in low energy density of SICs.

$$1/C_{SIC} = 1/C_{cathode} + 1/C_{anode} \tag{5.8}$$

Comparatively, the rocking-chair SICs could overcome this bottleneck. Various sodium-storage cathode materials are used in the field of SICs.

5.4.1 $NaMn_{1/3}Co_{1/3}Ni_{1/3}PO_4$ Cathodes

Sodium metal phosphates ($NaMPO_4$, where M represents Ni, Mn, Co, and Fe) have potential as SIC cathodes due to the thermal stability attributable to the maricite structure and high working voltage owing to the inductive effect. Senthilkumar et al. [110] prepared $NaMPO_4$ (M = Mn, Co, Ni) via solution combustion processes in 2014. These three sodium metal phosphates show different specific capacitances of 368 F g^{-1} ($NaNiPO_4$), 249 F g^{-1} ($NaCoPO_4$), and 163 F g^{-1} ($NaMnPO_4$) at the current density of 2 mA cm^{-2} in the NaOH aqueous solution. In addition, a SIC is assembled by AC anode and $NaNiPO_4$ cathode ($NaNiPO_4$//AC), which delivered energy and power densities of 20 W h kg^{-1} and 1358 W kg^{-1}.

Among sodium metal phosphates, iron-based phosphates have attracted extensive attention due to their low cost and environmental friendliness. Among them, $NaFePO_4$ stands out for its theoretical specific capacity of 154 mAh g^{-1} and suitable working potential. However, the thermodynamically stable $NaFePO_4$

phase lacks fast sodium ion transport channels and has low electrochemical activity [111, 112]. In addition, the intrinsic conductivity of $NaFePO_4$ is poor, and the crystal structure changes obviously after the intercalation of Na^+, which is not conducive to the rate performance and cycling stability. Therefore, it is often necessary to use other forms of ion doping or composites with other functional materials to improve the electrochemical performance. According to the conductivity sequence of U_{Fe} (3.71 eV) < U_{Mn} (3.92 eV) < U_{Co} (5.05 eV) < U_{Ni} (5.26 eV) in the standard Hubbard model (DFT + U method). The mixed sodium metal phosphates ($NaMn_{1/3}Co_{1/3}Ni_{1/3}PO_4$) have attracted much attention due to their narrow band gap and good electrical conductivity. In 2015, Sundaram et al. [113] proposed to replace the hybrid metal cations in $NaFePO_4$ with other metal elements to improve the electrical conductivity of iron-containing phosphate materials (Figure 5.18a). $NaMn_{1/3}Co_{1/3}Ni_{1/3}PO_4$ was successfully prepared by Sundaram et al. and used as cathode material for SICs. According to the results of DFT calculations (Figure 5.18b), the conduction band comes from the 3D electrons of the doped transition metal (Fe, Mn, Co, and Ni) atoms. The Ni-3D electrons play an essential role on the edge of the conduction band as well as the valence band, which is mostly dominated by the 2p-electrons of O and has hybridized O-2p and Mn-3D characters at the edge of the valence band. The observed smaller E_g (2.24–2.32 eV) of $NaMn_{1/3}Co_{1/3}Ni_{1/3}PO_4$ suggests that the electrical conductivity is improved after mixing with various transition metal such as Mn, Co, and Ni. Moreover, at the near core energy states, small peaks from Ni, Co, and Mn atoms, corresponding to their 3p-states, appear in the energy order of Ni < Co < Mn. Compared with $NaMPO_4$, the higher conductivity of $NaMn_{1/3}Co_{1/3}Ni_{1/3}PO_4$ is beneficial to the rate and cycle performance of the electrode and is more suitable for the field of SICs.

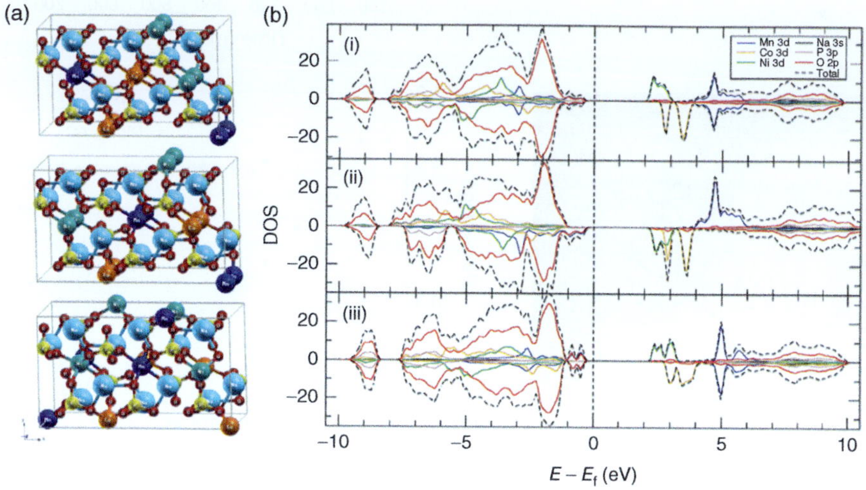

Figure 5.18 (a) The structures of $NaMn_{1/3}Co_{1/3}Ni_{1/3}PO_4$. (b) The total DOS (dash line) and orbital-decomposed DOS for each element (solid line) for the configuration of $NaMn_{1/3}Co_{1/3}Ni_{1/3}PO_4$ obtained from GGA + U calculations. Source: Reproduced with permission Sundaram et al. [113]. Copyright 2015, The Royal Society of Chemistry.

The reaction proposed for the charge and discharge processes of $NaMn_{1/3}Co_{1/3}Ni_{1/3}PO_4$ is as follows:

$$NaMn_{1/3}Co_{1/3}Ni_{1/3}PO_4 \leftrightarrow Na_{1-x}Mn_{1/3}Co_{1/3}Ni_{1/3}PO_4 + xNa^+ + xe^- \quad (5.9)$$

The SIC assembled with $NaMn_{1/3}Co_{1/3}Ni_{1/3}PO_4$ cathode and AC anode delivers an energy density of 50 Wh kg^{-1} at a power density of 180 W kg^{-1}. After 1000 cycles, the capacity retention rates and coulombic efficiency of the SICs devices were maintained at 95% and 92%, respectively, which suggested good electrochemical reversibility.

5.4.2 $Na_3V_2(PO_4)_3$ Cathodes

$Na_3V_2(PO4)_3$ (NVP) is a typical cathode material with NASICON structures for SIBs, which facilitates the insertion and extraction of Na^+ (Figure 5.19a) [117]. It features a highly covalent 3D framework that creates large interstitial spaces through which Na^+ can diffuse. One Na ion can easily insert into, or two Na ions can extract from the

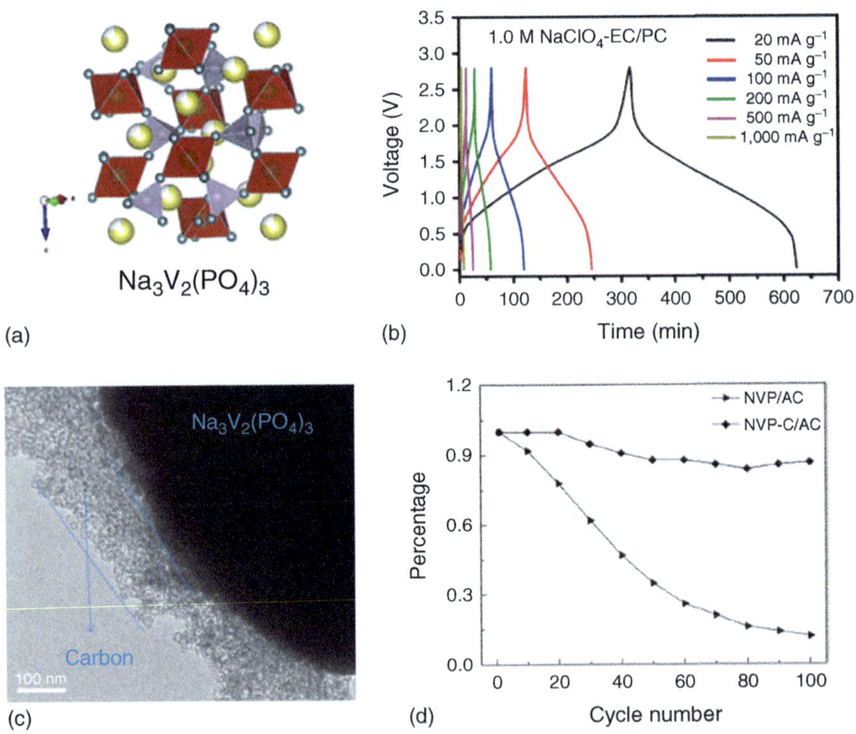

Figure 5.19 (a) Structure of NPF. Source: Reproduced with permission Yang et al. [114]. Copyright 2020, Elsevier. (b) Charge and discharge curves of the NPF//AC SIC employing 1.0 M $NaClO_4$-EC/PC at various current densities. Source: Reproduced with permission Park et al. [115]. Copyright 2018, WILEY-VCH. (c) TEM of NVP/C. (d) Cycling performance of NVP/AC and NVP-C/AC SIC at charge/discharge current density of 5 mA cm^{-1}. Source: Wang et al. [116]/Reproduced with permission from Springer Nature.

NVP host with two distinctive plateaus of 1.6 and 3.4 V, respectively. NVP can exhibit excellent structural stability, high working potential (about 3.4 V vs. Na^+/Na), good thermal stability, high theoretical energy density (about 394 Wh kg^{-1}), and open 3D framework structure. And the reaction mechanism of NVP cathode could be written as follows:

$$Na_3V_2(PO_4)_3 \leftrightarrow Na_{3-x}V_2(PO_4)_3 + xNa^+ + xe^- \quad (5.10)$$

Park et al. [115] assembled a SIC with AC anode and $Na_3V_2(PO_4)_3$ cathode. The SIC used ethylene carbonate (EC)/propylene carbonate with 1.0 M $NaClO_4$ as the electrolytes, showing good rate performances (Figure 5.19b). When the current density is 1 A g^{-1}, the $Na_3V_2(PO_4)_3$//AC has a good discharge capacity of 63.1 mAh g^{-1}. And at a current density of 0.5 A g^{-1}, the capacity retention rate is 90.8% after 1000 cycles. To further improve the energy density of SICs, pseudocapacitive capacity can be increased in anode materials with high specific surface area. Chen et al. [118] assembled a SIC by Zn-doped mesoporous hollow carbon spheres (MHCS–Zn) anode and $Na_3V_2(PO_4)_3$ cathode. MHCS–Zn displays an exceptional sodium storage kinetics and excellent long cycling life, which can effectively improve the energy density of SIC. The $Na_3V_2(PO_4)_3$//MHCS–Zn SIC exhibits a good cycling with a remained capacity of 84.2 mAh g^{-1} after 1000 cycles at 2 A g^{-1}. The SIC exhibits a high energy density of 145 Wh Kg^{-1} at a very high power of 8009 W kg^{-1} and a 168 Wh Kg^{-1} at 1694 W kg^{-1}.

However, NVP also suffers from the defect of low electronic conductivity (10^{-9} S cm^{-1}), which greatly limits the further practical applications. To address this issue, many studies have been devoted to the modification of NVPs with highly conductive materials such as carbon materials to improve electrochemical performance. Wang et al. [116] synthesized NVP/C by a modified sol–gel method in 2015. The transmission electron microscopy (TEM) of NVP/C is shown in the Figure 5.19c, NVP material is covered with porous amorphous carbon layer on the particles surface. The electronic conductivity and structure stability of NVP/C cathode are improved. The as-prepared NVP/C exhibits a specific capacitance as high as 51.0 F g^{-1} at a current density of 0.65 A g^{-1}, and the capacity retention rate exceeds 80% after 100 charge–discharge cycles. And the SIC assembled by NVP/C and AC delivers more stable cycling performance than NVP//AC (Figure 5.19d).

5.4.3 $Na_3V_2O_2(PO_4)_2F$ Cathodes

The polyanion-type fluorophosphates are always constructed by MO_xF_{6-x} (M represents the transition metals) octahedral and PO_4 tetrahedral units, thus delivering high structure and thermal stability. Moreover, the polyanion-type fluorophosphates have high Na^+ diffusion coefficient due to the layers along the *ab*-plane in crystal structure, which provide sites for sodium ion storage and transport, such as $Na_{1.5}VPO_{4.8}F_{0.7}$, $Na_3V_2(PO_4)_2F_3$, $NaVPO_4F$, and $Na_3V_2O_2(PO_4)_2F$ (NVOPF).

Among these compounds, NVOPF is a typical cathode material (Figure 5.20a). Like NVP, two Na^+ can be extracted from NVOPF with a theoretical specific capacity of 127 mAh g^{-1}. The structural and electrochemical properties of NVOPF are

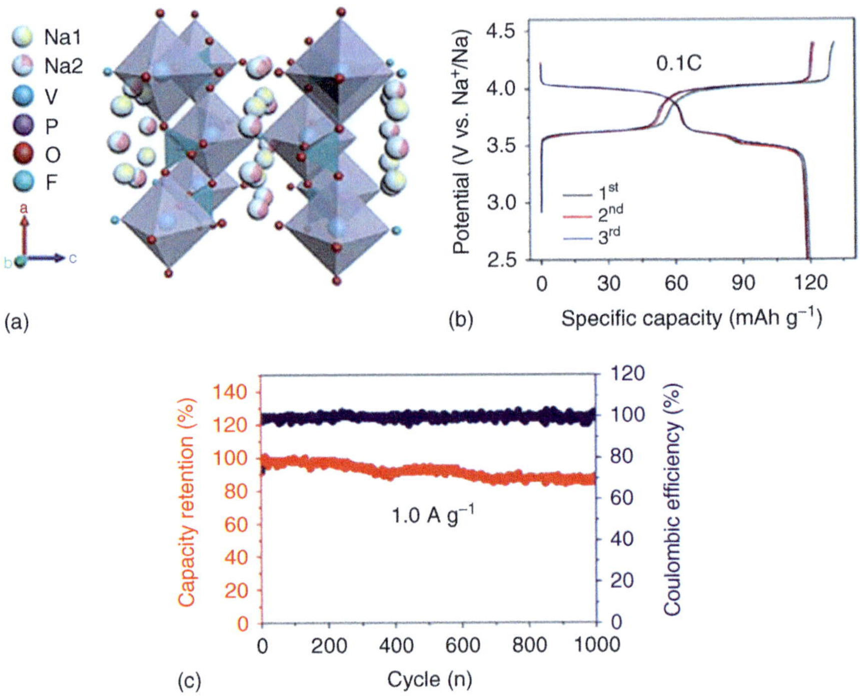

Figure 5.20 (a) Structure of NVOPF. (b) GCD curves of the NVOPF@PEDOT at 0.1 C. (c) Cycling performance of the NVOPF@ PEDOT//AC SIC at 1 A g^{-1}. Source: Reproduced with permission Wu et al. [119]. Copyright 2019, The Royal Society of Chemistry.

changed after the partial substitution of PO_4^{3-} by F^- anion. Two higher plateaus of 3.6 and 4.0 V appear in the charge and discharge process, which is beneficial for achieving high energy density (Figure 5.20b). The corresponding two step desodiation/sodiation processes:

$$Na_3V_2O_2(PO_4)_2F \leftrightarrow Na_2V_2O_2(PO_4)_2F + Na^+ + e^- \quad (5.11)$$

$$Na_2V_2O_2(PO_4)_2F \leftrightarrow NaV_2O_2(PO_4)_2F + Na^+ + e^- \quad (5.12)$$

However, the rate capability of NVOPF cannot meet actual requirements. The electronic conductivity of NVOPF (1.8×10^{-7} S cm^{-1}) is poor. To improve this issue, many studies have been devoted to constructing composite material systems. Goodenough et al. reported NVOPF/graphene sandwich nanostructure prepared by a solvothermal method that delivers a discharge capacity of about 110 mAh g^{-1} at 1 C. Wu et al. [119] proposed a facile low-temperature-polymerization method to synthesize conductive polymer PEDOT coated NVOPF nanorods (NVOPF@ PEDOT) as cathode materials for SIC. The NVOPF@PEDOT cathode exhibits a capacity of 120 mAh g^{-1} at 0.01 A g^{-1}. When the current density improves to 2 A g^{-1}, the specific capacity is still 50 mAh g^{-1}. The SIC assembled by NVOPF@ PEDOT cathode and AC anode (NVOPF@PEDOT//AC) delivers an energy density of 158 Wh kg^{-1} at power density of 230 W kg^{-1}. Even at high power density of 7000 W kg^{-1}, the energy density still retains up to 25 Wh kg^{-1} and has a good cycling life (85.1% capacity retention after 1000 cycles at 1 A g^{-1}) (Figure 5.20c). In addition, the use of carbon materials combined with NVOPE can also effectively improve the conductivity of the cathode material. Qiu et al. [120] used electrostatic spinning technology to composite NVPF and carbon fiber (CNF). The CNF@NVPF architecture can provide a stable Na^+ transfer channel. Therefore, CNF@NVPF has a high cycling stability (96% retention after 2000 cycles). A SIC is assembled by CNF@NVPF cathode and ZnO-activated porous CNFs anodes. The SIC can achieve a high energy density of 182.8 Wh kg^{-1} and a power density of 340 W kg^{-1}. In addition to CNTs, graphene is also used to improve the electrochemical properties of NVOPF based materials cathode. Dong et al. [121] prepared $Na_3(VO)_2(PO_4)_2F$/rGO (NVOPF/rGO) cathode and assembled SIC with $FeVO_4{\cdot}0.6H_2O$ anode. The NVOPF/rGO//$FeVO_4{\cdot}0.6H_2O$ SIC delivers a high specific energy of 88 Wh kg^{-1} at power density of 95 W kg^{-1}, 7.9 kW kg^{-1} at 35 Wh kg^{-1}, and superior cycling stability (5000 cycles)

5.4.4 Sodium Transition Metal Oxides Cathodes

Up to now, sodium transitional metal oxides with layered structures have been widely used in the field of sodium-ion battery cathodes, and there can be various choices of the transitional metals, such as Co, Ni, Mn, Fe, Cr, and V. Numerous modifications by metals doping improve the performances of layered sodium oxide.

5.4.4.1 $Na_{0.67}(Mn_{0.75}Al_{0.25})O_2$

Among these sodium transition metal oxides, Mn-containing layered oxides have advantages in cost and are more environmental friendly, which constitute the most prominent family of Na^+ storage electrode materials. However, the bigger size of Na^+ increases the lattice energy and retards its transport. Therefore, the charge–discharge curves of these oxides show many small potential plateaus, representing multiphase changes. These phase changes will cause continuous deformation and strain on the crystal structures of the oxide, thereby decreasing the electrochemical performances. One of the effective ways to smooth these potential plateaus is to replace part of manganese with other metals such as Co, Ni, Fe, and Al.

Al can often be incorporated into sodium manganese oxides as an inert element, which can form strong bonds with oxygen and stabilize the layered structure of oxides. In 2016, Tian et al. [122] first synthesized the cathode material of $Na_{0.67}(Mn_{0.75}Al_{0.25})O_2$ and found that aluminum and other inert elements could straighten the charge-discharge curves of the cathode materials and stabilize the structures of the cathode materials (Figure 5.21a,b). The reaction proposed for the charge–discharge processes of $Na_{0.67}(Mn_{0.75}Al_{0.25})O_2$ is as follows:

$$Na_{0.67}(Mn_{0.75}Al_{0.25})O_2 \leftrightarrow Na_{0.67-x}(Mn_{0.75}Al_{0.25})O_2 + xNa^+ + e^- \quad (5.13)$$

Al as an inert component, doping inside the material will inevitably reduce the specific capacity of the cathode material. However, most of the existing layered

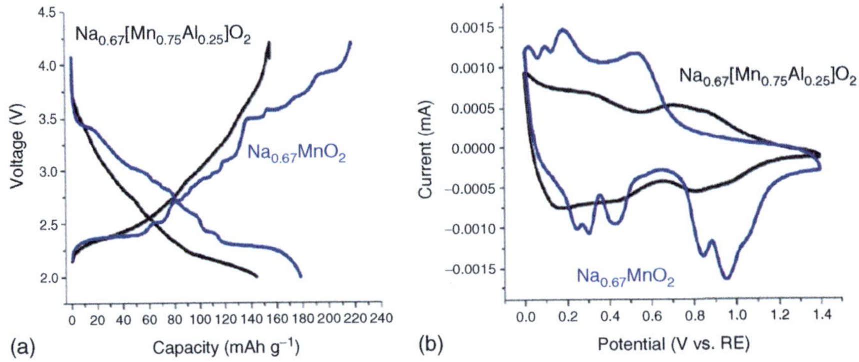

Figure 5.21 Electrochemical characterizations of $Na_{0.67}(Mn_{0.75}Al_{0.25})O_2$ and $Na_{0.67}MnO_2$ (a) the GCD curves. (b) The CV curves. Source: Reproduced with permission Tian et al. [122]. Copyright 2016, Elsevier.

sodium oxide cathode materials had many small platforms, which were not conducive to practical applications. The advantage of $Na_{0.67}(Mn_{0.75}Al_{0.25})O_2$ is mainly reflected in the cycling performances. The SIC assembled by $Na_{0.67}(Mn_{0.75}Al_{0.25})O_2$ and AC can operate stably in the voltage range of 2.7 V and retain 90% of the capacity after 1000 cycles.

5.4.4.2 $Na_{0.67}Co_{0.5}Mn_{0.5}O_2$

The coordination forms of alkali metals in layered TMOs can be represented by the mobile coordination P or by the octahedral O, and the numbers after these letters indicate the number of layers of transition metals in the stacked repeating unit. P2 type materials are suitable for SICs cathodes because Na^+ ions transport directly between two face-sharing trigonal prismatic sites, which leads to fast sodium diffusion and high rate performance and then enables high energy density at high power delivery. Moreover, due to their larger diameter, sodium ions are less prone to deep intercalate into the bulk, especially at higher rates. Therefore, the structural advantage of the P2 phase and the larger size of sodium ions can endow the electrode with fast kinetics of sodium ion storage. $Na_{0.67}Co_{0.5}Mn_{0.5}O_2$ as a typical P2 type oxides are also used in SICs. Gu et al. [123] selected P2-$Na_{0.67}Co_{0.5}Mn_{0.5}O_2$ (P2-NCM) as the cathode and zeolitic imidazolate framework-8 (ZIF-8) zeolitic derived carbon (ZDC) as the anode to construct SICs (Figure 5.22a). The sodium storage mechanism of P2-NCM cathode could be written as follows:

$$Na_{0.67}Co_{0.5}Mn_{0.5}O_2 \leftrightarrow Na_{0.67-x}Co_{0.5}Mn_{0.5}O_2 + xNa^+ + xe^- \quad (5.14)$$

The SIC assembled by P2-NCM cathode and ZDC anode P2-NCM//ZDC can work stably within the voltage window of 3 V. The charge–discharge curve is well symmetrical (Figure 5.22b). And the SIC could deliver an energy density of 18.8 Wh kg^{-1} at a high power density of 12.75 kW kg^{-1}. In addition to the fast kinetics, the P2-NCM//ZDC has a long working life at 5 A g^{-1}, high capacity retention after 2000 cycles with optimal coulombic efficiency (Figure 5.22c).

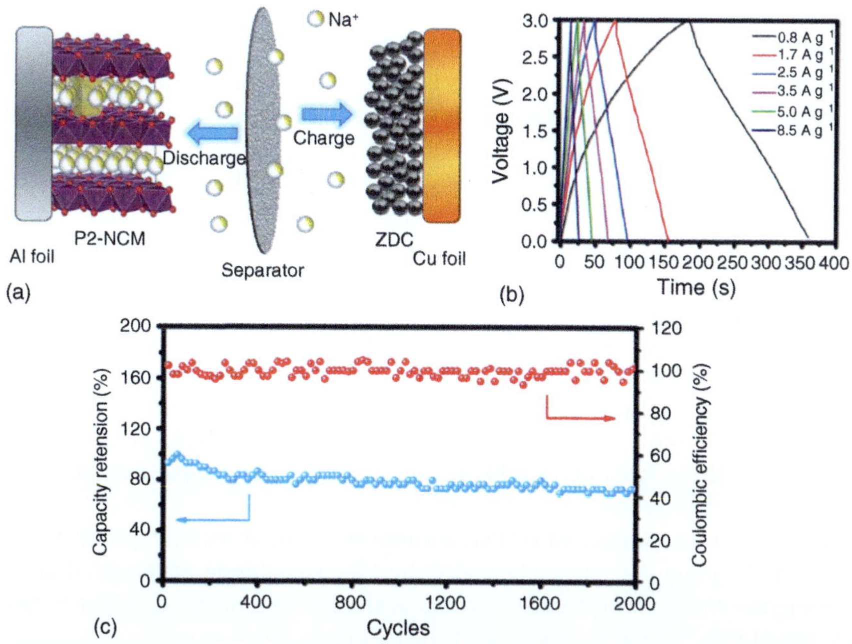

Figure 5.22 (a) Schematic of the coin-type ZDC//P2-NCM SIC; (b) the GCD curves of SIC; (c) Cycle performance at 5 A g^{-1} with Coulombic efficiency. Source: Reproduced with permission Gu et al. [123]. Copyright 2019, Elsevier.

5.4.4.3 $Na_{0.5}Mn_{0.5}Co_{0.48}Mg_{0.02}O_2$

A tunnel-type Na_xMnO_2 structure is formed when the composition of Na is between 0.22 and 0.44. This is an attractive candidate for NIC applications owing to its unique 3D structure composed of MnO_5 pyramids and MnO_6 octahedra arranged in a large S-shaped tunnel and a smaller pentagon tunnel with maximum stress tolerance during charge–discharge (Figure 5.23). Moreover, P2-type oxides provide more flexibility in design, allowing two or more transition-metal ions in the structure for better performance.

Kim et al. [124] first reported a novel composite layered tunnel-type $Na_{0.5}Mn_{0.5}Co_{0.48}Mg_{0.02}O_2$ cathode in 2020, and the possible reaction mechanism of the fabricated cathode can be written as follows:

$$Na_{0.5}Mn_{0.5}Co_{0.5}Mg_{0.02}O_2 \leftrightarrow Na_{0.5-x}Mn_{0.5}Co_{0.5}Mg_{0.02}O_2 + xNa^+ + xe^- \quad (5.15)$$

Figure 5.23 Schematic of the P2-layered and tunnel structure. Source: Reproduced with permission Kim et al. [124]. Copyright 2020, American Chemical Society.

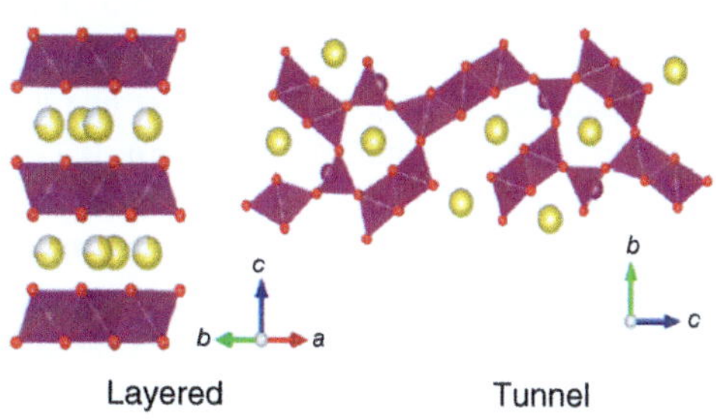

Combining tunnel structure, layered oxides, and the presence of Mg^{2+} ions, a high capacity of 134 mAh g^{-1} can be achieved at 0.085 A g^{-1}. When the current density improves to 1.7 A g^{-1}, the specific capacity is 41 mAh g^{-1}. In the composite layer tunnel structures, $Na_{0.5}Mn_{0.5}Co_{0.5}Mg_{0.02}O_2$ is utilized as the cathode and AC is utilized as the anode to construct SICs. The voltage range of $Na_{0.5}Mn_{0.5}Co_{0.5}Mg_{0.02}O_2$//AC SIC is fixed at 0–3 V to avoid electrolyte decomposition at higher voltages, The maximum energy density is 35 Wh kg^{-1} at a power density of 150 W kg^{-1}, which delivers good stability and can maintain 72% of the initial energy density after 3000 cycles.

5.4.4.4 $Na_{0.66}Mn_{0.54}Ni_{0.13}Co_{0.13}O_2$

In SICs intercalation cathodes, P2-layered materials exhibit high discharge capacity and rate performances. However, the P2-type materials display poor cycling stability. Therefore, Kaliyappan and Chen [125] used atomic layer deposition (ALD) to coat ultrathin (1 nm) Al_2O_3 on the $Na_{0.66}Mn_{0.54}Ni_{0.13}Co_{0.13}O_2$ (NMNC) surface to stabilize the electrolyte interfacial layer and construct extremely stable SICs with AC anode.

Through TEM, the surface of NMNC is uniformly coated with a layer of Al_2O_3, which can eliminate the phase change of the P2-type cathode materials, thereby improving the sodium-ion storage capacity of the layered materials. The reaction mechanism of the NMCM cathodes could be written as follows:

$$Na_{0.66}Mn_{0.54}Ni_{0.13}Co_{0.13}O_2 \leftrightarrow Na_{0.66-x}Mn_{0.54}Ni_{0.13}Co_{0.13}O_2 + xNa^+ + xe^- \quad (5.16)$$

The SIC assembled by the Al_2O_3-coated NMCM (NMNC-Al) cathode and AC anode exhibits an energy density of 63 Wh kg^{-1} and a power density of 6.6 kW kg^{-1}. The utilization of ALD to cover an ultrathin Al_2O_3 layer on the surface of the NMNC electrodes can overcome the kinetics of the slow response of the NMNC electrode under high current density, exhibiting extremely high electrochemical stability (98% capacity retention after 10 000 cycles). In contrast, the capacity retention is only 78% for original NMNC.

References

1 Kim, J., Choi, M.S., Shin, K.H. et al. (2019). *Adv. Mater.* 31: 1803444.
2 Dong, S., Lv, N., Wu, Y. et al. (2021). *Adv. Funct. Mater.* 31: 2100455.
3 Fang, S., Bresser, D., and Passerini, S. (2020). *Adv. Energy Mater.* 10: 1902485.
4 Kim, H., Cho, M.-Y., Kim, M.-H. et al. (2013). *Adv. Energy Mater.* 3: 1500.
5 Zhang, F., Zhang, T., Yang, X. et al. (2013). *Energy Environ. Sci.* 6: 1623.
6 Simon, P., Gogotsi, Y., and Dunn, B. (2014). *Science* 343: 1210.
7 Ding, J., Hu, W., Paek, E., and Mitlin, D. (2018). *Chem. Rev.* 118: 6457.
8 Shao, H., Wu, Y.-C., Lin, Z. et al. (2020). *Chem. Soc. Rev.* 49: 3005.
9 Park, S.K., Kwon, S.H., Lee, S.G. et al. (2018). *ACS Energy Lett.* 3: 724.
10 Lin, Z., Goikolea, E., Balducci, A. et al. (2018). *Mater. Today* 21: 419.
11 Zhang, Z., Gu, Z., Zhang, C. et al. (2021). *Batteries & Supercaps* 4: 1680.

12 Jia, R., Shen, G., and Chen, D. (2020). *Sci. China Mater.* 63: 185.
13 Deng, X., Zou, K., Cai, P. et al. (2020). *Small Methods* 4: 2000401.
14 Zhang, Y., Yuan, C., Ye, K. et al. (2014). *Electrochim. Acta* 148: 237.
15 Hu, Z., Zhu, Z., Cheng, F. et al. (2015). *Energy Environ. Sci.* 8: 1309.
16 Luo, J., Fang, C., Jin, C. et al. (2018). *J. Mater. Chem. A* 6: 7794.
17 Wang, H., Zhu, C., Chao, D. et al. (2017). *Adv. Mater.* 29: 1702093.
18 Kuratani, K., Yao, M., Senoh, H. et al. (2012). *Electrochim. Acta* 76: 320.
19 Zhai, Y., Dou, Y., Zhao, D. et al. (2011). *Adv. Mater.* 23: 4828.
20 Aravindan, V., Ulaganathan, M., and Madhavi, S. (2016). *J. Mater. Chem. A* 4: 7538.
21 Simon, P. and Gogotsi, Y. (2020). *Nat. Mater.* 19: 1151.
22 Zhang, L.L. and Zhao, X.S. (2009). *Chem. Soc. Rev.* 38: 2520.
23 Han, J., Li, H., Kong, D. et al. (2020). *ACS Energy Lett.* 1986: 5.
24 Beguin, F., Presser, V., Balducci, A., and Frackowiak, E. (2014). *Adv. Mater.* 26: 2219.
25 Wang, G., Zhang, L., and Zhang, J. (2012). *Chem. Soc. Rev.* 41: 797.
26 Liu, Y., Roy, S., Sarkar, S. et al. (2021). *Carbon Energy* 3: 795.
27 Liu, X.-M., Huang, Z.D., Oh, S.W. et al. (2012). *Compos. Sci. Technol.* 72: 121.
28 Rao, R., Pint, C.L., Islam, A.E. et al. (2018). *ACS Nano* 12: 11756.
29 Iijima, S. (1991). *Nature* 354: 56.
30 Soni, S.K., Thomas, B., and Kar, V.R. (2020). *Mater. Today Commun.* 25: 101546.
31 Lu, G., Wang, H., Zheng, Y. et al. (2019). *Electrochim. Acta* 319: 541.
32 Ma, H.M., Zeng, J.J., Realff, M.L. et al. (2003). *Compos. Sci. Technol.* 63: 1617.
33 De Jong, K.P. and Geus, J.W. (2000). *Catal. Rev.* 42: 481.
34 Inagaki, M., Yang, Y., and Kang, F. (2012). *Adv. Mater.* 24: 2547.
35 Xu, J., Liu, Z., Zhang, F. et al. (2020). *RSC Adv.* 10: 7780.
36 Bonaccorso, F., Colombo, L., Yu, G. et al. (2015). *Science* 347: 1246501.
37 Raccichini, R., Varzi, A., Passerini, S., and Scrosati, B. (2015). *Nat. Mater.* 14: 271.
38 El-Kady, M.F., Shao, Y., and Kaner, R.B. (2016). *Nat. Rev. Mater.* 1: 16033.
39 Li, C., Zhang, X., Sun, C. et al. (2019). *J. Phys. D: Appl. Phys.* 52: 143001.
40 Zhang, Y., Wan, Q., and Yang, N. (2019). *Small* 15: 1903780.
41 Hummers, W.S. and Offeman, R.E. (1958). *J. Am. Chem. Soc.* 80: 1339.
42 Zhao, Q., Yang, D., Zhang, C. et al. (2018). *ACS Appl. Mater. Interfaces* 10: 43730.
43 Thangavel, R., Moorthy, B., Kim, D.K., and Lee, Y.-S. (2017). *Adv. Energy Mater.* 7: 1602654.
44 Wang, F., Wang, X., Chang, Z. et al. (2015). *Adv. Mater.* 27: 6962.
45 Dong, S., Xu, Y., Wu, L. et al. (2018). *Energy Storage Mater.* 11: 8.
46 Chen, J., Zhou, X., Mei, C. et al. (2017). *J. Power Sources* 342: 48.
47 Surendran, V., Arya, R.S., Vineesh, T.V. et al. (2021). *J. Energy Storage* 35: 102340.
48 Lukatskaya, M.R., Dunn, B., and Gogotsi, Y. (2016). *Nat. Commun.* 7: 12647.
49 Xiao, J., Han, J., Zhang, C. et al. (2022). *Adv. Energy Mater.* 12: 2100775.

50 Kang, J., Zhang, S., and Zhang, Z. (2017). *Adv. Mater.* 29: 1700515.
51 Bae, S.-E., Kim, K.-J., Choi, I.-H., and Huh, S. (2016). *Carbon* 99: 8.
52 Gao, S., Chen, Y., Fan, H. et al. (2014). *J. Mater. Chem. A* 2: 3317.
53 Li, Y., Hu, Y.-S., Titirici, M.-M. et al. (2016). *Adv. Energy Mater.* 6: 1600659.
54 Wang, S., Wang, R., Zhang, Y. et al. (2018). *J. Power Sources* 379: 33.
55 Thangavel, R., Kannan, A.G., Ponraj, R. et al. (2020). *Energy Storage Mater.* 25: 702.
56 Yin, L., Feng, J., Zhang, X. et al. (2019). *J. Mater. Sci.* 54: 4124.
57 Dong, G., Wang, H., Liu, W. et al. (2018). *ACS Appl. Energy Mater.* 1: 5636.
58 Han, P., Han, X., Yao, J. et al. (2015). *J. Power Sources* 297: 457.
59 Wang, H., Mitlin, D., Ding, J. et al. (2016). *J. Mater. Chem. A* 4: 5149.
60 Liu, M., Niu, J., Zhang, Z. et al. (2019). *J. Power Sources* 414: 68.
61 Sevilla, M. and Mokaya, R. (2014). *Energy Environ. Sci.* 7: 1250.
62 Chen, Z., Augustyn, V., Jia, X. et al. (2012). *ACS Nano* 6: 4319.
63 Huang, L., Wang, S., Zhang, Y. et al. (2021). *New Carbon Mater.* 36: 1128.
64 Sun, F., Liu, X., Wu, H.B. et al. (2018). *Nano Lett.* 18: 3368.
65 Liu, X., Wang, H., Cui, Y. et al. (2018). *J. Mater. Sci.* 53: 6763.
66 Zou, K., Cai, P., Liu, C. et al. (2019). *J. Mater. Chem. A* 7: 13540.
67 Ding, J., Wang, H., Li, Z. et al. (2015). *Energy Environ. Sci.* 8: 941.
68 Ding, J., Li, Z., Cui, K. et al. (2016). *Nano Energy* 23: 129.
69 Phattharasupakun, N., Wutthiprom, J., Ma, N. et al. (2018). *Electrochim. Acta* 286: 55.
70 Zhang, J., Lv, W., Zheng, D. et al. (2018). *Adv. Energy Mater.* 8: 1702395.
71 Liu, W., Zhang, X., Xu, Y. et al. (2021). *Batteries Supercaps* 4: 407.
72 Wang, X., Liu, L., and Niu, Z. (2019). *Mater. Chem. Front.* 3: 1265.
73 Han, J., Li, H., and Yang, Q.-H. (2021). *Mater. Today* 51: 552.
74 Li, B., Wang, Y., Jiang, N. et al. (2020). *Nano Energy* 72: 104727.
75 Choi, C., Ashby, D.S., Butts, D.M. et al. (2020). *Nat. Rev. Mater.* 5: 5.
76 Fleischmann, S., Mitchell, J.B., Wang, R. et al. (2020). *Chem. Rev.* 120: 6738.
77 Zheng, J.P., Cygan, P.J., and Jow, T.R. (1995). *J. Electrochem. Soc.* 142: 2699.
78 Roberto, G., Jose, M.O., Bernabe, A.-R., and Juan, M.F. (2004). *J. Phys. Chem. B* 108: 228.
79 Augustyn, V., Simon, P., and Dunn, B. (2014). *Energy Environ. Sci.* 7: 1597.
80 Jiang, Y., Shen, Y., Dong, J. et al. (2019). *Adv. Energy Mater.* 9: 1900967.
81 Conwsay, B.E. and Gileadi, E. (1962). *Trans. Faraday Soc.* 58: 2493.
82 Lee, H.Y. and Goodenough, J.B. (1999). *J. Solid State Chem.* 144: 220.
83 Conwsay, B.E. (1993). *Electrochim. Acta* 38: 1249.
84 Levi, M.D. and Aurbach, D. (1999). *Electrochim. Acta* 45: 167.
85 Hu, X., Wang, G., Li, J. et al. (2021). *Energy Environ. Sci.* 14: 4564.
86 Kim, H., Kwon, J.E., Lee, B. et al. (2015). *Chem. Mater.* 27: 7258.
87 Kim, H., Kim, H., Ding, Z. et al. (2016). *Adv. Energy Mater.* 6: 1600943.
88 Wan, F., Wu, X.-L., Guo, J.-Z. et al. (2015). *Nano Energy* 13: 450.
89 Thangavel, R., Kaliyappan, K., Kim, D.-U. et al. (2017). *Chem. Mater.* 29: 7122.
90 Wei, Q., Liu, J., Feng, W. et al. (2015). *J. Mater. Chem. A* 3: 8070.

91 De jesus, L.R., Andrews, J.L., Parija, A., and Banerjee, S. (2018). *ACS Energy Lett.* 3: 915.

92 Mai, L., Dong, Y., Xu, L., and Han, C. (2010). *Nano Lett.* 10: 4273.

93 Dong, J., Jiang, Y., Wei, Q. et al. (2019). *Small* 15: 1900379.

94 Wei, Q., Jiang, Y., Qian, X. et al. (2018). *iScience* 6: 212.

95 Zhang, X., Wang, L., Liu, W. et al. (2020). *ACS Omega* 5: 75.

96 Yi, S., Wang, L., Zhang, X. et al. (2021). *Sci. Bull.* 66: 914.

97 Naguib, M., Kurtoglu, M., Presser, V. et al. (2011). *Adv. Mater.* 23: 4248.

98 Gao, L., Bao, W., Kuklin, A.V. et al. (2021). *Adv. Mater.* 33: 2004129.

99 Li, Y., Shao, H., Lin, Z. et al. (2020). *Nat. Mater.* 19: 894.

100 Wang, L., Zhang, X., Xu, Y. et al. (2021). *Adv. Funct. Mater.* 31: 2104286.

101 Abdolhosseinzadeh, S., Jiang, X., Zhang, H. et al. (2021). *Mater. Today* 48: 214.

102 Anasori, B.A., Lukatskaya, M.R., and Gogotsi, Y. (2017). *Nat. Rev. Mater.* 2: 16098.

103 Wang, X., Mathis, T.S., Li, K. et al. (2019). *Nat. Energy* 4: 241.

104 Okubo, M., Sugahara, A., Kajiyama, S., and Yamada, A. (2018). *Acc. Chem. Res.* 51: 591.

105 Dall'Agnese, Y., Taberna, P.L., Gogotsi, Y., and Simon, P. (2015). *J. Phys. Chem. Lett.* 6: 2305.

106 Zhu, K., Zhang, H., Ye, K. et al. (2017). *ChemElectroChem* 4: 3018.

107 Cai, P., Zou, K., Deng, X. et al. (2021). *Adv. Energy Mater.* 11: 2203804.

108 Fei, R., Wang, H., Wang, Q. et al. (2020). *Adv. Energy Mater.* 10: 2002741.

109 Li, Y., Wang, H., Wang, L. et al. (2019). *Energy Storage Mater.* 23: 95.

110 Senthilkumar, B., Sankar, K.V., Vasylechko, L. et al. (2014). *RSC Adv.* 4: 53192.

111 Lao, M., Zhang, Y., Luo, W. et al. (2017). *Adv. Mater.* 29: 1700622.

112 Tan, H., Feng, Y., Rui, X. et al. (2019). *Small Methods* 4: 1900563.

113 Sundaram, M.M., Watcharatharapong, T., Chakraborty, S. et al. (2015). *Dalton Trans.* 44: 20108.

114 Yang, Z., Li, G., Sun, J. et al. (2020). *Energy Storage Mater.* 25: 724.

115 Park, M.S., Veerasubramani, G.K., Thangavel, R. et al. (2018). *ChemElectroChem* 6: 653.

116 Wang, S., Zhao, J., Wang, L. et al. (2015). *Ionics* 21: 2633.

117 Ali, Z., Zhang, T., Asif, M. et al. (2020). *Mater. Today* 35: 131.

118 Chen, Y., Zhao, S., Yu, Y. et al. (2021). *Small* 18: 2106513.

119 Wu, L., Dong, S., Pang, G. et al. (2019). *J. Mater. Chem. A* 7: 1030.

120 Qiu, R., Fei, R., Guo, J.-Z. et al. (2020). *J. Power Sources* 466: 228249.

121 Dong, J., He, Y., Jiang, Y. et al. (2020). *Nano Energy* 73: 104838.

122 Tian, S., Qi, L., and Wang, H. (2016). *Solid State Ionics* 289: 194.

123 Gu, H., Kong, L., Cui, H. et al. (2019). *J. Energy Chem.* 28: 79.

124 Kim, H.J., Ramasamy, H.V., Jeong, G.H. et al. (2020). *ACS Appl. Mater. Interfaces* 12: 10268.

125 Kaliyappan, K. and Chen, Z. (2018). *Nano Energy* 48: 107.

6

Anode Materials for Sodium-Ion Capacitors

Kangyu Zou[1,2], *Wentao Deng*[1], *Hongshuai Hou*[1], *Xiaobo Ji*[1], *and Guoqiang Zou*[1]

[1] *Central South University, College of Chemistry and Chemical Engineering, No. 932, Lushan South Road, YueLu District, Changsha 410083, China*
[2] *Changsha University of Science and Technology, School of Materials Science and Engineering, Changsha 410114, China*

It is well known that anodes are a very important component in sodium-based energy storage systems [1]. Different from SIBs, sodium-ion capacitors (SICs) possess various mechanism configurations; thus, the anode materials of SICs can be divided into two major types: capacitor-type anode and battery-type anode [2]. Notably, capacitor-type anode is depended on fast surface-induced behavior [3]. Meanwhile, battery-type anode is determined by the bulk phase transportation of Na^+ [4]. At present, there are many anode materials have been explored for construction the high-performance SICs. Moreover, some new materials have also been exploited, which are regarded as the promising anode candidates of SICs. The commonly desirable anode species have been summarized in Figure 6.1 [5]. It has been found that required demands of anode are low potential, high capacity, and excellent structure stability [6]. Therefore, the specific criteria for evaluating the high-performance anode materials should be listed.

Firstly, the working potential serves a vital role in energy density of cell. Moreover, the cell voltage is mainly decided by the affinity of Na^+ ions of anode/cathode and the electrochemical stability window of electrolyte [7]. Thus, desired anode candidates should exhibit electrochemical activity at low potential, which is conducive to enlarging the potential difference between the cathode and anode, thus leading to the high energy density of the entire system [8]. However, the anode potential is incapable of closing to the sodium plating potential, which potentially causes the notorious dendritic growth accompanied by safety hazard [9]. For example, due to the excellent redox activity at low-voltage region, carbon materials have attracted enormous attentions [10]. Nevertheless, the problems of pronounced electrolyte reductive decomposition and instability of the solid electrolyte interphase (SEI) should be urgently resolved [11]. In terms of Ti-based compounds, a relatively high

Sodium-Ion Capacitors: Mechanisms, Materials, and Technologies, First Edition.
Edited by Guoqiang Zou, Xiaobo Ji, and Hongshuai Hou.

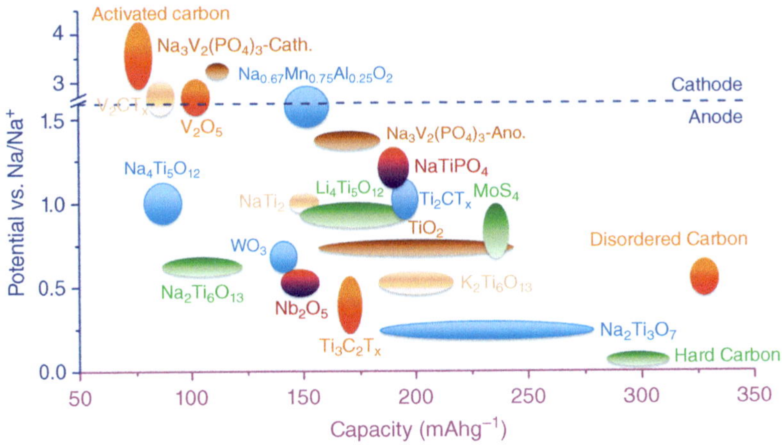

Figure 6.1 Summary of the average working potential and practical capacity of representative electrode materials of SICs. Source: Reproduced with permission Ding et al. [5]. Copyright 2018, American Chemical Society.

operating potential triggered by insertion mechanism could ensure the desired safety, but the decreased cell voltage would cause the undesired energy density [12]. Meanwhile, the specific capacity is also a critical parameter for anode materials. Notably, the theoretical capacity of anode material is directly determined by its ability to store sodium ions [13]. Thus, the stronger storage ability of Na^+ ion represents the larger capacity. Generally, the key factors for high storage capacity of anode are depended on the available space for accommodating ions, the ability to change valence state, and the degree of reversibility of the reactions [14]. It is noticed that the irreversible capacity loss of anode occurs in the first sodiation process, which is mainly attributed to the formation of SEI film and other parasitic side reactions [15]. Unfortunately, the irreversible capacity loss could lead to the low Coulombic efficiency, which restricts the practical application [16]. Moreover, the unstable SEI film in Na-based system, continuously consumes electrolyte and deteriorates structure, which brings out the fast capacity decaying [17]. As a result, a stable structure of anode material could ensure the high Coulombic efficiency and superb cycling durability. In consideration of the raw material employed in Na-based systems is lower cost, such as the price of Na_2CO_3 is 25–30 times cheaper than that of Li_2CO_3. Thus, the cost and sustainability are identified as the important selection criterion for anode materials, which preferably highlight the merits and significances of Na-based systems [18].

There are two kinds of capacitor-type anodes: EDLC and pseudocapacitive materials [19]. Porous carbon material could adequately adsorb the Na^+ ions in the electroyte based on the EDLC mechanism, which is resulted from the high specific surface area and abundant porosity [20]. Meanwhile, pseudocapacitive materials (such as MXene) could ensure the rapid surface redox reaction accompanied with the adsorption/desorption of Na^+ ions [21]. It is noticed that the most anode materials could exhibit the insertion of Na^+ ions in the relatively low potential [22]. Thus, it is required that capacitor-type anode only presents the surface adsorption/desorption at high working potential region, which could effectively avoid the unwished bulk redox behavior. Unfortunately, due to the relatively high potential of capacitor-type anode, the working voltage interval of SIC is inevitably decreased, thus leading to the reduced energy density. Inspiringly, in the capacitor-type anode//battery-type cathode cell configuration, the Na-containing cathode could provide the Na source in the entire system and the pre-sodiation treatment is unnecessary, greatly simplifying the assemble steps.

In terms of battery-type anode, the related reaction mechanism can be classified into three categories: insertion reaction, alloying reaction, and conversion reaction (Figure 6.2) [23]. Insertion type anodes (such as carbon and TiO_2) can allow Na^+ ions to be inserted into the material matrix with a small and reversible structural change [24]. However, the insertion type anodes usually show a relatively low theoretical capacity, which is attributed to the limited amount of Na^+ insertion. Notably, rational utilization of insertion type anode with optimized mass loading can certainly ensure the substantial increase of energy density in the practical application. Alloy-type anodes refered to semi-metal or metallic elements (such as Si, Sn, and Bi) could exhibit the ultrahigh theoretical capacity through the sodiation process [25]. In addition, the alloy-type material is regarded as a promising anode due to the abundant resources. Unfortunately, the issue of dramatic volume expansion seriously leads to the structural deterioration, thus hindering the utilization and development. Simarily, conversion-type mechanism could provide a high

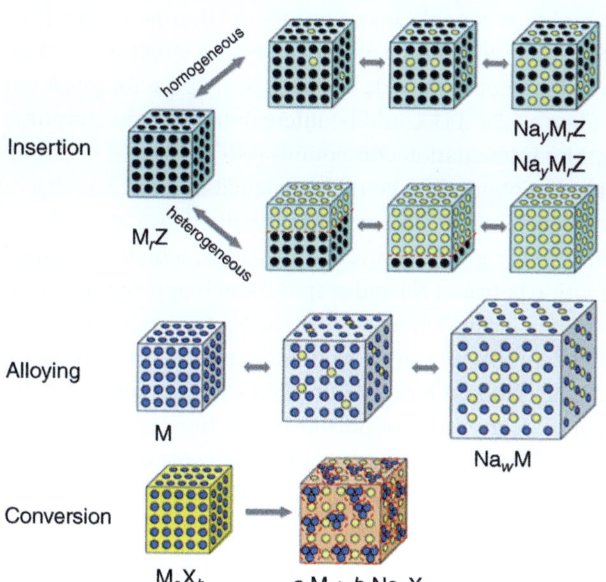

Figure 6.2 Schematic of the different reaction mechanisms of insertion-type anodes in SICs. Source: Reproduced with permission Rosa Palacín [23]. Copyright 2009, Royal Society of Chemistry.

capacity triggered by the multiple electron reactions, but large volume variation and high redox potential are the troublesome problems [26]. It is noteworthy that the SEI could formed on the surface/interface of battery-type anode, which largely consumes the active sodium source, causing the irreversible capacity loss. Moreover, the unstable SEI would deteriorate the material structure. Limited by the sluggish kinetic of battery-type anode, the dynamics mismatch should be resolved in the battery-type anode//capacitor-type cathode cell configuration [27]. As a result, in views of battery-type anode, the fast Na^+ transportation is the critical factor in the constrction of high-performance SICs instead of high capacity. In the above Chapters 2–4, the electrochemical storage mechanisms of EDLC and pseudocapacitive behavior have been systematically described and compared. Therefore, the detailed electrochemical storage processed and differences in the battery-type anode mechanism should be adequately discussed in this chapter, which is better understanding the battery-type anode of SICs.

In alkali metal ion system, the physicochemical properties of alkali metal element are discrepant, thus bringing out the different storage behavior [28]. For example, in terms of the ion radius, the ion radius of Na^+ is 1.02 Å, which is larger than that of Li^+, thus leading to the relatively difficult insertion behavior for anode materials [29]. Moreover, the much larger ion size of Na^+ avoidly brings out a troublesome issue for designing electrode materials with high storage capacity and fast ion transport. In sodium-based storage system, hard carbon (HC) has been regarded as the most common insertion-type anode materials [30]. Taking HC as example, different from graphite, HC possesses the graphitic layer with a large interlayer distance, which is beneficial for the highly reversible insertion/deinsertion processes [31]. It is well known that graphite material exhibits a unique layered structure with an interlayer spacing of 3.35 Å, which could provide the sufficient space for guest ion intercalation. As shown in Figure 6.3a, Li^+ could be intercalated into the graphitic layer for generating the graphite intercalation compounds (GICs) [32]. In addition, the theoretical Li storage capacity of graphite could be reached up to 372 mAh g^{-1}, corresponding to the LiC_6 configuration [34]. However, it is verified the electrochemical intercalation of Na^+ ion into the graphite layer is infeasible, which is resulted from the weak binding interaction between Na and graphite and inappropriate interlayer spacing. In view of this issue, Cao et al. have studied the Na-ion insertion behavior into carbon as a function of carbon interlayer by the theoretical simulations on the energy cost [33]. And the calculation is based on the balance of attractive van der Waals interactions between carbon layers and repulsive interactions between Na or Li ions and carbon. As shown in Figure 6.3b, when the interlayer spacing of graphite is 0.335 nm, the energy cost for Li-ion insertion is 0.03 eV, whereas energy cost for Na-ion insertion is 0.12 eV. This obviously large energy barrier is hard to overcome, which illustrates that Na^+ ions are incapable of intercalating into the graphite layers. Significantly, when the layer spacings is increased to 0.37 nm, the energy barrier for Na ion insertion is markedly decreased to 0.053 eV, which is low enough for feasible Na-storage capacity. Furthermore, the equilibrium interplanar distance for NaC_6 is about 0.45 nm.

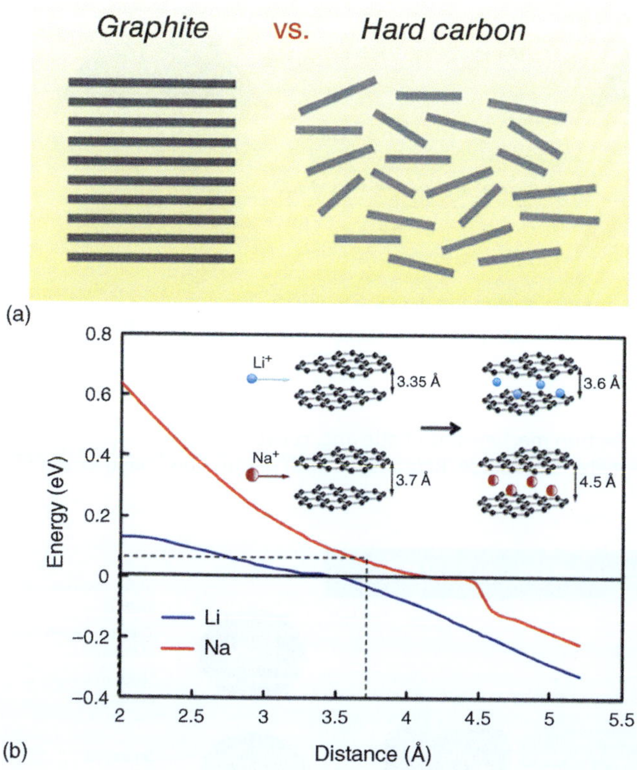

Figure 6.3 (a) Comparison of structure models of graphite and amorphous hard carbon. Source: Reproduced with permission Sun et al. [32]. Copyright 2022 Wiley-VCH. (b) Theoretical energy cost for Na^+ and Li^+ ions' insertion into carbon as a function of carbon interlayer distance. Source: Reproduced with permission Cao et al. [33]. Copyright 2012 American Chemical Society.

In terms of alloying- and conversion-type anode materials, the high theoretical specific capacities are the most attractive advantage, which is resulted from the multi-electron reactions with Na^+ ions during the charge and discharge (Figure 6.4) [35]. Generally, alloying-type materials, such as tin (Sn) and antimony (Sb), can form binary alloys with sodium, providing a high theoretical capacity triggered by the multiple electron exchange [36]. Moreover, conversion-type materials undergo phase transition during the uptake and extraction of Na^+ accompanied by the breaking and formation of bonds [37]. However, some key issues for alloying- and conversion-type materials should be solved before their practical applications [38]. Typically, a large volume variation would lead to the aggregation and pulverization of active materials, further causing the weak electrical contact of pulverized integrated electrode and the slow kinetics. In addition, voltage hysteresis issue during charge and discharge could lower the round-trip energy efficiency with a consequent heat generation. The formed thick SEI and/or insulating layers would largely retard the charge transfer and lead to sharp capacity decay.

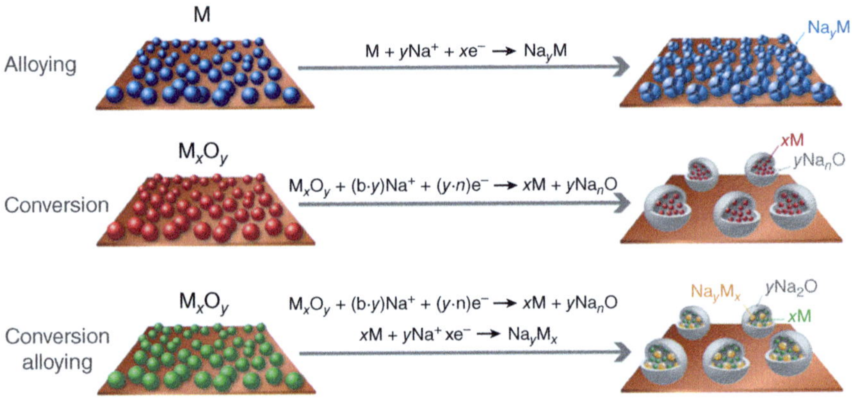

Figure 6.4 Schematic of reaction mechanisms of alloying, conversion, and conversion-alloying anode materials. Source: Reproduced with permission Zhang et al. [35]. Copyright 2018 WileY-VCH.

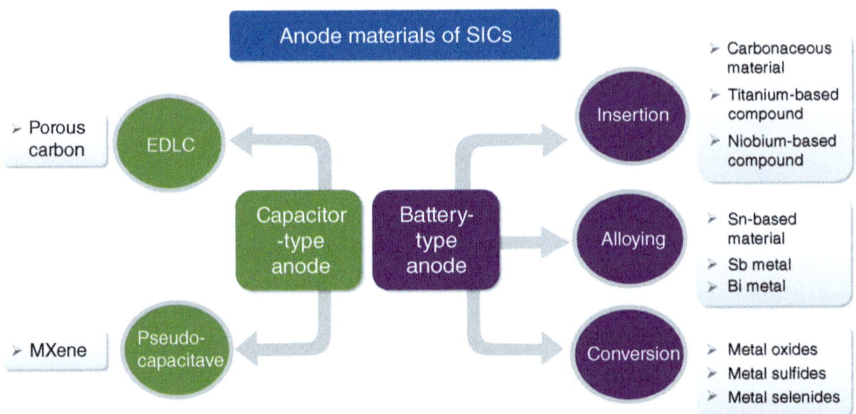

Figure 6.5 Schematic of classification of the various anodes of SICs.

In the past time, many anode materials have been explored for high-performance SICs. Moreover, the various modified strategies have been developed, which is aimed to resolve confronting problem. In this chapter, different kinds of anode of SICs have been summarized (Figure 6.5). Moreover, the related physicochemical property of anode materials and the structure-performance correlation have been detailed discussed.

6.1 EDLC Anode Materials

Porous carbon anode: Generally, porous carbon materials have been defined as the EDLC anode of SICs paired with battery-type cathode. In terms of the physical adsorption/desorption behavior of capacitive nature, the porous carbon anode should possess the large specific surface area and satisfying porous size distribution [39]. It is noticed that activated carbon (AC) has been utilized as the typical EDLC anode, which exhibits the distorted rectangular CV shape and not strictly linear GCD curve in the selected potential range [40]. In consideration of the ubiquitous Na^+ insertion reaction of carbon anode in the low potential, the relatively high potential range should be specifically chosen, which only remain the occurrence of the Na^+ ions of adsorption/desorption. Moreover, the large specific surface area of AC anode could give rise to the notorious side reactions in the low potential region, seriously bringing out the dissatisfied electrochemical performance [41]. Notably, the possible energy storage mechanism of AC anode could be presented as follows:

$$AC + xe^- + xNa^+ \leftrightarrow AC(xe^-)//Na^+ \text{ (// stands for double layer)} \tag{6.1}$$

As shown in Figure 6.6a, the different charge/discharge curves of AC and $Na_3V_2(PO_4)_3$ (NVP) electrodes have been observed in the corresponding potential range [42]. The plateaus of NVP electrode around 3.4 V vs. Na/Na^+ have been presented, which are resulted from the reversible phase transformation between $NaV_2(PO_4)_3$ and $Na_3V_2(PO_4)_3$ triggered by the redox reactions of V species. Moreover, the AC electrode exhibits an obviously linear charge/discharge behavior in

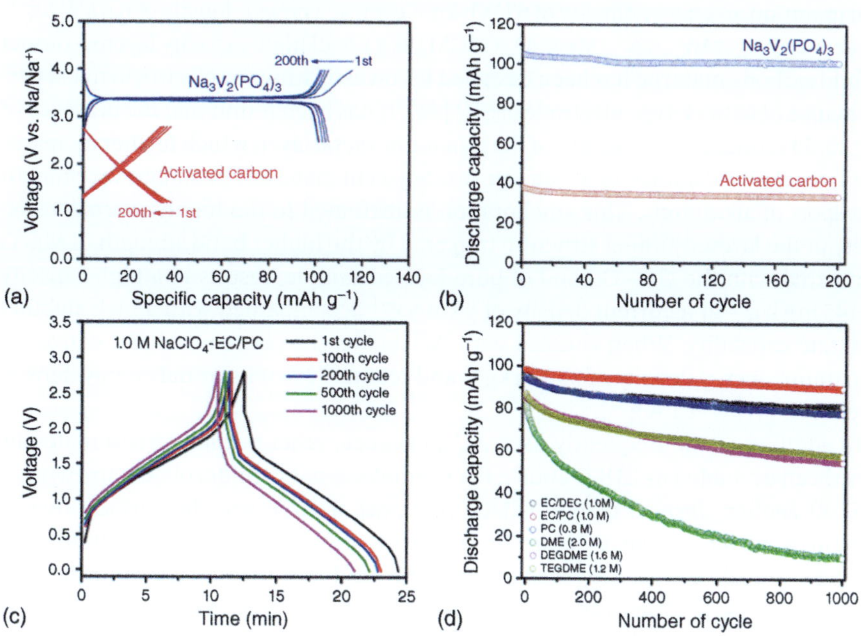

Figure 6.6 (a) Charge and discharge curves and (b) cycle performance of the AC and NVP electrodes in 1.0 M $NaClO_4$_EC/PC electrolyte at a current density of 25 mA g^{-1}. (c) Charge and discharge curves of AC//NVP SIC in 1.0 M $NaClO_4$_EC/PC at a current density of 500 mA g^{-1} and (d) cycling performance of assembled SICs with different electrolytes at a current density of 500 mA g^{-1}. Source: Reproduced with permission Park et al. [42]. Copyright 2018, Wiley.

the 1.2–2.8 V, which delivers the discharge capacities around 36.0 mAh g^{-1} with excellent cycling stability, illustrating the adsorption/desorption processes of ions without any faradaic reactions are highly reversible (Figure 6.6b). Significantly, the different electrolytes have been further utilized for evaluating the electrochemical performance of AC//NVP SICs. As shown in Figure 6.6c,d, it is noticed that the carbonate-based SIC cells all exhibit the initial coulombic efficiencies larger than 94.0%. However, the employed ether-based electrolytes bring out the relatively low initial coulombic efficiencies of 87.1%, 88.9%, and 89.7% for dimethyl ether (DME), diethylene glycol dimethyl ether (DEGDME), and tetraethylene glycol dimethyl ether (TEGDME) in assembled SICs, respectively. Notably, the AC//NVP SIC cell with 1.0 M $NaClO_4$–ethylene carbonate (EC)/propylene carbonate (PC) electrolyte shows the most stable cycling behavior with the best capacity retention of 90.8% after 1000 cycles. Unfortunately, the DME-based SIC cell exhibit a discharge capacity of 10.0 mAh g^{-1} at 1000th cycle, which could remain capacity retention of 11.8%.

Meanwhile, stabilizing surface strategy by the atomic layer deposition (ALD) of Al_2O_3 is aimed at boosting the rate stability of P2-type $Na_{0.66}Mn_{0.54}Ni_{0.13}Co_{0.13}O_2$ (NMNC) material for further fabricating high-performance SIC [43]. Moreover, the used commercial AC anode presents a discharge capacitance of 68 F g^{-1} at 0.35 A g^{-1} and shows extremely high ~98% retention after 10 000 cycles. The NMNC-Al/AC SIC delivers an energy density of 82 Wh kg^{-1} at a power density of 1350 W kg^{-1} and could still maintain an energy density of 63 Wh kg^{-1} even at a power density of 6.6 kW kg^{-1}. Moreover, $Na_{0.5}Mn_{0.5}Co_{0.48}Mg_{0.02}O_2$ (MCM) as a novel high-capacity layered-tunnel hybrid cathode material has been designed for investigating the electrochemical performance of battery-type electrode of SIC [44]. It has been found that the presence of Mg could reduce the slab length of the transition-metal layer, which further increases the interplanar distance between the two adjacent slabs and facilitate the smooth transport of alkali ions. This amelioration is attributed to the highly reversible stability of the layered-tunnel structure triggered by the higher bond strength of Mg–O compared with the Co—O bond of pure layered sample, results in a high capacity of 145 mAh g^{-1} at a current density of 85 mA g^{-1} accompanied with a high stability and rate capability. When coupled with AC anode, the constructed SIC delivers a maximum energy density of 35 Wh kg^{-1} and retains 72% of its initial energy density after 3000 cycles at 1 A g^{-1}.

In addition to the frequently used AC substance, other porous carbon materials have been evaluated as EDLC anode. For example, zeolitic imidazolate framework-8 (ZIF-8) zeolitic derived carbon (ZDC) possesses a large specific surface area of 1259 m^2 g^{-1} and a high exterior surface area, providing more adsorption sites for ions, which delivers the desired capacity and the excellent electrochemical reversibility [45]. Moreover, P2-$Na_{0.67}Co_{0.5}Mn_{0.5}O_2$ (P2-NCM) has been chosen as the battery-type cathode material to pair with ZDC capacitor-type anode for fabricating the SIC [46]. Endowed by the improved kinetic match and high-rate performance of both electrodes, the ZDC//P2-NCM SIC delivers an energy output of 18.8 Wh kg^{-1} at a high-power density of 12.75 kW kg^{-1}. Interestingly, peanut shell-derived carbon as a kind of lab-synthesized AC cathode has been utilized for constructing SIC by pairing with the poly(3,4-ethylenedioxythiophene)

(PEDOT)-coated $Na_3V_2O_2(PO_4)_2F$ cathode [47]. Benefiting from the conductive polymer PEDOT coating layer, $Na_3V_2O_2(PO_4)_2F$@PEDOT electrode exhibits enhanced electrochemical properties with outstanding rate capability and ultralong cycling stability. Moreover, the as-assembled SIC efficiently improves the energy density by minimizing the amount of electrolyte because Na-ion is de-intercalated from the cathode while it is adsorbed to the anode during charging. Notably, the $Na_3V_2O_2(PO_4)_2F$@PEDOT//AC (peanut shell derived carbon) SIC delivers a high energy density of ~158 Wh kg^{-1} and a large power density of 7000 W kg^{-1} in the voltage window of 1.0–4.2 V.

6.2 Pseudocapacitive Anode Materials

MXene: MXenes not only serve as the capacitive cathode, but also can be regarded as a typical pseudocapacitive anode for fabricating the SIC coupled with battery-type or EDLC cathode [48].

The Ti_2CT_x MXene nanosheets have been firstly employed as pseudocapacitance anode of SIC in 2015, which is prepared from the MAX phase Ti_2AlC through the hydrofluoric acid treatment (Figure 6.7) [49]. It has been noticed that the interlayer distance of the Ti_2CT_x nanosheets has been expanded from 7.7 to 10.1 Å during the first cathodic process triggered by the bare Na^+ intercalation. Moreover, the activated Ti_2CT_x can ensure reversible Na^+ intercalation/de-intercalation into the

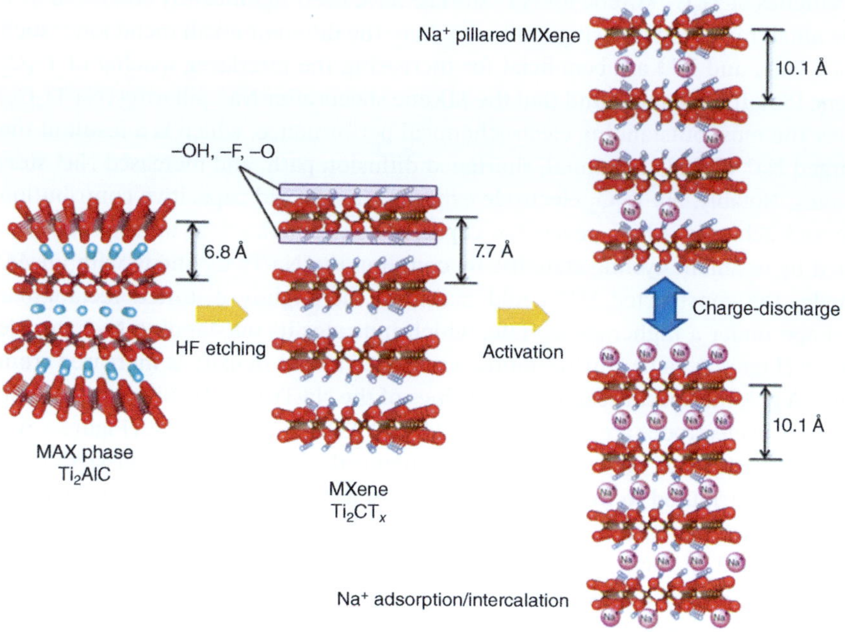

Figure 6.7 Schematic illustration of the reaction mechanism of Ti_2CT_x by electrochemical activation. Source: Wang et al. [49]/Springer Nature/Licensed under CC BY 4.0.

interlayer space as well as reversible Na^+ adsorption/desorption onto the surface of each layer/sheet, dramatically delivering a higher specific capacity relative to double-layer capacitor electrodes and a higher rate capability relative to ion intercalation electrodes. By utilizing an alluaudite $Na_2Fe_2(SO_4)_3$ as a cathode, the SIC cell consisting of pseudocapacitance Ti_2CT_x MXene anode delivers a high specific energy of 260 Wh kg^{-1} at a high specific power of 1.4 kW kg^{-1} (based on the weight of Ti_2CT_x). Meanwhile, the porous Ti_3C_2 MXene/CNT composite anode has been employed with battery-type $Na_{0.44}MnO_2$ cathode to fabricate a prototype SIC in 2016, which delivers a volumetric discharge capacity of 286 mAh cm^{-3} and could power a 2.5 V light-emitting diode (LED) for approximately 25 minutes [50]. Notably, a prototype SIC has been assembled by combining the bistacked MXene anode with a capacitive AC cathode in a nonaqueous 1 M $NaClO_4$ electrolyte [51]. In this SIC configuration, AC electrode exhibits a typical rectangular shape due to reversible adsorption–desorption of ClO_4^- anions in the potential window of 3–4.5 V (vs. Na^+/Na) and the pseudocapacitive bistacked MXene in the potential range of 0.01–3 V presents a mixed battery/capacitor behavior. The hybrid SIC system shows the combined energy storage behavior of the redox and double-layer mechanisms. It has been found that the Ti_3C_2//AC SIC delivers 84.2% capacitance retention after 4000 cycles. Moreover, the as-constructed Ti_3C_2//AC SIC device achieves an energy density of 39 Wh kg^{-1} in the operating voltage window of 3.4 V, which could successfully light a green LED.

In order to improve the pseudocapacitive feature of MXene anode, the compositional and structural amelioration have been exploited. For instance, the capacity and kinetics of Ti_3C_2 MXene for Na^+ storage have been significantly enhanced by a facile alkali metal ion pillaring method, where the different alkali metal ions (such as Li^+, Na^+, and K^+) are beneficial for increasing the interlayer spacing of Ti_3C_2 MXene [52]. It has been found that the MXene sheets after Na^+ pillaring (Na-Ti_3C_2) deliver the most outstanding electrochemical performance, which is a result of the enlarged Na^+ diffusion channel, shortened diffusion path, and increased Na^+ storage sites. Notably, Na-Ti_3C_2 electrode exhibits an enlarged capacitive contribution ratio and achieves a high reversible capacity of 175 mAh g^{-1} at 0.1 A g^{-1} accompanied by excellent cycling stability. By coupling the Na-Ti_3C_2 anode with an AC cathode, the constructed SIC could maintain the slightly distorted rectangular CV shape under a higher sweep rate, which indicates its outstanding Na^+ storage kinetics (Figure 6.8a,b). Furthermore, when the current density is increased from 0.1 to 3 A g^{-1}, the specific capacitance values of the Na-Ti_3C_2//AC SIC based on the total mass of electrodes are 32, 30, 28, 25, 24, and 23 F g^{-1}, respectively (Figure 6.8c). It is noteworthy that the excellent electrochemical performance of Na-Ti_3C_2//AC SIC can be attributed to the enhanced Na^+ storage sites and kinetics of Na-Ti_3C_2 anode. Meanwhile, the S atoms intercalated Ti_3C_2 MXene (CT-S@Ti_3C_2) has been fabricated through the effective cetyltrimethylammonium bromide (CTAB) pretreatment and S atoms intercalation methods [54]. Interestingly, elemental S could easily intercalate into interlayer of the CTAB-pretreated Ti_3C_2 and the intercalated elemental S can interact with the interfacial Ti atoms to form Ti—S bond, which further brings out the expanded interlayer spacing. Inspired by

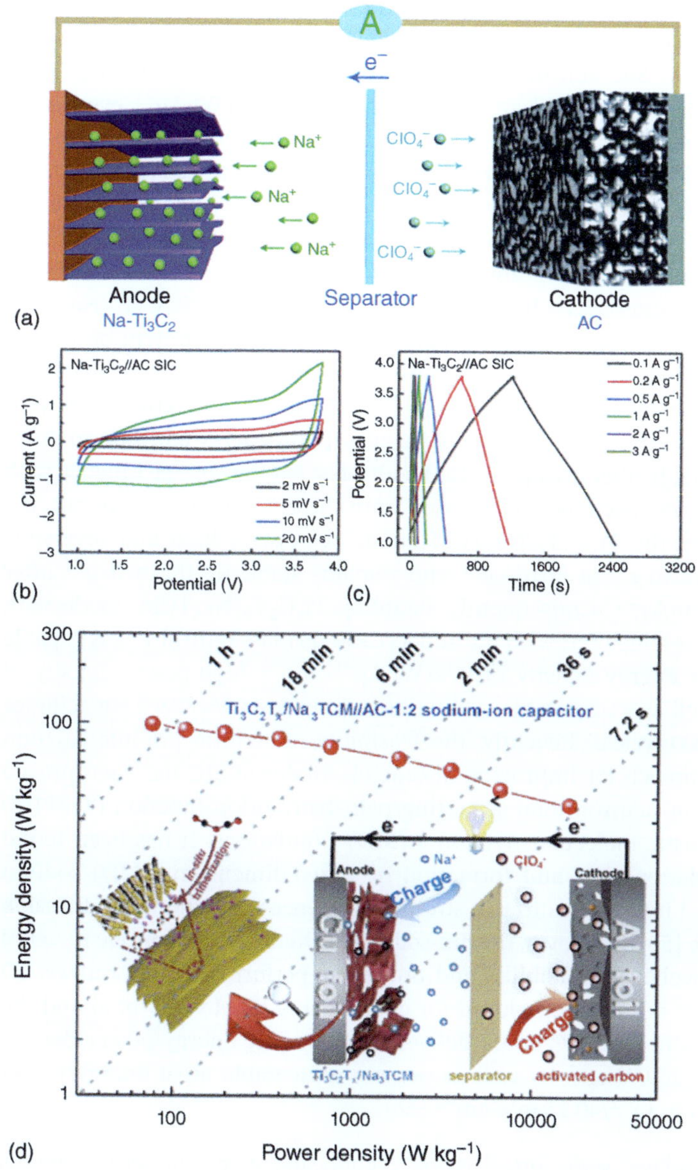

Figure 6.8 (a) Schematic illustration of the Na-Ti_3C_2//AC SIC device. (b) CV curves and (c) galvanostatic charge/discharge plots of the assembled Na-Ti_3C_2//AC SIC. Source: Reproduced with permission Luo et al. [52]. Copyright 2018, The Royal Society of Chemistry. (d) The detailed electrochemical performance of the $Ti_3C_2T_x$/Na_3TCM//AC SIC. Source: Liu et al. [53]/Springer Nature/Licensed under CC BY 4.0.

these above-mentioned structural features, the CT-S@Ti_3C_2-450 possesses fast Na-ion storage kinetics and incremental storage sites. Besides, the assembled CT-S@Ti_3C_2-450//AC SIC delivers a high energy density of 114.03 Wh kg^{-1} at a power density of 237 W kg^{-1} and an energy density of 52.64 Wh kg^{-1} even at an ultrahigh power density of 8240 W kg^{-1}, respectively. It is noteworthy that in situ polymeric strategy of sodium dicyanamide (Na-dca) has been utilized for fabricating a novel nitrogen-doped $Ti_3C_2T_x$ MXene by Liu et al. (Figure 6.8d) [53]. Acted as intercalant and N-source, the Na-dca can intercalate into the interlayers of $Ti_3C_2T_x$ nanosheets and simultaneously form sodium tricyanomelaminate (Na_3TCM) between the $Ti_3C_2T_x$ interlayers. It has been found that modified $Ti_3C_2T_x$/Na_3TCM shows a large interlayer spacing of 12.6 Å and a high N-doping content of 5.6 at.% originated from strong Ti—N bonding and stabilized triazine ring structure. According to the density functional theory (DFT) calculations and kinetic analysis, the N element could effectively reduce the Na adsorption energy and Na-ion diffusion kinetic barriers, further giving rise to the ultrafast Na-ion storage kinetics with the increased pseudocapacitance. Benefiting from the enlarged interlayer spacing and N-rich surface, the $Ti_3C_2T_x$/Na_3TCM anode achieves a high and reversible capacity of 243.4 mAh g^{-1} at 20 mA g^{-1} and remains stable at 182 mAh g^{-1} after 1000 cycles at 100 mA g^{-1}. Consequently, coupling $Ti_3C_2T_x$/Na_3TCM anode with AC cathode, the assembled SIC with the anode/cathode mass ratio of 1 : 2 is capable to deliver a larger energy density of 97.6 Wh kg^{-1} and a high-power density of 16.5 kW kg^{-1} as well as a ultrastable cycling life within 0–4.0 V (82.6% capacitance retention after 8000 cycles). Recently, the feasibility of sodium pre-intercalation as a promising approach for improving MXene electrodes of SIC has been proved by a combination of neutron total scattering/pair distribution function (TS/PDF) analysis and atomistic molecular dynamics (MD) simulations. It has been found that multiple sodiation sites and formation of a two-dimensional (2D) sodium domain structure at interfaces/surfaces strongly influence the sodium ion diffusion behavior of MXene [55]. Moreover, the pre-sodiated MXene as an electrode material in a SIC shows excellent reversibility and promising performance. Nowadays, 3D printing technology has been employed for constructing SIC by Fan et al and the 3D-printed N-$Ti_3C_2T_x$//AC SIC could deliver a gravimetric energy/power density of 101.6 Wh kg^{-1}/3269 W kg^{-1} as well as possess a favorable areal energy/power density of 1.18 mWh cm^{-2}/40.15 mW cm^{-2} [56].

Na_2CoPO_4F: Nowadays, some novel redox pseudocapacitive (surface) materials have been exploited as the next-generation anodes of SIC. For example, the Na_2CoPO_4F (NCPF) species as pseudocapacitive anodes exhibit no defined peaks in the 0–3 V region, which suggests the charge-storage mechanism is primarily based on redox pseudocapacitance on the surface of the active material accompanied with the adsorption/desorption of Na^+ ions. Moreover, NCPF anode with a pseudocapacitive charge-storage mechanism is beneficial for matching the fast adsorption-based kinetics of the AC cathode [57]. As a result, the assembled SIC delivers an outstanding cycle stability and rate performance in 0–3 V, retaining a 93% capacitance retention over 100 000 cycles.

Layered iron vanadate: Layered iron vanadate $Fe_5V_{15}O_{39}(OH)_9{\cdot}9H_2O$ ultrathin nanosheets (FeVO UNSs) as a novel pseudocapacitance anode have been explored for rapid and reversible sodium-ion storage [58]. Significantly, the sodium-ion storage mechanism and the interface reaction chemistry of FeVO UNSs anode have been carefully analyzed according to the in situ and ex situ characterization. It has been found that FeVO UNSs anode can manifest the highly reversible sodium ion intercalation with single-phase changes by limiting the cutoff potential at 0.8 V (Figure 6.9a). Moreover, this Na^+ intercalation process is pseudocapacitive dominated, enabling a high ICE and stable interface. However, when the sodiation potential is discharged to 0.01 V, the conversion reaction of FeVO anode can be observed, which brings out the low ICE and poor cycling stability. Thus, the optimized potential is operated above 0.8 V vs. Na^+/Na to ensure the highly reversible pseudocapacitive Na^+ intercalation behavior. A novel

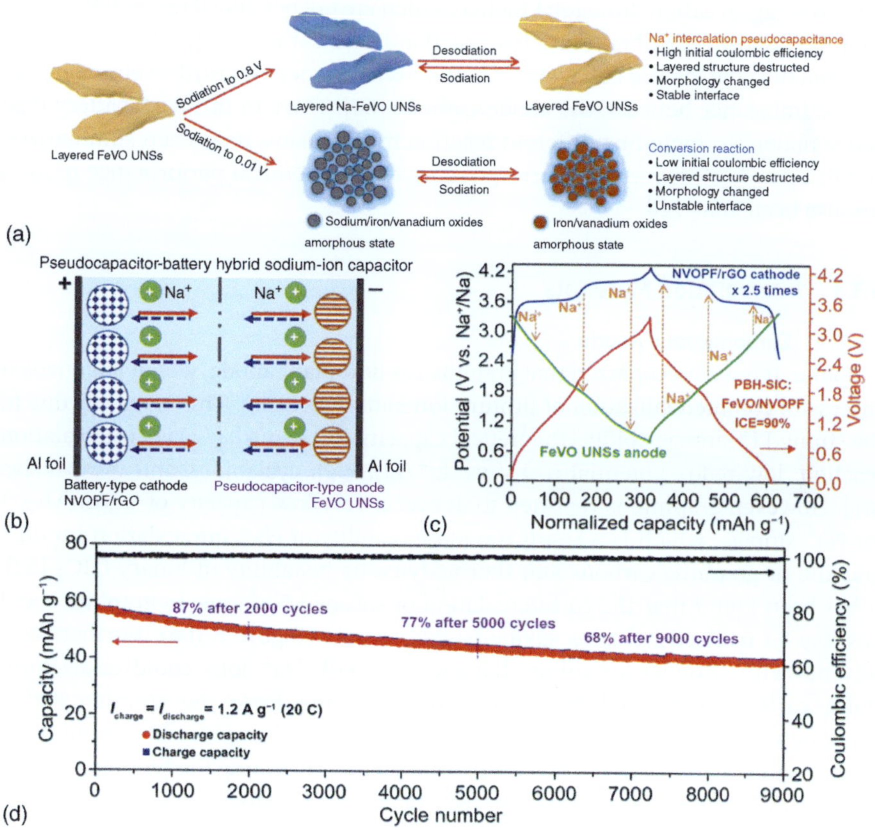

Figure 6.9 (a) Illustration sodium-ion storage mechanism for FeVO UNSs anode in different potential ranges. (b) Schematic of PBH-SIC configuration composed of pseudocapacitive FeVO UNSs anode and battery-type NVOPF/rGO cathode. (c) GCD profiles of FeVO//NVOPF PBH-SIC accompanied with the corresponding anode/cathode. (d) Long-term cycling stability of the PBH-SIC at 1.2 A g^{-1}. Source: Wei et al. [58]/Springer Nature/Licensed under CC BY 4.0.

pseudocapacitor-battery hybrid sodium-ion capacitor (PBH-SIC) prototype has been proposed by employing the pseudocapacitive FeVO anode and Na-rich battery-type $Na_3(VO)_2(PO_4)_2F$ (NVOPF) cathode (Figure 6.9b). As shown in Figure 6.9c, the GCD curves of FeVO//NVOPF PBH-SIC display the capacitive quasi-linear slopes in the voltage 0–3.3 V. Remarkably, the as-assembled FeVO//NVOPF PBH-SIC can achieve the maximum specific energy and power densities of 126 Wh kg^{-1} and 7.6 kW kg^{-1}, respectively. Moreover, the capacity retention of 68% after 9000 cycles at 1.2 A g^{-1} can be obtained, resulted from the stable electrode–electrolyte interface (Figure 6.9d).

6.3 Battery-Type Anode Materials

It is noticed that battery-type anodes, unlike EDLC and pseudocapacitive anodes, possess sluggish cation diffusion kinetics, which are not beneficial for matching well with the fast kinetics of non-faradaic reaction-based cathode [59]. The major issue is to enhance the kinetic behaviors of battery-type anodes and further eliminate the kinetic imbalance between the anode/cathode [60]. Thus, in this part, battery-type anode materials, including different reaction mechanisms, have been summarized and discussed. Moreover, the corresponding electrochemical performance of SICs has also been analyzed.

6.3.1 Intercalation Materials

6.3.1.1 Carbonaceous Anode

Graphite: It is well recognized that graphite is a dominant anode, which has enabled successful commercialization of lithium-ion batteries (LIBs). This is mainly due to the strong Li^+ storage ability (theoretical capacity of 372 mAh g^{-1}) via intercalation reaction, low redox potential (0.1 V vs. Li^+/Li), high cyclability, and inexpensive [61]. However, graphite is reported to deliver a very low capacity of <20 mAh g^{-1} for Na^+ storage, which is a result the inaccessibility of Na^+ intercalation into the graphite or graphitic carbons and thermodynamic instability of binary GICs [62]. It has been found that the co-intercalation of solvated Na^+ ions in graphite could possibly be realized by using glyme-based solvents (Figure 6.10a). Furthermore, solvents with stronger solvation characterstics with Na^+ ions could easily form stable carbon-intercalated structure [63]. Benefiting from the co-intercalation phenomenon of 0.5 M $NaPF_6$ in TEGDME, the recovered graphite (RG) from spent LIBs as anode active material has been explored for fabricating the high-energy SIC paired with commercial AC cathode [64]. Notably, RG anode possesses moderate electrochemical performance in half-cell assembly with 0.5 M $NaPF_6$ TEGDME electrolyte. As shown in Figure 6.10b, RG anode exhibits the highly reversible oxidative and reductive peaks originating from sodium storage. Moreover, Figure 6.10c shows a discharge capacity of approximately 220 mAh g^{-1} with initial coulombic efficiency of approximately 50%. It is noticed that RG electrode possesses discharge capacities in the range of 125–52 mAh g^{-1} at various current rates (0.025–5 A g^{-1}),

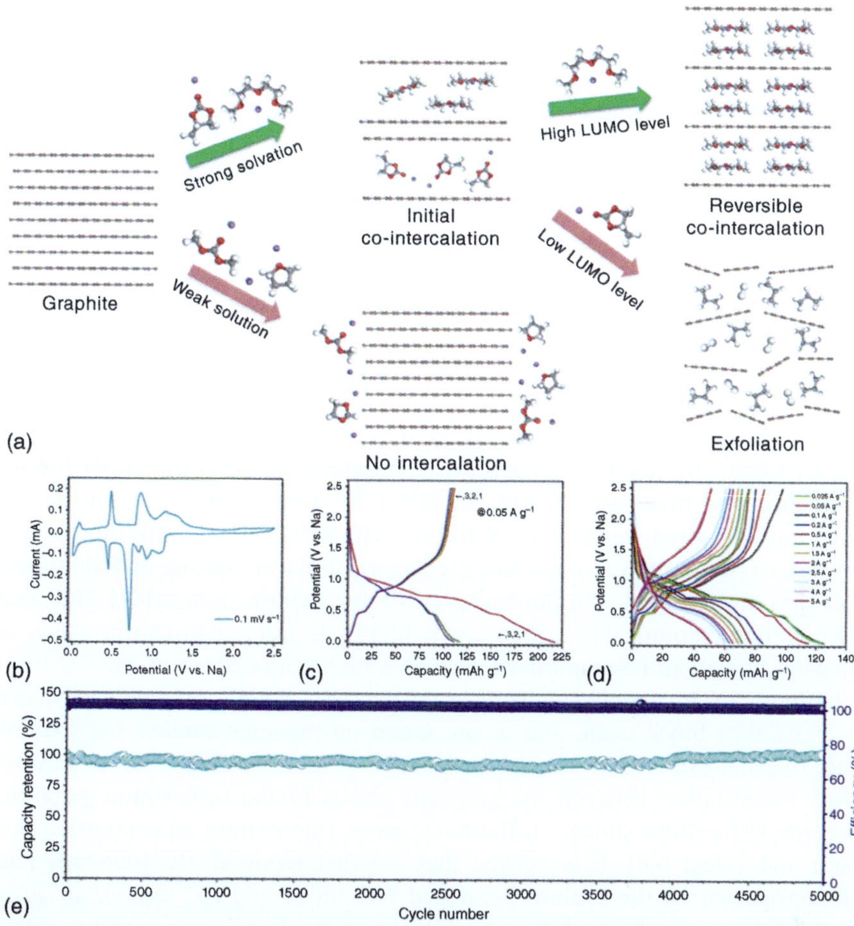

Figure 6.10 (a) Schematic illustration of the solvent influence on realizing Na^+ intercalation in graphite. Source: Reproduced with permission Yoon et al. [63]. Copyright 2016, Wiley. (b) CV curve at 0.1 mV s^{-1}, (c) galvanostatic charge/discharge plot at 0.05 A g^{-1} and (d) galvanostatic charge/discharge profiles at different current rates of RG anodes in half-cell with 0.5 m $NaPF_6$ TEGDME electrolyte. (e) Long-term cyclability of assembled pre-sodiated RG/AC SIC at 0.5 A g^{-1} in the potential window of 1–3.75 V. Source: Reproduced with permission Divya et al. [64]. Copyright 2020, Wiley.

which implies the excellent reversibility of the insertion host along with the solvent co-intercalation (Figure 6.10d). Subsequently, the fabricated RG (pre-sodiated)/AC SIC delivered a maximum energy density of 59.93 Wh kg^{-1} and a maximum power density of 6.8 kW kg^{-1} with superior cycling stability (approximately 98% capacity retention even after 5000 cycles at ambient temperature conditions) (Figure 6.10e).

In order to achieve the commercial application of low-cost graphite in SIC, graphitic mesocarbon microbead (MCMB) anode considered as new artificial graphite has been utilized for constructing a high energy density SIC through co-intercalation mechanism in diglyme-based electrolyte system. The energy densities of this SIC system are 93.5 and 86.5 Wh kg^{-1} at 573 and 2832 W kg^{-1} in the voltage window at 1–4 V, respectively. Notably, the capacity retention ratio is as high as 98.3% after 3000 cycles at 20 C (1 C equals 50 mA g^{-1}), resulted from the reversible intercalated/de-intercalated reaction of sodium ions and the formation of ternary GICs in diglyme-based electrolyte [65]. An electrochemically exfoliated graphite (EEG) has been developed as the anode for high-performance SIC. Notably, the 2D ultrathin nanosheets of EEG can boost the Na-ion reaction kinetics in a diglyme-based electrolyte. Furthermore, the well-stacked architecture with open void effectively alleviates the volume expansion in co-intercalation process. Compared with the pristine graphite, EEG anode presents the broader redox peaks and more sloping charge/discharge profiles with a less discharge plateau. Moreover, the EEG anode delivers a high reversible capacity of 109 mAh g^{-1} and retains a good capacity retention of 90% after 1000 cycles. Endowed by the above-mentioned merits, the assembled EEG//AC SIC could achieve a maximum energy density of 90 Wh kg^{-1} and an ultrahigh power density of 17 500 W kg^{-1} [66].

Hard carbon: In view of the appropriate interlayer spacing, the larger tunnel of carbonaceous material is more beneficial for intercalation/de-intercalation of Na^+ ions. Inspired this structural feature, HC with low graphitization and a short-range order structure possesses the larger interlayer spacing and abundant defect sites are regarded as a promising anode for sodium storage [67]. HC, also known as non-graphitizable carbon, owns highly disordered structure, which is extremely difficult to be graphitized even at a high temperature of 2500 °C [68]. Furthermore, HC is generally prepared by the pyrolysis of high-molecular-polymers and crosslinked block resin, and so on. Based on previous studies, the sodium storage mechanism of HC can be proposed as the following three processes: (i) Na^+ intercalation between the graphene sheets in the turbostratic graphitic structures; (ii) sodium storage in the micropores; (iii) sodium adsorption on the surface and defect [69]. It is noticed that the first reported SIC prototype has been constructed by the sodium pre-doped HC anode and AC cathode in 2012 [70]. With the successful exploitation of HC as an anode material for SICs, many carbonaceous materials derived from natural biomaterials have been widely investigated as electrodes. Furthermore, the biomass is a green and economical source for preparing HC materials, promoting the commercialization of SICs [71]. For example, a SIC entirely based on electrodes designed from recycled olive pit bio-waste-derived carbon materials has been designed by Ajuria et al. [72]. In this scenario, a HC anode pyrolyzed by olive nucleus possesses a low specific surface area. On the other hand, the same HC has been chemically activated to achieve the AC material with a high specific surface area, which is employed as the capacitor-type cathode of SIC. The resulting SIC demonstrates good cycling stability up to 5000 cycles with 70% capacity retention. Meanwhile, HC derived from mulberry tree stems has been explored for SIC by Xu et al. [73]. When paired with a biomass-derived AC cathode, the assembled SIC delivers a maximum energy density of 170 Wh kg^{-1} at 2800 W kg^{-1} and remains at an energy density of Wh kg^{-1} at 67 200 W kg^{-1}, which could sustain 10 000 cycles with 70% capacity retention. Notably, nanotube-like hard carbon (NTHC) anode material has been prepared by high-temperature carbonization (1150 °C) of polyaniline (PANI) nanotubes for high-performance SICs (Figure 6.11a) [74]. As shown in Figure 6.11b,c, three small

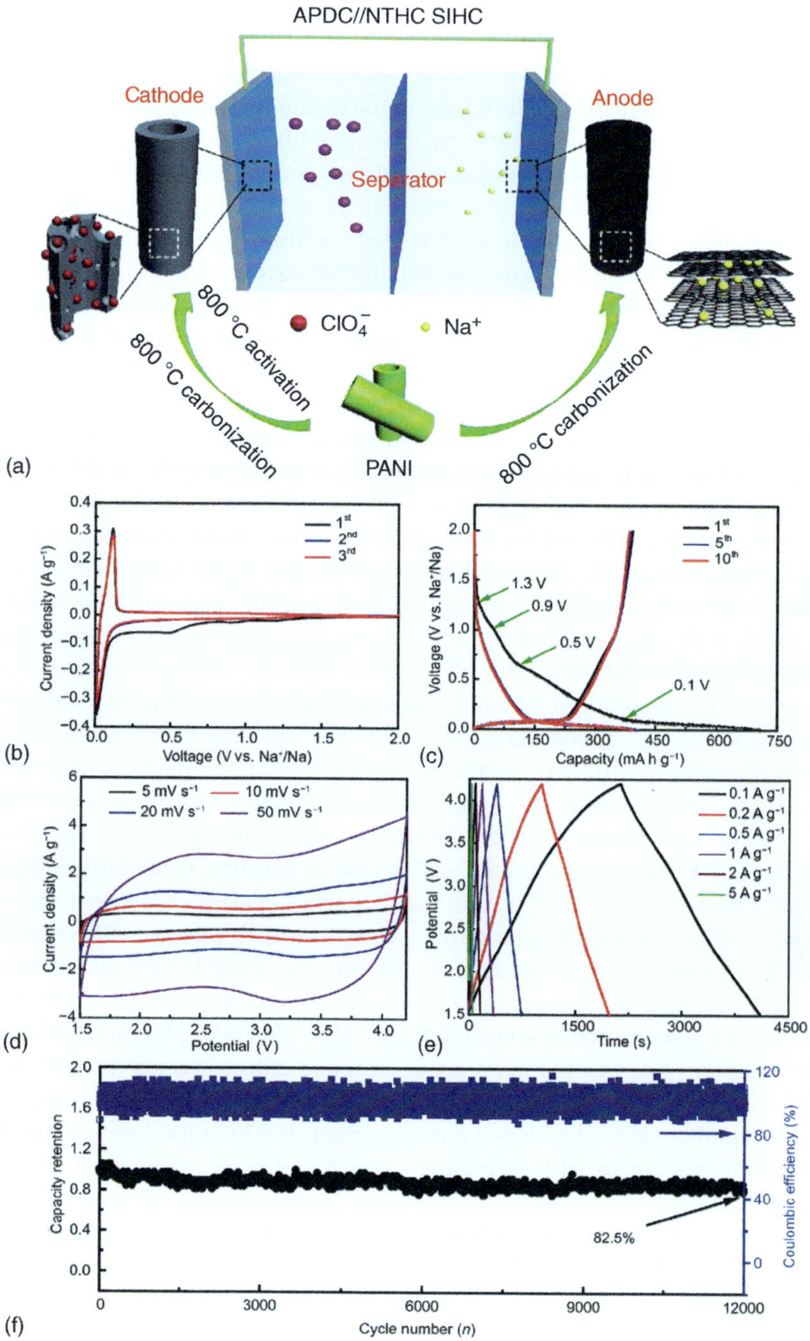

Figure 6.11 (a) Illustration of dual-carbon NTHC//APDC SIC configuration. (b) First three CV curves at 0.1 mV^{-1} and (c) GCD curves at 50 mA g^{-1} of NTHC-1150 anode. (d) CV profiles, (e) GCD curves and (f) cycle performance of NTHC-1150//APDC SIC. Source: Reproduced with permission Ding et al. [74]. Copyright 2018, Springer.

reduction peaks located at 1.3, 0.9, and 0.5 V (vs. Na^+/Na) appear only in the first cathodic scan, which illustrates that the formation of SEI film associated with decomposition of electrolyte on the electrode surface. Importantly, a pair of distinct redox peaks about 0.114/0.01 V in the low potential region is resulted from reversible insertion/extraction of Na^+ ions between quasi-graphite layers. Coupled with a high-capacity activated carbon (APDC) as cathode, a dual-carbon SIC with the optimized mass ratio shows a high energy density of 133.0 Wh kg^{-1} at 2850 W kg^{-1} and still remains 100.9 Wh kg^{-1} at 14 250 W kg^{-1} within the potential range of 1.5–3.5 V. Moreover, the as-fabricated SIC exhibits an outstanding cycling stability (the capacity retention of 82.5%) at 2 A g^{-1} even after 12 000 cycles (Figure 6.11d–f). By virtue of green and abundantly, a superior garlic-derived SIC has been designed by Liu et al., in which a garlic-derived hard carbon (GDHC) as a cation-intercalation anode and garlic-derived porous carbon (GDPC) as an anion-adsorption cathode [75]. The obtained SIC displays an excellent energy density of 156 Wh kg^{-1} at 355 W kg^{-1} and maintain 73% of its initial capacity after 10 000 cycles in 1.5–4.2 V. Interestingly, hydrogen-bonded melamine and cyanuric acid complex (MCA) is employed as a reactive template for regulating the physicochemical characteristics of the biomass-derived HC anode, which could realize triple functionalities of (i) structure-directing agent, (ii) nitrogen source, and (iii) porogen [76]. It has been found that the mass ratio of MCA to glucose precursor can well control the interlayer spacing, lateral size, ratio of graphitic content, nitrogen content and porosity of HC. At an optimum ratio of 1, the resulting HC anode has an interlayer spacing of 0.408 nm, graphene lateral size of 21 nm, stacking height of three to five layers, N content of 10.6 wt%, a specific surface area of 347 m^2 g^{-1}, and a pore volume of 0.29 cm^3 g^{-1}, achieving a specific capacity of 380 mAh g^{-1} at a current density of 250 mA g^{-1} after 100 cycles. Combined with a commercial microporous carbon fiber (CC) cathode, the full SIC cell is able to deliver a volumetric energy density of 2.89 mWh cm^{-3} and a volumetric power density of 160 mW cm^{-3}, which shows an amazing cycling performance with >97% Coulombic efficiency in 10 000 cycles. A novel NH_3 plasma strategy has been proposed to well manipulate the doped N element of advanced bamboo-derived carbon materials. [77] The bamboo-derived N-doped hard carbon (NHC-5) anode after NH_3 plasma treatment with five minutes possesses the more active sites and defects, which delivers the improved rate capability originated from the enhanced charge transport kinetics. Meanwhile, bamboo-derived N-doped porous carbon (NPC-60) has been prepared by NH_3 plasma treatment with 60 minutes, which has been investigated as the cathode of SIC. It has been found that NH_3 plasma treatment as an efficient strategy could give rise to the appropriate aperture and large specific surface area of NPC-60 cathode, dramatically facilitating the adsorption of ClO_4^- ions. Importantly, pyridinic-N and C=O bond on the surface of the anode/cathode are capable for introducing extra pseudocapacitance during the charging/discharging processes, which has been verified by the combination analysis of in situ/ex situ characterizations and theoretical calculations. Notably, the designed bamboo-derived dual-carbon SIC achieves an energy density of 107 Wh kg^{-1} at a power density of 200 W kg^{-1}.

Graphene: Graphene is a kind of 2D material consisting of carbon atoms densely arranged in six-membered rings. Peculiarly, graphene has a large theoretical specific surface area of $2630\,m^2\,g^{-1}$, an outstanding fracture strength of 125 GPa, a high Young's modulus (about 1100 GPa), a good thermal conductivity (about $5000\,Wm^{-1}\,K^{-1}$), and fast heterogeneous electron transfer at the edges ($>15\,000\,cm^2\,V^{-1}\,s^{-1}$) [78]. Endowed by these unique characteristics, graphene as a promising anode has been widely investigated in various electrochemical energy storage systems. However, in terms of SIC, the Na^+ storage capacity of pristine graphene is limited triggered by the unsatisfied interlayer spacing similar to graphite. Thus, the ameliorated strategy of graphene should be implemented, which is aimed at dramatically improving its electrochemical performance. For example, graphene hollow nanospheres (NS-GHNSs) have been rationally designed by using a sacrificial template method with an amino-functionalized silica sphere [79]. By virtues of the advantages of hollow 3D architecture with high porosity, large surface area, and a dual-heteroatom doping, GHNS exhibits enhanced electrolyte compatibility as well as improved electrochemical reaction kinetics. Therefore, all NS-GHNS-based SIC delivers a high energy density of $121\,Wh\,kg^{-1}$ at the power density of $100\,W\,kg^{-1}$. Furthermore, the SIC could remain the remarkable cycling stability of 85% retention after 10 000 cycles as well as maintaining the high-Coulombic efficiency of >99% cycle.

Graphdiyne: Graphdiyne (GDY) as a new carbon allotrope has been successfully designed by Li et al. through a cross-coupling reaction method in 2010 [80]. It is noticed that GDY is composed by sp^2- and sp- hybridized carbon atoms, which features high π-conjugation degree and uniformly distributed pores. Importantly, function tunability of GDY can be easily manipulated by the substitution of special groups, which exhibits the promising applications in the electrochemical energy store systems. In 2017, 3D-architected graphdiyne nanosheet (GDY-NS) was prepared via a modified Glaser–Hay coupling reaction by Wang et al. [81]. It is observed that GDY-NS is arranged along the vertical direction on copper substrate, further forming a large-scale 3D structure with sub-micrometer voids (Figure 6.12a). In terms of Na^+ storage, GDY-NS as anode displays a stable reversible capacity of approximately $400\,mAh\,g^{-1}$ over 1000 cycles at a current density of $1\,A\,g^{-1}$ accompanied with a good rate and excellent cycle performance. Figure 6.12b shows SIC configuration composed of the AC cathode and pre-sodiated GDY-NS anode by direct attachment of extra sodium. The assembled SIC could exhibit a capacitance more than $200\,F\,g^{-1}$ over 3000 cycles at $1\,A\,g^{-1}$ and deliver an initial specific energy as high as $182.3\,Wh\,kg^{-1}$ at a power density of $300\,W\,kg^{-1}$ (Figure 6.12c). Nitrogen-doped graphdiyne (N-GDY) has been prepared through the nitriding of GDY under ammonia, which features fast diffusion and excellent storage ability of sodium ions triggered by its high conductivity, wide interlayer distance, and unique 3D porous structure and associated defects from nitrogen doping. Compared with pure GDY, the energy density of N-GDY/AC SIC can be enlarged by 58.7% ($119\,Wh\,kg^{-1}$) even at $22\,500\,W\,kg^{-1}$ [82].

1D nanostructure carbon: Different from the 2D plane architecture, the unique 1D nanostructure provides an enhanced surface/volume ratio, shortened transport

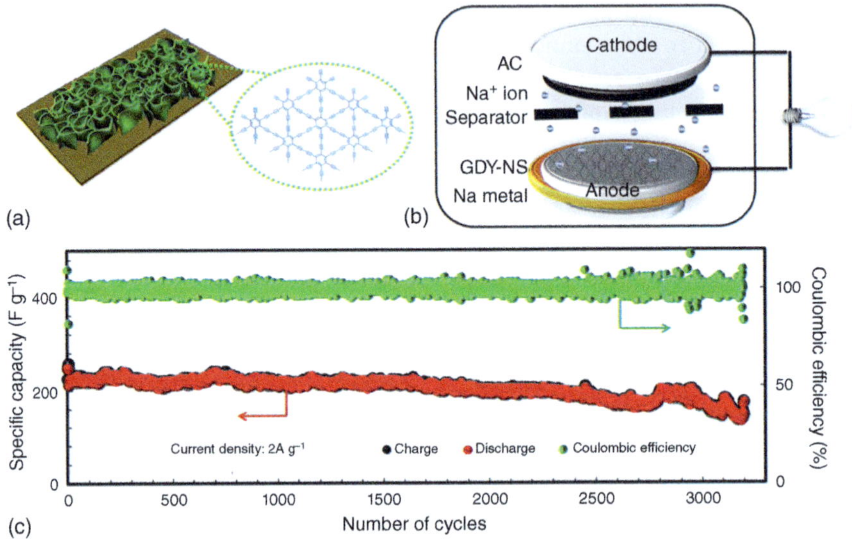

Figure 6.12 (a) Illustration of the morphology and molecular structure of GDY-NS on the Cu substrate. (b) Schematic diagrams of the SIC configuration consisted of the AC cathode and GDY-NS anode. (c) Cycle performance of the GDY-NS//AC SIC at the current density of $1\,A\,g^{-1}$. Source: Reproduced with permission Wang et al. [81]. Copyright 2020, American Chemical Society.

lengths for ionic transport, and efficient 1D electron transport along the longitudinal direction, which offer many advantages for improving the electrochemical properties of carbon anodes in SICs [83]. For instance, nitrogen-doped hollow carbon nanowires (N-HCNWs) with a large specific surface area and high capacitive contribution have been prepared by Li et al., which display excellent rate capability and cyclic performance. A SIC prototype consisting of a N-HCNW anode and a commercial AC cathode can achieve a maximum energy density of $108\,Wh\,kg^{-1}$ and a maximum power density of $9\,kW\,kg^{-1}$, respectively [84]. Interestingly, nitrogen/sulfur co-doped hierarchical hollow carbon nanofibers (N/S-HCNFs) by a facile and controllable method (Figure 6.13a) [85]. It is certified that the S, N dual-heteroatom dopant could enlarge the interlayer distance and provide more exposed active sites accompanied with increased surface area, further bringing out the fast kinetics of N/S-HCNFs and outstanding sodium-ion storage properties. Then, a SIC device is constructed by using N/SHCNFs anode and commercial AC cathode in 1 M $NaClO_4$ in EC-DMC electrolyte solution, where ClO_4^- ions are absorbed in the porous structure of AC and Na^+ ions are inserted into N/S-HCNFs during the charge process (Figure 6.13b). At the mass ratio of 3 : 1 (AC to N/S-HCNFs), the as-prepared SIC device could manifest the optimized electrochemical performance in a relatively large voltage window between 0 and 4.0 V (Figure 6.13c,d). It is calculated that the corresponding specific capacities (based on the total mass of active materials) are 58.2, 42.5, 39.5, 36.2, 33.9, 28.2, and $24.1\,mAh\,g^{-1}$ at 0.1, 0.2, 0.5, 1, 2, 5, and $10\,A\,g^{-1}$, respectively. Moreover, the excellent cycling stability of 81% capacity retention is achieved after 3000 cycles

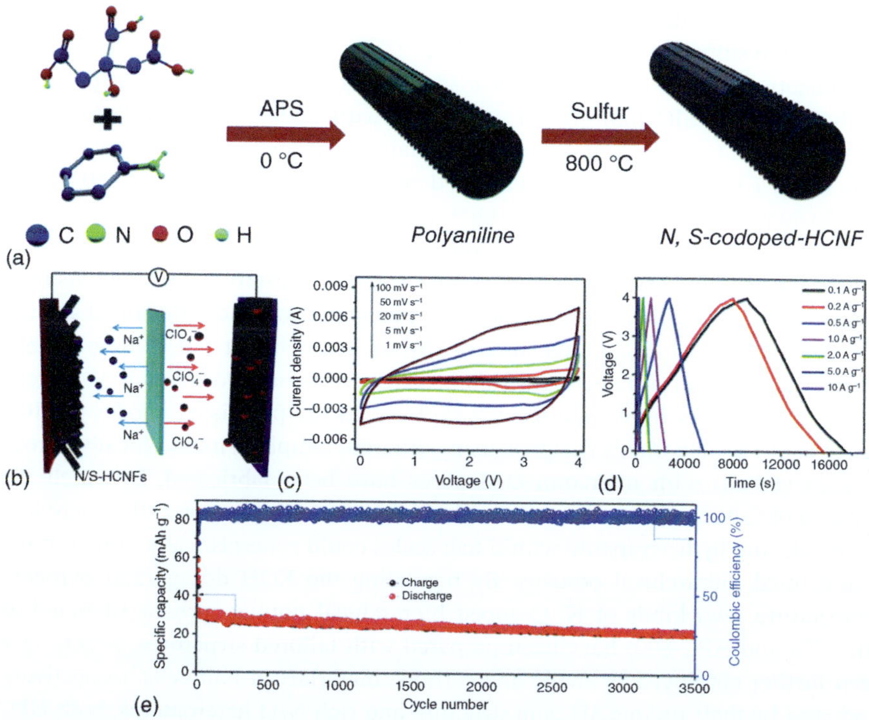

Figure 6.13 (a) Schematic illustration of the preparation of N/S-HCNFs. (b) Illustration of energy storage mechanism of N/S-HCNFs//AC SIC. (c) CV curves at various scan rates, (d) GCD curves at different current densities and (e) cycling performance at a current density of 2 A g^{-1} of N/S-HCNFs//AC SIC. Source: Reproduced with permission Liao et al. [85]. Copyright 2019, The Royal Society of Chemistry.

at a current density of 2 A g^{-1}, corresponding to a high-Coulombic efficiency of nearly 100% (Figure 6.13e). Meanwhile, the sulfur and nitrogen dual-doped carbon nanofiber (S-NCNF) membrane has been prepared as the anode of SIC [86]. The interlayer distance of S-NCNF has been enlarged to 0.38 nm after the introduction of S atoms, which is beneficial to Na^+ ion insertion and the electron transport. As a result, the S-NCNF electrode achieves a high sodium-storage capacity (1888 mAh g^{-1} at 5 A g^{-1}), which is much larger than NCNF without S doping (89 mAh g^{-1} at 5 A g^{-1}). Coupled with commercial AC as the cathode, the SIC achieves a high energy density of 97 Wh kg^{-1} and a large power density of 17 kW kg^{-1} as well as long cycle life (97% capacitance retention after 3000 cycles). In addition, self-supported N/P dual-doped carbon nanofiber film (NP-CNF) anode has been synthesized via the electrospinning method assisted by pre-oxidization and carbonization [87]. It is demonstrated that the N and P contents of the NP-CNF are optimized to be 12.8 and 4.1 at%, resulting in enhanced specific capacity triggered by increased defects and boosted rate performance originating from the larger interlayer distance. As a result, the advanced SIC device paired with AC cathode displays an energy density of 95.6 Wh kg^{-1}, a power density of 20.0 kW kg^{-1}, and a good capacity retention of 77.8% after 10 000 cycles at 2.0 A g^{-1}. On the other hand, nitrogen and sulfur co-doped nanotube-like carbon (NS-NTC) prepared by a simple carbonization process of high sulfur-loaded PANI nanotubes has been proposed as the anode of SIC [88]. When assembled with activated polyaniline-derived carbon (APDC) as the cathode, the as-obtained SIC with optimal active mass ratio of 1 : 1 could manifest the energy densities of 100.9 Wh kg^{-1} at 250 W kg^{-1} and 50.69 Wh kg^{-1} even at 12 500 W kg^{-1}, respectively.

3D porous carbon: In general, 3D porous structures can feature the merits of interconnected networks, large contact area, abundant reactive sites, and outstanding mechanical stability, ensuring the rapid transportation and diffusion of sodium ions [89]. Moreover, 3D architecture of carbon could buffer the volume change during the ion insertion and extraction processes. It is noteworthy that 3D porous carbon frameworks have been usually explored as the promising anodes of SICs, which could be elaborately designed using different template. hierarchically porous carbons (HPCs) with interconnected pores have been fabricated, in which the collagen of fish scale is served as the carbon precursor [90]. Importantly, potassium hydroxide and hydroxyapatite within fish scales could synergistically bring out the well-defined hierarchical porosity. By regulating the KOH dosage and pyrolysis temperature, two kinds of N, O-doped hierarchical porous carbons (denoted as HPC-550 and HPC-800) have been prepared with tailored structures, which have been further employed as anode and cathode materials in half cells, respectively. Endowed by their unique 3D pore structure and rich N/O heteroatoms, both HPC electrodes exhibit superior electrochemical performance. As a result, the assembled HPC-550//HPC-800 SIC can achieve a large energy density (103.2 Wh kg^{-1}), an ultrahigh power density (15.9 kW kg^{-1}), and a long cycling lifetime (81.1% of the initial capacitance over 2500 cycles). Similarly, direct pyrolysis of organometallic salts is regarded as a novel approach for the preparation of 3D porous carbon frameworks, in which the in-situ generated metallic compound could serve as the template. For example, the large-scale 3D framework carbon (3DFC) prepared by a direct calcination of sodium citrate without introduction of any extra carbon source, template, or catalyst [91]. Notably, the Na element in sodium citrate molecules could be in situ transformed into hexagonal Na_2CO_3 species. Furthermore, the Na_2CO_3 crystals can act as catalyst and hard template for the formation of 3D framework. It has been found that the DFT calculations demonstrate the curved structural unit which formed in self-growth process is more beneficial for Na ion insertion/extraction. The optimal 3DFC-700 anode could manifest the superior sodium storage performance, particularly reflected in rate capability and long-term cycling stability, resulted from the integrated merits of 3D interconnected structure, large graphitic interlayer spacing as well as curved interface, and high electrical conductivity. On this basis, a dual-carbon SIC composed of 3DFC as battery-type anode and 3DFC-derived nanoporous carbon as capacitor-type cathode has been constructed, which successfully delivers the high energy- and power-density feature as well as outstanding cycling stability in the potential range of 0–4 V (Figure 6.14a). Meanwhile, the 3D nitrogen-doped hierarchical porous carbons (NHPCs) have been prepared by thermolysis of tetrasodium ethylenediaminetetraacetate tetrahydrate

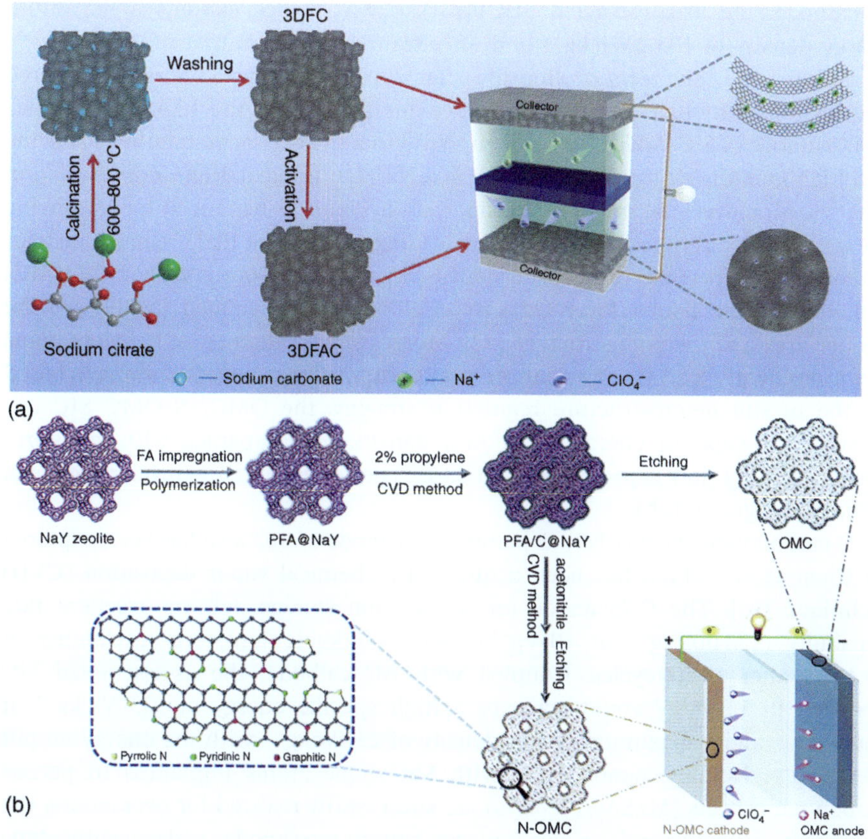

Figure 6.14 (a) Schematic illustration for preparation of a dual-carbon SIC consisted of 3DFC as battery-type anode and 3DFAC as capacitor-type cathode. Source: Reproduced with permission Yang et al. [91]. Copyright 2018, Wiley. (b) Schematics of the design route of OMC//N-OMC SIC. Source: Reproduced with permission Li et al. [92]. Copyright 2021, The Royal Society of Chemistry.

(EDTA-4Na·4H_2O), which present the enlarged interlayer distance and suitable porous feature, further leading to the strong pseudocapacitance characteristics and remarkable electrochemical sodium-storage performance [93]. In consideration of kinetic-match strategy about similar material system, nitrogen-doped hierarchical porous activated carbon (NHPAC) has been employed as the SIC cathode. As a consequence, the assembled NHPC-800//NHPAC SIC manifest a high energy density of 115 Wh kg^{-1} and excellent cycling stability in the potential range of 0–4.0 V. Novelty, a new strategy to synthesis the 3D carbon framework (3DCF) as unique anode candidate of SIC is proposed, in which the 3DCFs are prepared by using electrolysis based on ethanol/sodium hydroxide electrolyte combined with subsequent calcination process [94]. Moreover, the resulting 3DCFs feature an interconnected porous structure and defects, greatly boosting the rate capability and long-term cyclability. Paired with sodium alginate derived activated carbon (SDAC) as cathode, the as-constructed 3DCF//SDAC SIC device exhibits a maximum energy density of 133.2 Wh kg^{-1} and an ultrahigh power output of 20.0 kW kg^{-1} as well as good long-term cycling life after 4000 cycles. In addition, the ordered microporous structure has been designed for fabricating the high-performance carbon anode of SIC device. It has been found that the expanded interlayer spacing and high graphite degree can facilitate fast Na^+ ion intercalation/de-intercalation kinetics. Moreover, the ordered microporous structure is beneficial for improving the ion Na^+ adsorption/desorption, thus leading to the extra EDLC capacitance for carbon anode. In terms of nitrogen-doped ordered microporous carbon (N-OMC) cathode, electric double-layer capacity could be well regulated by the specific surface area and the pore structure [92]. Moreover, introduction of N heteroatoms can prominently lead to the enhanced pseudocapacitive storage. As a result, based on the in situ heterostructure regulation strategy, the OMC//N-OMC SIC cell presents the superior energy and power densities accompanied with long-term cycling stability (low capacity decay of 0.015% per cycle after 1800 cycles) between 0 and 4 V (Figure 6.14b).

3D carbon nanosheet (CNS) as battery-type anode of SIC also has been explored by Chen et al., which has been acquired by chemical vapor deposition (CVD) technique [95]. The CNS anode for sodium ion storage delivers excellent rate capability (138 mAh g^{-1} at 10 A g^{-1}) and remarkable long-term cyclability at 10 A g^{-1} after 12 500 cycles. Coupled with AC cathode, the as-assembled SIC operated in 1.0–4.0 V could achieves a high specific energy of 135 Wh kg^{-1} at 250 W kg^{-1} and a maximum power density of 25 kW kg^{-1} with the energy output 69 Wh kg^{-1}. Notably, nanohybrids with Mn single atoms implanted in porous carbon nanosheets (MnSAs/NF-CNs) are successfully realized for overcoming the mismatch of kinetics and capacity between battery-type anodes and capacitive-type cathodes (Figure 6.15a) [96]. The experimental study coupled with theoretical calculations verifies that N-coordinated Mn atoms (Mn-N_4) can act as sites not only for reversible Na^+ storage with reduced energy barrier but also for improving the pseudocapacitance, thus greatly accelerating the reaction kinetics of the SIC anode (Figure 6.15b). Furthermore, Mn-N_4 active sites also could lead to stronger capacitive adsorption ability of ClO_4^- ions for MnSAs/NF-CNs as SIC cathode. As a result, MnSAs/NF-CNs//MnSAs/NF-CNs SIC delivers a maximum energy/power density feature (197 Wh kg^{-1}/9350 W kg^{-1}) and ultralong cycling life (85.2% capacity retention after 10 000 cycles at 1 A g^{-1}), which is more superior to the previously reported metal-ion capacitors (Figure 6.15c,d). A new kind of 3D nitrogen functionalized carbon nanocages (NFCNs) have been successfully synthesized by a modified in situ MgO template method using acetonitrile as the precursor [97]. It has been noticed that the as-prepared NFCNs manifest the hierarchical micro/meso/macroporous structures, effectively ensuring the fast pathways for sodium in diffusion, which further gives rises to the excellent rate capability. Benefited from the enhanced kinetics feature triggered by the unique architecture, the as-assembled NFCNs-800//NFCNs-800 SIC could achieve an energy density of 102.5 Wh kg^{-1} at the power density of 331 W kg^{-1} and a remarkable cycle life of 100 000 cycles with 74.2% of the capacity retention. On the other

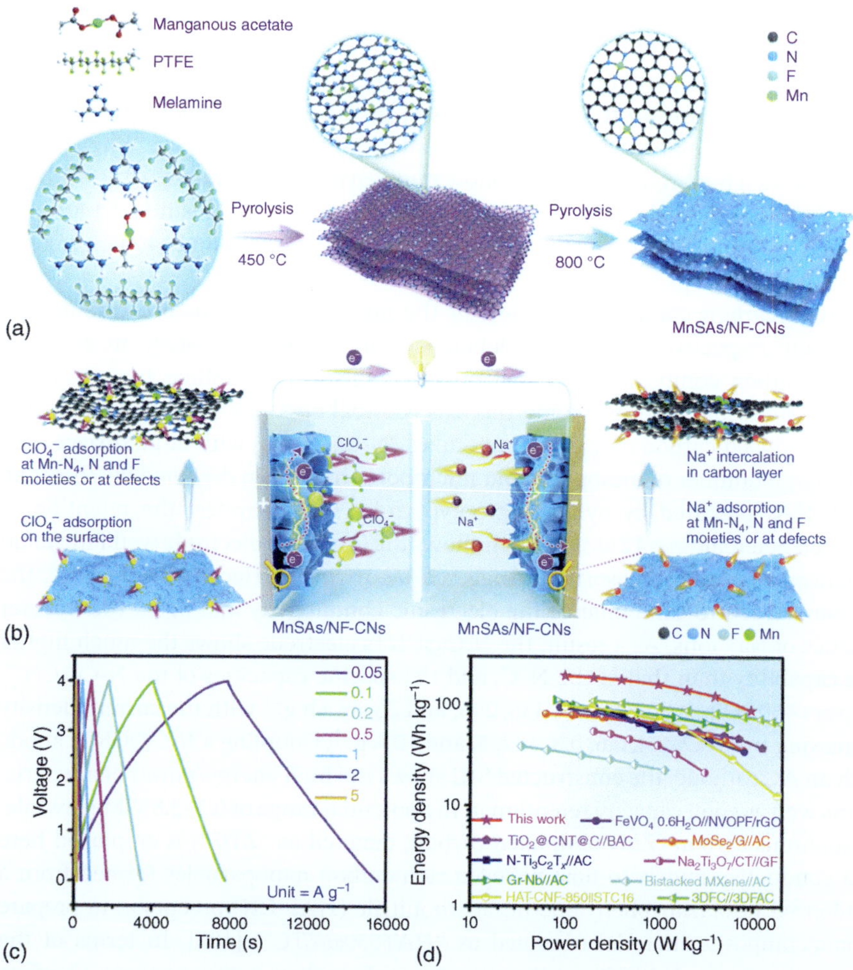

Figure 6.15 (a) Schematic diagram of preparation process of MnSAs/NF-CNs. (b) Schematic illustration of the MnSAs/NF-CNs//MnSAs/NF-CNs SIC device. (c) GCD curves at different current density and (d) Ragone plots of the SIC device compared with previously reported SICs. Source: Reproduced with permission Hu et al. [96]. Copyright 2021, The Royal Society of Chemistry.

hand, hierarchical porous hollow carbon sphere (HPCS) has been as a novel anode for ameliorating the sluggish kinetics of SICs. When evaluated in Na half-cell, HPCS electrode displays an outstanding rate capability and long-life capability. Interestingly, hierarchical porous hollow carbon bowl (HPCB) as the cathode has been fabricated by chemical activation of HPCS, which exhibits the excellent capacitive performance [98]. By virtues of superior properties and porous structures of the anode/cathode, the designed dual-carbon SIC manifest the maximum energy/power density ($128.5\,Wh\,kg^{-1}/11.9\,kW\,kg^{-1}$). In addition, hierarchical porous hollow multicavity carbon spheres (HMCSs) as a novel intercalated anode of SIC have been investigated, which are prepared from aniline and pyrrole through a one-step carbonization with F-127 as soft template [99]. It has been noted that the hollow multi-cavities of HMCSs provide abundant electrolyte reservoirs. Moreover, the interconnected pores on the shells and on the walls of inner cavities are served as shortcuts for fast ion transport. Endowed by the above-mentioned merits, the HMCSs deliver a high reversible capacity and good rate performance. When paired with activated hierarchical porous hollow multicavity carbon spheres (AHMCSs) as cathode, the as-assembled SIC manifests the superior energy densities of 140.1 and 87.1 Wh kg^{-1} under the corresponding power densities of 380.0 and 14 249.7 W kg^{-1}, respectively.

Carbon/carbon composite: In view of the different nanostructure carbon possess their respective merits and deficiency, the elaborate strategy to engineer carbon/carbon composite could not only synergistically combine the advantage of both sides, but also overcome the existed weakness [100]. For example, rich nitrogen-doped carbon on carbon nanotubes (N-C@CNTs) with large surface area and a large number of mesopores and micropores have been designed by Zhu et al. [101]. N–C prepared by pyrolyzing polypyrrole (PPy) manifest the retention of nitrogen heteroatoms (12 at.%), which shows the improved electrode wettability and enhanced sodium ion absorption triggered by extrinsic defects. Furthermore, the 1D nanostructure can enhance the electronic conductivity and boost the transfer kinetics of Na^+ ions. As a result, the N-C@CNTs electrode shows the much higher rate capability than that of the N–C, and the specific capacities of the N-C@CNTs delivers 480, 432, 408, 376, 368, 336, 296, and 264 mAh g^{-1} with the current density increasing from 0.2, 0.4, 0.6, 0.8, 1, 2, 5, and 10 A g^{-1}. Coupling a N-C@CNTs anode with an AC cathode, the constructed SIC shows the high energy/power characteristics as well as good capacity retention in the potential range of 0.3–3.8 V Meanwhile, mesoporous carbon (ZnO templated carbon, denoted as "ZTC") is employed here as a conductive matrix to host nitrogen-rich carbon nanoparticles formed from a condensed hexaazatriphenylene-hexacarbonitrile (HAT-CN) precursor, to prepare a nanocomposite material (denoted as "HAT550@ZTC") [102]. In terms of the structural merits, HAT550 as the nitrogen-rich carbon component could offers the abundant stabilization sites to storage sodium, while the ZTC matrix ensures the good electrical conductivity. It has been noticed that this ingenious architecture further affords the strong contact and short transport pathways between the less electron conducting HAT550 and the highly conductive ZTC, greatly boosting the electron transport inside the whole framework. Thus, the HAT550@ZTC anode could deliver a reversible capacity of 343 mAh g^{-1} at 0.1 A g^{-1}. Moreover, a SIC employing HAT550@ZTC anode and carbon cathode shows the remarkable energy and power density in the voltage range of 1.0–4.0 V.

Interestingly, a complex carbon anode consisting of N and O dual-doped carbon nanofibers (N, O-CNFs) and CNTs has been purposefully synthesized, in which modified N,O dual-functional group effectively regulates the surface activity of CNFs while the integration of CNTs can dramatically increase the graphitization extent and electrical conductivity (Figure 6.16a) [103]. When paired with the activation of CNFs (A-CNFs) as cathode, the as-assembled SIC achieves a maximum energy density of 59.2 Wh kg^{-1} at a power density of 275 W kg^{-1} as well as

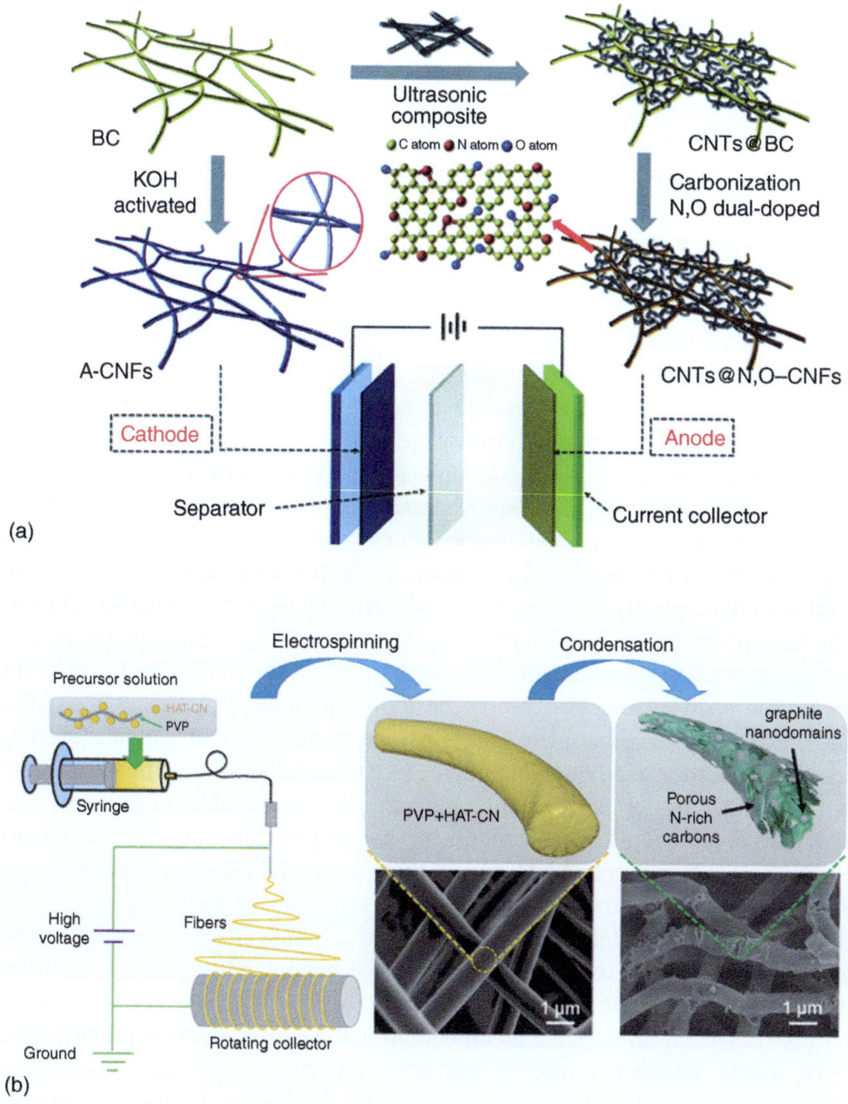

Figure 6.16 (a) Schematic illustration of the synthesis and related construction of CNTs@N,O-CNFs//A-CNFs SIC. Source: Xu et al. [103]/Royal Society of Chemistry/Licensed under CC BY 3.0. (b) Schematic illustration of the preparation procedures of porous HAT-CNF. Source: Yan et al. [104]/Reproduced with permission from John Wiley & Sons.

a high energy density of 38.7 Wh kg^{-1} at a power density of 5500 W kg^{-1} in the voltage range of 1.5–4.0 V. A free-standing carbon fiber composite film (denoted as HAT-CNFs) has been prepared by HAT-CN and polyvinylpyrrolidone (PVP) as precursors (Figure 6.16b) [104]. Notably, the high electrical conductivity and efficient ion transport could be improved by the interconnected 1D structure of the fibers, and the high nitrogen content is effectively introduced by condensation of

HAT-CN. Moreover, the sodium storage properties of HAT-CNFs could be adjusted by the ratio of nitrogen-doped carbon and graphitic carbon domains through different condensation temperatures. A SIC prototype combining with HAT-CNF anode and porous salt-templated carbon (STC) as cathode displays the excellent electrochemical performance in the voltage range of 0.5–4.0 V (95 Wh kg^{-1} at 0.19 kW kg^{-1} and 18 Wh kg^{-1} at 13 kW kg^{-1}).

6.3.1.2 Titanium-Based Compound

TiO_2: Featured with the natural abundance, nontoxicity, low-cost and good stability, and titanium dioxide (TiO_2) has been regarded as a promising anode material in SIB [105]. Commonly, TiO_2 with electrochemical activity can be divided into five polymorphs: amorphous, anatase, rutile, TiO_2 bronze (TiO_2-B), and TiO_2 hollandite (TiO_2-H). It has been noticed that the above-mentioned four crystalline types of TiO_2 are composed of TiO_6 octahedra, sharing edges, and corners but with different regular arrangements of the TiO_6 octahedra units [106]. In terms of anatase, the stacking of TiO_6 octahedra gives rise to the empty zigzag channels in the structure for providing interstitial sodium ion storage sites. Rutile TiO_2 possesses a tetragonal $P42/mnm$ structure including corner-shared octahedra. It is well-known that rutile TiO_2 is the most thermodynamically stable structure, which could be irreversibly formed from the anatase phase by the heating process. Meanwhile, TiO_2-B presents the monoclinic $C2/m$ structure, in which the edge-sharing and corner-sharing TiO_6 octahedra form a more open architecture. Moreover, the open channel of TiO_2-B is parallel to the b-axis. The tetragonal $I4/m$ TiO_2-H structure has the double chains of TiO_6 octahedra along the c-axis. And the unique (2×2) channels running the [001] direction are obtained by corner cross linking the double chains. More importantly, the different arrangements of TiO_6 octahedra in TiO_2 polymorphs can bring out the different Na^+ migration processes. Moreover, these structural differences would cause the different mass densities and electronic band structures of TiO_2 species.

On account of the deficiency of d electrons, pure TiO_2 as one of n-type semi-conductors manifests poor electronic/ionic conductivity, giving rise to the sluggish chemical kinetics during electrochemical reactions. Hence, a variety of strategies have been investigated for ameliorating the electrochemical performance of TiO_2 anode, which is aimed to fabricating the high-performance SICs. It is verified that doping other elements as a practicable method could effective enhance the electronic and ionic conductivities of TiO_2. Typically, nitrogen-doped mesoporous titanium dioxide (NMTiO_2) structures are successfully achieved through the controlled pyrolysis of NH_2-MIL-125 (Ti) [107]. It has been found that NMTiO_2 possesses a lower Ohmic resistance and charge transfer resistance values than those of the bare anatase TiO_2 sample, which demonstrates that the electrical conductivity of pristine TiO_2 electrode can be effectively enlarged by N-doping. Benefiting from the interconnected open mesopores architecture with heterogeneous nitrogen dopants, the MTiO_2 delivers the enhanced fast ion transport during repeated charge–discharge cycles. Subsequently, the designed SIC device composed of MTiO_2 anode and nitrogen-doped graphene (NG) as a cathode has been fabricated (Figure 6.17a). As shown in Figure 6.17b,c, the related curves and

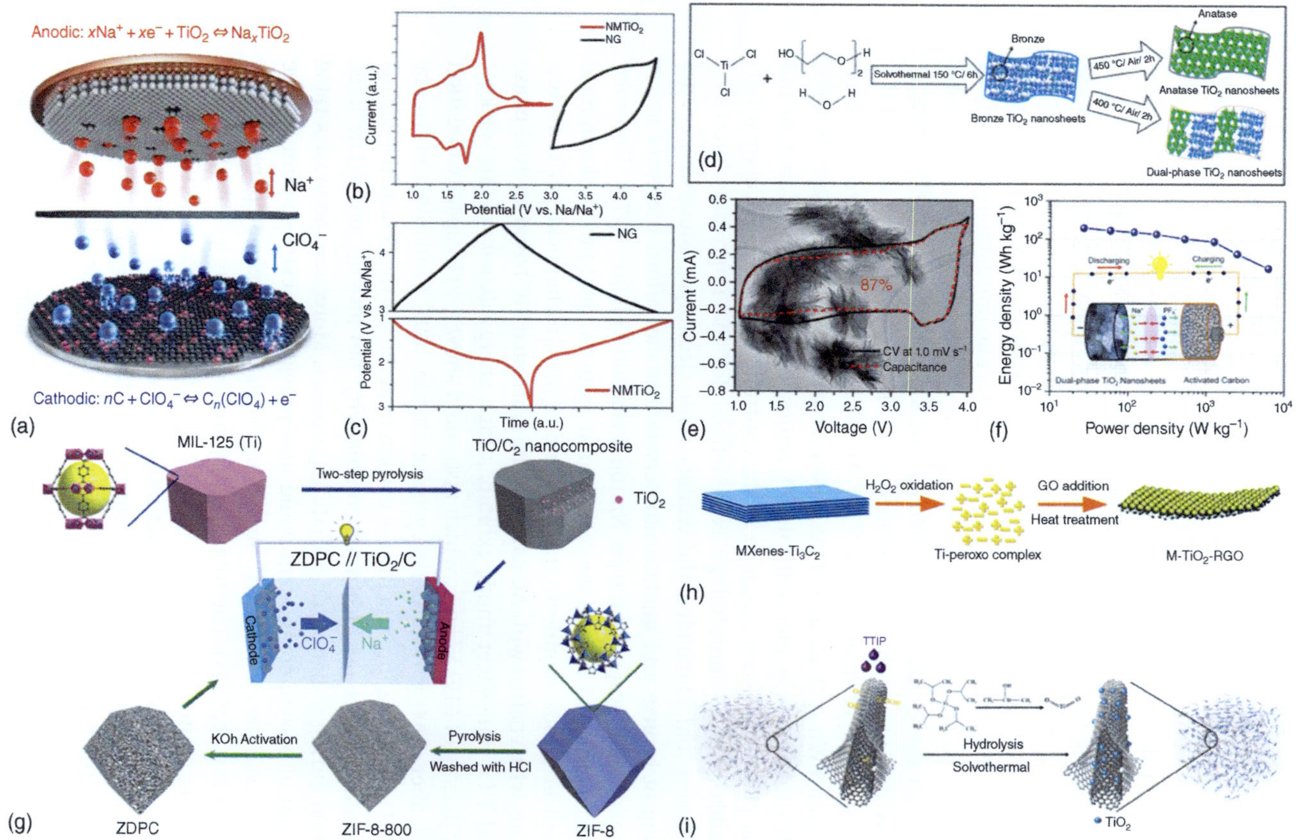

Figure 6.17 (a) Schematic illustration of $NMTiO_2$//NG SIC configuration. (b) CV curves and (c) GCD profiles of $NMTiO_2$ anode and NG cathode in half-cell configurations. Source: Lee and Kang [107]/John Wiley & Sons/CC BY 4.0. (d) Illustration of the preparation of single- and dual-phase TiO_2 nanosheets. (e) Morphology of dual-phase hierarchical TiO_2 nanosheets with enhanced pseudocapacitive feature. (f) Ragone plot of dual-phase TiO_2 nanosheets//AC SIC. Source: Feng et al. [108]/Reproduced with permission from John Wiley & Sons/CC BY 4.0. (g) Schematic illustration of preparation process of a SIC device consisted of TiO_2/C anode and ZIF-8 derived nanoporous carbon (ZDPC) cathode. Source: Reproduced with permission Li et al. [109]. Copyright 2018, Wiley. (h) Schematic illustration of synthetic process of M-TiO_2-rGO composite derived from MXenes-Ti_3C_2. Source: Reproduced with permission Wang et al. [110]. Copyright 2018, The Royal Society of Chemistry. (i) Illustration of the formation process of 3D TiO_2/SCNT composite. Source: Reproduced with permission Luo et al. [111]. Copyright 2020, Wiley.

corresponding GCD profiles for anode and cathode demonstrate that the $NMTiO_2$ anode displays the Faradaic reaction curve triggered by Na-ion insertion/desertion, while the NG cathode manifests the capacitive mechanism. Furthermore, the assembled $NMTiO_2$//NG SIC shows a maximum energy density of 90.5 Wh kg^{-1} at the power density of 91.1 W kg^{-1}. Meanwhile, nitrogen-doped anatase TiO_2 (N-TiO_2) nanospheres with enhanced electronic conductivity deliver a reversible specific capacity of 162 mAh g^{-1} over 1000 cycles at 1 A g^{-1} and a superior rate performance [112]. Coupled with the AC cathode, the constructed SIC device could achieve the high energy/power characteristic (80.3 Wh kg^{-1}/12 500 W kg^{-1}) and excellent long-term cycling stability. Similarly, metal-ion doped anatase ($Mo_{0.1}Ti_{0.9}O_2$ and $Nb_{0.25}Ti_{0.75}O_2$) materials also manifest significantly higher specific discharge capacities than that of undoped anatase [113]. When paired with the AC cathode, the $Mo_{0.1}Ti_{0.9}O_2$//AC SIC delivers a maximum energy density of 60 Wh kg^{-1} at a power density of 350 W kg^{-1}. Moreover, the $Nb_{0.25}Ti_{0.75}O_2$//AC SIC achieves an energy density of 21 Wh kg^{-1} at a power density of 6000 W kg^{-1}. It is obvious that electrochemical performance of SICs comprising doped TiO_2 anode are much larger than that of TiO_2/AC SIC (relatively low energy densities of 31 and 3 Wh kg^{-1} at moderate power densities of 322 and 2350 W kg^{-1}, respectively). Interestingly, hierarchical TiO_2 nanosheets consisting of anatase and bronze dual-phase have been designed by the controlled calcination (Figure 6.17d) [108]. Notably, the 2D morphology is beneficial for the improving interfacial Na-ion diffusion and enhancing the contact with electrolyte. Moreover, crystal mismatching of bronze and anatase nanocrystallites can afford the extra active sites for rapid pseudocapacitive-type Na-ion insertion (Figure 6.17e). Endowed by the above-mentioned unique microstructural features, dual-phase hierarchical TiO_2 nanosheet anode manifests the outstanding electrochemical performance. Furthermore, when paired with AC cathode, the fabricated SIC based on dual-phase hierarchical TiO_2 nanosheet anode exhibits a high energy density of 200 Wh kg^{-1} and power density of 6191 W kg^{-1}, which also could show a 80% capacity retention at 1.0 A g^{-1} after 5000 galvanostatic cycles with the 100% Coulombic efficiency (Figure 6.17f). Typically, introducing the carbon materials as the conductive agents not only could enhance the electronic conductivity, but also accommodate large volume strain during sodiation/de-sodiation processes. Moreover, the incorporated small amount of carbon component could also provide the extra sodium storage capacity. For example, the TiO_2/C anode has been elaborately prepared by the thermolysis of MIL-125 (Ti). The robust architecture comprising a continuous conductive network of the TiO_2/C nanocomposite leads to the pseudocapacitive charge storage with fast kinetics and high structural stability, thus delivering a high capacity of 215 mAh g^{-1} at 0.1 A g^{-1} and remains a reversible capacity of 120 mAh g^{-1} at 5 A g^{-1} after 10 000 cycles [109]. When assembled with the capacitor-type carbon cathode derived from ZIF-8, the SIC achieves high energy and power densities and outstanding cyclability (Figure 6.17g). Meanwhile, graphene-supported TiO_2 (M-TiO_2-RGO) nanocomposite with highly porous nano-/meso-hybrid architecture has been prepared by a facile stepwise transformation from accordion-like Ti_3C_2 MXenes precursor (Figure 6.17h) [110]. It has been found that the unique hybrid architecture gives rise to the pseudocapacitive behavior in Na^+ charge storage with high reversibility, fast kinetics, long cyclability, and negligible degradation. Moreover, the as-built SIC using M-TiO_2-RGO anode together with a porous carbon cathode is capable of operating at a high voltage of 4.0 V and delivering a maximum energy density of 94.7 Wh kg^{-1}. The single-crystal-like anatase TiO_2 mesocages anchored on graphene sheets as a sodium storage anode have been acquired by Le et al. Importantly, the mesoporous texture TiO_2 accompanied with graphene conducting network allows the fast Na^+ insertion/deinsertion and short diffusion distance [114]. By coupling with commercial AC cathode, the constructed SIC displays a high energy density of 64.2 Wh kg^{-1} at 56.3 W kg^{-1} and 25.8 Wh kg^{-1} at 1357 W kg^{-1} as well as an ultralong lifespan of 10 000 cycles with over 90% of capacity retention. A unique architecture consisting of TiO_2 nanoparticles anchored on a sheared-carbon nanotubes backbone (TiO_2/SCNTs) has been designed (Figure 6.17i) [111]. It is noticed that the SCNT framework leads to the enhanced Na^+ ion diffusivity and electrical conductivity of TiO_2/SCNT composite, over other carbon architectures commonly used. Thus, the TiO_2/SCNT electrode presents the superior electrochemical properties in a half-cell. When combined with AC cathode in SIC prototype, the maximum energy density and power density of 54.9 Wh kg^{-1} and 1410 W kg^{-1} are successfully achieved, respectively. Interestingly, the advanced architecture of TiO_2@CNT@C nanorods have been engineered, in which the stable external carbon skeleton and internal 1D nanoarchitecture would be beneficial for electrochemical performance [115]. Employed with the biomass-derived AC cathode, the assembled SIC presents an exceptionally high energy and power density as well as long cycling stability in the voltage window of 1.0–4.0 V. On the other hand, a defective TiO_2@reduced graphene oxide (M-TiO_2@rGO) self-supporting foam electrode has been constructed via a facile MXene decomposition and graphene oxide self-assembling process [116]. It is confirmed that a large number of lattice defects and oxygen vacancies are in situ generated during the conversion process from Ti_2CT_x-MXene to M-TiO_2 by consuming the –OH in the H_2O and self-doping the F functional group, which fundamentally improves the electrons/ions conductivity. Moreover, the rGO could afford a highly conductive pathway for accelerated ion diffusion.

Titanate: Generally, layered titanates with the $M_2Ti_nO_{2n+1}$ formula (M is mainly H, Li, Na, and K) crystallize into a monoclinic structure, which are composed of edge sharing and strongly distorted TiO_6 octahedra, yielding ribbons whose thicknesses depend on the n value [22]. Typically, sodium trititanate $Na_2Ti_3O_7$ presents a monoclinic $P2_1/m$ space group. Three Ti atoms are located at nonequivalent positions and each one is connected by six oxygens as first neighbors, yielding a strongly distorted edge sharing TiO_6 octahedra. As shown in Figure 6.18a, $\{Ti_3O_7\}^{-2}$ ribbons generated from these octahedra are linked with the oxygens in the equatorial vertices, further leading to the stepped-like layers. Moreover, the layers with a zigzag format are extended in [010] direction, where the sodium atoms are located between the layers (Figure 6.18b) [117]. Most importantly, $Na_2Ti_3O_7$ as anode for sodium storage could manifest the relative low Na^+ insertion voltage and accommodate 3.5 Na ions with capacity of 310 mAh g^{-1}. For instance,

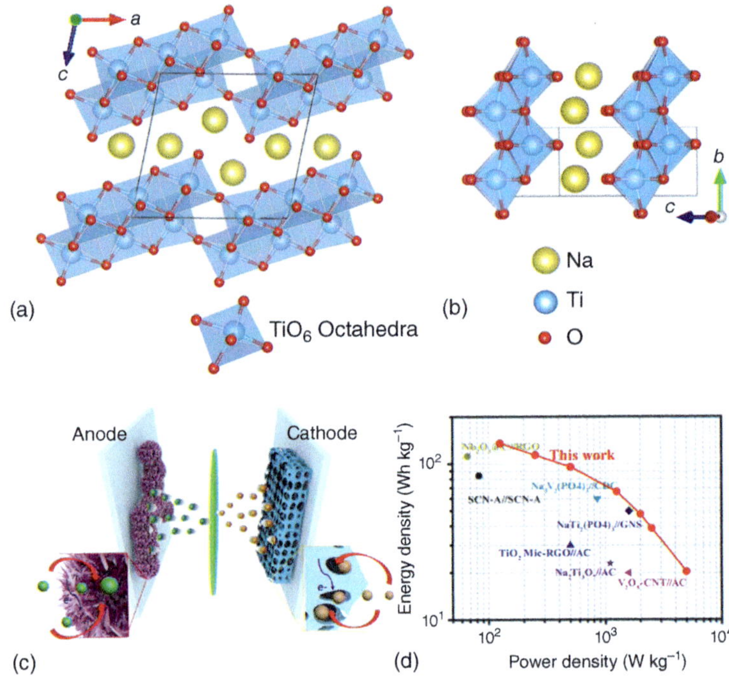

Figure 6.18 (a) Stepped layer and lamellar format in [010] zone axis and (b) zigzag format in *bc*-plane of $Na_2Ti_3O_7$ crystal structure. Source: Reproduced with permission Silva et al. [117]. Copyright 2018, Wiley. (c) Structure schematic and (d) Ragone plot of the as-assembled $Na_2Ti_3O_7$-IcUH//AC SIC. Source: Reproduced with permission Chen et al. [118]. Copyright 2019, American Chemical Society.

interconnected urchin-like hollow $Na_2Ti_3O_7$ ($Na_2Ti_3O_7$-IcUH) chains have been elaborately designed by the assistance of H_2O_2 reagent [118]. It has been discovered that the H_2O_2 cannot only trigger the 1D crystal-oriented growth of $Na_2Ti_3O_7$ but also induce the (001) facet preferential growth of $Na_2Ti_3O_7$. Due to the more exposed (001) facet and shorter Na^+ diffusion path, $Na_2Ti_3O_7$-IcUH exhibits faster Na^+ storage characteristics than those of 2D $Na_2Ti_3O_7$ nanosheet. By coupling the $Na_2Ti_3O_7$-IcUH anode and a commercial AC cathode, the as-fabricated SIC achieves the high energy densities of 134.3 and 38.2 Wh kg^{-1} at power densities of 125 and 2500 W kg^{-1} (Figure 6.18c,d). A novel $Na_2Ti_3O_7$@CNT coaxial nanocable has been prepared by Dong et al., which is explored as an anode material for SICs. [119]. Featured with unique 1D nanostructure, high conductivity and extra pseudocapacitive behaviors, the $Na_2Ti_3O_7$@CNT anode presents the excellent rate capability and outstanding cycling stability. Moreover, the constructed SIC based on $Na_2Ti_3O_7$@CNT anode delivers superior electrochemical performance. Interestingly, sodium titanate nanotubes (Na-TNTs) as Na^+-insertion compounds present the diversified crystal phases of $Na_xH_{2-x}Ti_3O_7{\cdot}nH_2O$ ($0<x<2$, and $n<1.2$), $Na_xH_{2-x}Ti_3O_7$ ($x>1$) and $Na_2Ti_6O_{13}$ at different calcined temperatures [120]. Moreover, it is confirmed that $Na_xH_{2-x}Ti_3O_7$ possesses the superior electrochemical performance due to the larger interlayer spacing. The Na-TNT//AC SIC displays the desirable energy density/power density (34 Wh kg^{-1}/889 W kg^{-1}) as well as a long cycling life. In terms of $Na_2T_7O_{15}$ substance, a $Na_2T_7O_{15}$ (NTO) nanowire/graphene anode has been synthesized through an ALD seeding technique by Liu et al. [121]. It has been stated that fabric-like framework composed of cross-connected NTO nanowires and graphene nanosheets and strong bonding interaction between NTO and graphene promote the electron transport and sodium ion migration. Thanks to the above-mentioned virtues, the composite anode achieves a superior rate capability of 60 mAh g^{-1} even at an ultrahigh rate of 17.7 A g^{-1}, and an extraordinary cyclic stability (90% retention after 10 000 cycles at 8.85 A g^{-1}). The ALD NTO@graphene//AC SICs manifests high energy/power features (82.7 Wh kg^{-1}/97.5 W kg^{-1} and 16 Wh kg^{-1}/25 kW kg^{-1}) particularly at extremely fast charge/discharge condition. Moreover, about 90% capacity retention after 10 000 cycles at the current density of 3.2 A g^{-1} can be achieved. A new nanostructure $Na_2Ti_9O_{19}$ material, as a new Na-insertion electrode, exhibits excellent Na-ion kinetics with a capacitive nature of 86% at 1.0 mV s^{-1}, which is considered as a potential anode candidate for SICs.

It has been found that Ti–O distance in $Na_2Ti_9O_{19}$ structural framework is varied during the sodiation and desodiation processes, further bringing out the tunnel structural change. Meanwhile, the distribution of Na ions is also disturbed by the entering sodium atoms. Thus, the sodium insertion and extraction processes in $Na_2Ti_9O_{19}$ through a single-phase solid solution mechanism. And the related Na^+-insertion principle could be listed as follows:

$$Na_2Ti_9O_{19} + xNa^+ + xe^- \rightarrow Na_{2+x}Ti_9O_{19} \tag{6.2}$$

Notably, a SIC prototype employing the $Na_2Ti_9O_{19}$ anode and activated porous carbon cathode displays a maximum energy density of 54 Wh kg^{-1} and a maximum power density of 5 kW kg^{-1}, respectively. A novel hydrated layered sodium titanite $Na_2Ti_2O_4(OH)_2$ has been investigated as promising anode material for SIC. It is demonstrated that $Na_2Ti_2O_4(OH)_2$ shows an intercalation mechanism of Na^+ intercalation with a large pseudocapacitive contribution to the total charge storage according to the kinetics analysis [122]. Moreover, rice husk derived porous carbon (RHDPC-KOH) as cathode presents the adsorption/desorption process of PF_6^- anions originated from nearly rectangular CV curves. The assembled SIC device could show excellent electrochemical performance at the operated voltage of 1–4 V. Interestingly, spinel $Li_4Ti_5O_{12}$ (LTO) cannot only be investigated as an anode of LICs, but also reversibly intercalate/deintercalate sodium ions through the three-phase separation mechanism according to the following reaction: $2Li_4Ti_5O_{12} + 6Na^+ + 6e^- \leftrightarrow Li_7Ti_5O_{12} + Na_6LiTi_5O_{12}$. Moreover, LTO manifests a lower working potential plateau of 0.9 V vs. Na/Na^+ for Na^+ ion intercalation reaction (equivalent to 1.2 V vs. Li/Li^+) than that for Li^+ ion intercalation reaction (1.55 V vs. Li/Li^+). High-tap density $Li_4Ti_5O_{12}$/rGO microsphere composites as anodes have been firstly paired with AC cathode for constructing a high-performance SIC [123]. Notably, the SIC shows a large energy density of 83 Wh kg^{-1}, which is twice as high as the energy density of lithium-ion capacitor (42 Wh kg^{-1}).

$NaTi_2(PO_4)_3$: Sodium super ionic conductor (NASICON)-structured $NaTi_2(PO_4)_3$, as a typical representatives of polyanion-type electrode materials, can allow the reversible intercalation of two sodium ions (corresponding to a theoretical capacity of 133 mAh g^{-1}), which is resulted from a stable 3D framework consisted of corner- and edge-connected [PO_4] tetrahedra and [TiO_6] octahedra [124]. Moreover, this skeleton with 3D diffusion pathways for Na ions can ensure the controllable lattice expansion below 8%, indicating their good cycling stability. For example, quasi-cubic mesoporous $NaTi_2(PO_4)_3$ (NTP) nanocages has been explored as anode of SIC, which shows a large ICE of 94% and outstanding rate capability of 98 mAh g^{-1} at 10 C (1 C = 133 mA g^{-1}) [125]. It has been demonstrated that the tiny nanostructure and abundant pores provide more active sites and shorten the ion transport path for the enhanced sodium-ion transport kinetics. Furthermore, the as-built NTP//AC SIC presents superb cycling performance at various charging/discharging rates. On the other hand, gadolinium ions (Gd^{3+}) with a half-full 4f electron shell are introduced to substitute Ti of mesoporous $NaTi_2(PO_4)_3$ (MNTP) nanocrystals, which is aimed to improve Na storage properties [126]. According to the microstructure characterization and kinetics analysis, it is confirmed that the superior electrochemical characteristics are attributed to larger cell size, optimized porous structure, faster intrinsic electron conductivity, and higher Na ion mobility, which is resulted from the introduction of Gd^{3+} ions. In terms of the SIC configuration composed of Gd^{3+} doped MNTP anode and AC cathode, a high energy density of 63 Wh kg^{-1} can be successfully achieved at a power density of 46 W kg^{-1}. Meanwhile, it is proved that the doping of Mn^{2+} ions also could regulate the intrinsic conductivity of NTP anode, which is beneficial for construction of high-performance SICs [127]. Interestingly, a uniform and controllable carbon coating strategy is designed for improving Na^+ and electron kinetics of the self-supported NTP nanorod arrays [128]. As a result, the as-fabricated NTP@C NR anode exhibits a high initial coulombic efficiency of 97% and a good rate performance (89.1 mAh g^{-1} at 100 C, 1 C = 133 mA g^{-1}). The NTP@C NR//AC SIC displays a high retention of ≈94.5% at 1 A g^{-1} over 7000 cycles. Moreover, an energy density of 37.1 Wh kg^{-1} at 0.4 A g^{-1} and a maximum power density of 9180.0 W kg^{-1} at 10 A g^{-1} are successfully obtained.

Meanwhile, porous $NaTi_2(PO_4)_3$/C nanofibers (NTP/CNFs) via an electrospinning method have been prepared by Wei et al., which manifests the excellent reversible capacity and a long cycling stability [129]. When assembled with AC cathode, the SIC prototype achieves a maximum specific energy density of 56 Wh kg^{-1} at a specific power density of 174 W kg^{-1}. Moreover, the rGO has been utilized for chemically bonding to $NaTi_2(PO_4)_3$ nanoparticles to improve their electrical conductivity. The interconnected nanoporous structures could ensure easy access of Na^+ ions into $NaTi_2(PO_4)_3$ active material [130]. Benefiting from the above-mentioned merits, the SIC using $NaTi_2(PO_4)_3$/rGO microsphere composite as an anode and commercial AC as a cathode shows excellent energy/power features.

6.3.1.3 Niobium-Based Compound

Nb_2O_5: Generally, Nb_2O_5 is classified into five types crystalline phases: pseudohexagonal Nb_2O_5 (TT-Nb_2O_5), orthorhombic Nb_2O_5 (T-Nb_2O_5), tetragonal

Nb_2O_5 (M-Nb_2O_5), and monoclinic Nb_2O_5 (H-Nb_2O_5), which can be manipulated by the different heat treatment temperatures [131]. Notably, T-Nb_2O_5 manifests the superior electrochemical performance among them, which is resulted from fast ion insertion–extraction mechanism and excellent cycling stability. In the T-Nb_2O_5 crystal structure, each Nb atom is coordinated with six or seven oxygen atoms, forming edge/corner-sharing distorted octahedral (NbO_6) or pentagonal bipyramids (NbO_7) [132]. According to the atomic arrangement of T-Nb_2O_5, the crystal structure can be regarded as two alternating atomic layers, including the loosely packed 4g layer and the denser 4h layer. Moreover, the atomic density of 4h layer is approximately 2.6 times higher than that of 4g layer. Due to the larger accommodation spaces, the 4g layer affords the more preferential sites for storage and transmission of alkali ions.

By virtue of these remarkable structural features, T-Nb_2O_5 has been widely studied as the promising anode materials for SICs. A novel self-assembled strategy has been utilized for fabricating various Nb_2O_5 morphologies triggered by different pH values (Figure 6.19a) [133]. It is noticed that the optimized Nb_2O_5 nanosheets achieve the excellent sodium ion storage property, showing a high capacity of 85.6 mAh g^{-1} at 500 mA g^{-1} (Figure 6.19b). A SIC prototype is constructed by a Nb_2O_5 nanosheet anode and peanut shell carbon (PSC) cathode, in which the Nb_2O_5 anode exhibits the reversible sodium ion insertion/de-insertion process and the PSC cathode undergoes the ClO_4^- adsorption/desorption behavior (Figure 6.19c). Moreover, the resulted SIC could achieve an energy density of 43.2 Wh kg^{-1} at a power density of 160 W kg^{-1}. Meanwhile, a unique nanocomposite comprising Nb_2O_5@carbon core–shell nanoparticles (Nb_2O_5@C NPs) and rGO has been investigated as a high-power anode material for SICs in 2016 [135]. Benefiting from the synergistic effects of nanosized Nb_2O_5@C core–shell structures and enhanced electron mobility of rGO, Nb_2O_5@C/rGO availably delivers a highly reversible capacity (285 mAh g^{-1} at 0.025 A g^{-1}). The designed SIC using an Nb_2O_5@C-50 anode and commercial AC cathode successfully obtains the maximum energy/power densities (76 Wh kg^{-1}/20 800 W kg^{-1}) and a stable cycle life in the potential range of 1.0–4.3 V. A mesoporous T-Nb_2O_5/carbon composite (signed as m-Nb_2O_5/C) has been prepared by using the sol–gel method, which is further studied as anode of SICs [136]. It has been found that besides the structure advantages of rich mesopores with a high specific surface, a highly electron-conductive niobium carbide (NbC) layer between the T-Nb_2O_5 phase and the carbon matrix has been observed, which significantly enlarging the electric conductivity of the integrated composite. Tested in Na half-cell, m-Nb_2O_5/C anode exhibits the fast charge transfer kinetics, which displays a reversible sodium-ion storage capacity of 252 mAh g^{-1} at 50 mA g^{-1} and a long cycling stability. When paired with AC cathode, the as-built SIC delivers a high energy density of ~72.9 Wh kg^{-1} with a power density of 250 W kg^{-1} at the current density of 0.1 A g^{-1}. Furthermore, excellent capacity retention with 100% Coulombic efficiency is obtained. Interestingly, an effective and simple method is designed to fabricate carbon-encapsulated T-Nb_2O_5 composites by exploiting the acid-catalyzed condensation reactions of acetophenone [134]. Originally, Nb_2O_5 species is generated via the aldol condensation reaction of acetophenone promoted

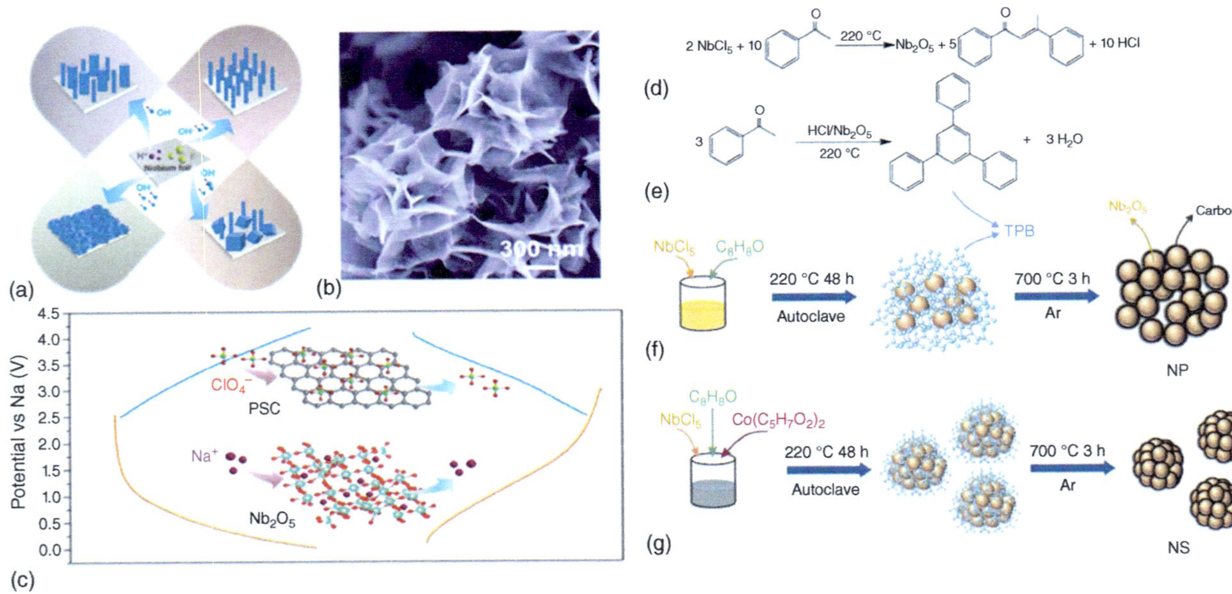

Figure 6.19 (a) Illustration of the various Nb_2O_5 morphologies prepared by different pH values. (b) FESEM images of the self-assembled Nb_2O_5 nanosheets with a flower-like structure. (c) Illustration of potential variations of electrodes in Nb_2O_5 nanosheets//PSC SIC. Source: Li et al. [133]/Reproduced with permission from American Chemical Society. (d) Formation mechanism of Nb_2O_5 in acetophenone. (e) Condensation reaction of acetophenone into TPB. Schematic illustration of the preparation processes of (f) NP and (g) NS composites. Source: Reproduced with permission Han et al. [134]. Copyright 2019, Wiley.

by $NbCl_5$ as Lewis acid (Figure 6.19d). Subsequently, Nb_2O_5 and HCl produced by the previous aldol condensation reaction synergistically catalyze the condensation of the acetophenone into 1,3,5-triphenylbenzene (TPB) (Figure 6.19e). Furthermore, thermal treatment of the Nb_2O_5/TPB mixtures at 700 °C under argon atmosphere is aimed to prepared T-Nb_2O_5/C composites with increased the crystallization degree. It is confirmed that the carbon content and size of the Nb_2O_5 nanocrystals can be controlled varying the Nb_2O_5/TPB ratio and the addition of promoting agents, further well manipulated the corresponding electrochemical properties. Specifically, small carbon-capsulated Nb_2O_5 nanocrystals (NP-X, where X is the carbon content) are synthesized by reaction of $NbCl_5$ with acetophenone (Figure 6.19f). At the same time, spherical structures (400–600 nm) consisting of carbon-coated Nb_2O_5 nanocrystals (NS-X, where X is the carbon content) are successfully obtained by introduction of extra Co^{2+} ions (Figure 6.19g). The resulted T-Nb_2O_5/C composite as anode manifest the remarkable performance in Na half-cell and SIC devices paired with AC cathode. Oxygen defected T-Nb_2O_{5-x} confined in necklace-like N-doped carbon fibers (T-Nb_2O_{5-x}@NC) has been acquired for optimizing the transmission dynamics and stability of T-Nb_2O_5 anode material [137]. In detail, the yolk-shell architecture can alleviate the volume expansion of Nb_2O_5 during Na^+ embedding and ensure good structural stability. Moreover, the presence of oxygen vacancies reduces the transmission barrier of Na^+ and enhances the electrical conductivity. As a result, the optimal T-Nb_2O_{5-x}@NC-7 is coupled with AC cathode to achieve the high-performance SIC in a broad potential range of 1–4 V. This SIC system could deliver a maximum energy density of ~83.0 Wh kg^{-1} at a power density of ~75.0 W kg^{-1}. A flexible free-standing sulfur doped Nb_2O_5 quantum dots/nitrogen and sulfur co-doped microporous carbon nanofiber (S-Nb_2O_5@NS-PCNF) film has been fabricated by She et al., which exhibits superior rate capability and excellent cycling stability [138]. The outstanding electrochemical performance is attributed to the synergistic effect of 3D interconnected network and sulfur doping. The S-Nb_2O_5@NS-PCNF//AC SIC shows a large energy density of 112 Wh kg^{-1} at a power density of 80 W kg^{-1} and a high-power density of 7949 W kg^{-1} at an energy density of 55 Wh kg^{-1} as well as a long-term cycling stability (81% capacity retention after 10 000 cycles at a current density of 1 A g^{-1}). On the other hand, fluorine substitution is employed to boost pseudocapacitive intercalation of Na^+ ions in T-Nb_2O_5. DFT calculations demonstrates that the fluorine-substituted provides more favorable sodiation sites and enable lower diffusion barriers compared to the pristine T-Nb_2O_5. Therefore, T-$Nb_2O_{5-x}F_y$ nanoslabs embedded in a carbon nanobelt matrix (T-$Nb_2O_{5-x}F_y$@C-NB) composite displays excellent electron conductivity and electron/ion transport [139]. Paired with AC cathode, the Swagelok-type SIC cell shows energy densities of 86.8 and 32.1 Wh kg^{-1} at power densities of 250 and 18 000 W kg^{-1}, respectively. Interestingly, the uniform sandwich-structure mesoporous Nb_2O_5/graphene/mesoporous Nb_2O_5 (G@mNb_2O_5) nanosheets has been synthesized, in which Nb_2O_5 layers are formed by stack of several nanometersized particles [140]. The G@mNb_2O_5 nanosheets anode delivers prominent sodium-storage properties triggered by the increased surface pseudocapacitive contribution, improved sodium-ion diffusion coefficient and short characteristic relaxation process. By employing the G@mNb_2O_5 nanosheets as anode and AC as cathode, the SIC prototype demonstrates the maximum energy/power densities (56.1 Wh kg^{-1}/7200 W kg^{-1}) and a stable capacitance retention of ~89% at 1 A g^{-1}.

$NbSe_2$: The $NbSe_2$ layer structure is formed via interactions between Se–Nb–Se and Se–Se, in which Nb atom is coordinated by six Se atoms by forming a trigonal prism, while each Se atom is linked with three Nb atoms in the layer. According to the previously reported results using first-principles calculations, $NbSe_2$ anode could possess a high theoretical specific capacity of 312 mAh g^{-1} and low average potential (0.22 V vs. Na/Na^+). In 2019, a SIC prototype consisted of $NbSe_2$ anode and AC cathode has been proposed [141]. It is confirmed that the intercalation process of Na^+ ions into the $NbSe_2$ interlayer spacing between the layers, which can be transformed into Na_xNbSe_2 phase. In addition, the as-constructed $NbSe_2$//AC SIC demonstrates an energy density of 41.7 Wh kg^{-1} at a power density of 45.0 W kg^{-1} and can maintain an energy density of 17.3 Wh kg^{-1} at a maximum power density of 599.6 W kg^{-1}.

6.3.1.4 Vanadium-Based Oxide

Vanadium-based Oxide: Vanadium (V) is a highly abundant element in the Earth. It is well known that vanadium oxides possess the multiple oxidation states (II–V) and numerous polymorphs. Generally, vanadium oxides including VO_2, V_2O_5, V_2O_3, Magnéli phases (V_nO_{2n+1}), and Wadsley phases (V_nO_{2n-1}). However, due to the sluggish diffusion kinetics, low ionic conductivity, dissatisfactory energy density, and unstable cycling performance, the applications of vanadium oxides are limited in energy conversion fields [142]. For example, the interpenetrating porous networks composed of layer-structured V_2O_5 nanowires and CNTs have been synthesized, which effectively provides a facile sodium insertion/extraction and fast electron transfer [143]. As a result, the V_2O_5/CNT nanocomposite shows better rate capability and charge storage compared to V_2O_5 electrode without CNTs in Na half-cells. Moreover, the electrochemical Na^+ insertion/extraction process in V_2O_5 electrode can be demonstrated by:

$$V_2O_5 + xNa^+ + xe^- \leftrightarrow Na_xV_2O_5 \tag{6.3}$$

A SIC configuration based on V_2O_5/CNT anode and AC cathode has been constructed, which is original prototype in the early stage of the development of SIC. Moreover, this as-assembled SIC can be operated at a maximum voltage of 2.8 V, achieving a maximum energy of 40 Wh kg^{-1}. Moreover, V_2O_5 nanorods anchored rGO nanocomposite as an anode has been explored for SIC [144]. The rGO serves as a backbone for supporting V_2O_5, which ensures enhanced electrical conductivity and chemical stability. A SIC comprising V_2O_5@rGO anode and honey derived activated carbon cathode delivers the energy density of 65 Wh kg^{-1} at a power density of ~72 W kg^{-1}. For the first time, a flexible free-standing nanohybrid with V_2O_3 nanoparticles embedded in the 1D multinanochannel nanofibers (V_2O_3@MCNFs) has been designed as the promising anode for SIC [145]. Thanks to the enhanced electron/mass transfer resulted from transportation the interconnected conducting network, the V_2O_3@MCNF//AC SIC successfully achieves the high energy densities of 96/76.8 Wh kg^{-1} at a power density of 250/7680 W kg^{-1}, respectively. Moreover, the as-fabricated device demonstrates an excellent cycling stability with a high capacity retention of 80.9% over 10 000 cycles at 1 A g^{-1}, which can light up a smiling face logo made up of 35 red LEDs.

6.3.1.5 Other New Intercalation Anodes

Zinc Vanadate: A rGO-$Zn_{0.25}V_2O_5{\cdot}nH_2O$ nanobelt (G-ZVO) composite is proposed as a high-power anode for SIC, which exhibits fast Na-ion intercalation kinetics, high electronic conductivity and small volume change, resulted from layered structure of ZVO and conducting network of rGO (Figure 6.20a) [146]. In Na half-cell, G-ZVO anode shows the outstanding rate capability and cycling stability. Coupled with an ordered mesoporous and microporous hard salt-templated carbon (OM-HSTC) as a high-performance capacitive cathode, the as-resulted SIC could afford the high energy/power features and impressive cycling performance.

WO_3: The ideal WO_3 crystal structure exhibits a cubic perovskite-like (ReO_3) structure. Due to the distortion from the cubic ReO_3 structure, WO_3 possesses different structural polymorphs [147]. This structure is formed by a cation surrounded by oxygen atoms in an octahedral arrangement, resulting in a 3D network of corner-sharing WO_6 octahedra. It is well known that WO_3 features a semiconducting nature, whereas nonstoichiometric tungsten oxides (WO_{3-x}) are electron conducting in nature triggered by the oxygen vacancies. Thus, mesoporous reduced tungsten oxide@N-doped macroporous graphene (m-WO_{3-x}@NM-rGO) has been rational designed by using an appropriate silica template [148]. Significantly, the ex situ analyses demonstrate that the electrochemical process of tungsten-based oxide is based on Na^+ intercalation/de-intercalation mechanism, which is different from the common conversion reaction in the Li^+ storage behavior of WO_3. By using commercial AC cathode, the as-built SIC delivers a maximum energy density of 67 Wh kg^{-1} at power density of ~21 000 W kg^{-1}.

Na_2FeSiO_4: A novel intercalation-type carbon-coated Na_2FeSiO_4 (CNFS) has been designed for combining with AC cathode to construct the high-performance SIC [149]. Notably, the carbon coating and stable Si—O bonding synergistically

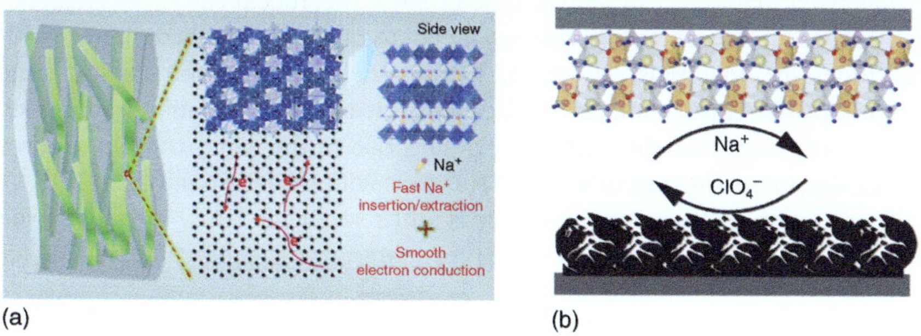

Figure 6.20 (a) Schematic of the advantages of G-ZVO as high power SIC anode. Source: Reproduced with permission Huang et al. [146]. Copyright 2018, Wiley. (b) Illustration of NCPF//AC SIC. Source: Reproduced with permission Or et al. [57]. Copyright 2020, Elsevier.

improves the conductivity and stability of CNFS anode. Moreover, the CNFS//AC SIC manifests the maximum energy and power densities of 70 Wh kg^{-1} and 5400 W kg^{-1}, respectively.

Na_2CoPO_4F: Sodium fluorophosphate materials have been proposed as potential cathode materials for SIBs, which shows a higher operating voltage than that of layered cathodes due to the higher ionicity of the M—F bond. Among them, NCPF exhibits a discharge plateau near 4.3 V (vs. Na/Na^+) with a high energy density of 525 Wh kg^{-1}. Surprisingly, NCPF has been firstly explored as a promising insertion anode for SIC [57]. The obtained NCPF//AC SIC can be operated at 3.25 V, achieving a 93% capacitance retention over 100 000 cycles at 1 kW kg^{-1} (Figure 6.20b).

VP: 3D vanadium monophosphide (VP) coupled with carbon nano-onions (CNOs) has been firstly prepared, which is further investigated as a novel host anode for SICs [150]. It is confirmed that VP anode undergoes the topotactic intercalation mechanism due to the no phase changes during intercalation/de-intercalation processes. Featured with the metallic feature and topotactic structure of VP with a spacious interstitial position for the accommodation of guest species, VP-CNO anode delivers 140 mAh g^{-1} at a current density of 2 A g^{-1} in Na half-cell. Moreover, optimized SIC paired with AC cathode can output the high energy/power densities s well as an ultrastable lifespan of 20 000 cycles.

6.3.2 Conversion Materials

Unlike the intercalation reaction, the conversion reactions involve the formation and breaking of new chemical bonds during the electrochemical processes. Notably, some transition metal oxide (TMO), transition metal sulfide (TMS), and transition metal phosphide (TMP) and other transition metallic compounds can storage Na^+ ions through conversion reactions [151]. Generally, the conventional type of conversion reaction can be written as follows:

$$M_aX_b + (b^*c)Na^+ + (b^*c)e^- \leftrightarrow aM + bNa_cX \tag{6.4}$$

where M represents the transition metal, such as Fe, Co, Ni, Cu, and Mn, while X demonstrates the anionic species (like O, S, Se, P, N, and F). It is easily known that conversion materials can deliver much higher theoretical capacities triggered by the full reduction of the transition metal to metallic state compared with intercalation-type materials. However, low ionic/electronic conductivity, large volume expansion/contraction, and undesirable electrolyte compatibility seriously causes the structural instability, further giving rise to the rapid capacity fading. Moreover, due to the sluggish mobility of Na^+ ions with the large ionic size (1.02 Å), the full utilization of theoretical capacity is still restrained. Thus, optimizing the electrochemical properties of conversion-type materials is a promising strategy for constructing high-performance SICs.

6.3.2.1 Metal Oxides

CoO: An interface engineering approach has been implemented for modifying CoO material, in which CoO nanoparticles are anchored on the N-doped reduced graphene oxide (NrGO) to from 3D architecture. It is noticed that Co—O—C bond prevents CoO nanoparticle agglomeration and rGO sheet restacking, providing the efficient electronic pathways for enhanced charge separation and interfacial charge storage. Hence, the 3D-CoO-NrGO anode has been employed to combined with AC cathode for assembling a SIC prototype with high energy/power peculiarity [152]. It is noteworthy that the conversion reaction mechanism of CoO ($3CoO + 6Na^+ + 6e^- \leftrightarrow 3Co + 3Na_2O$) has been confirmed by the ex situ XRD results.

MnO: Conversion-type MnO species embedded into interconnected hollow carbon nanoboxes (MnO@HCNb) have been prepared by a simple template-assisted CVD method [153]. As shown in Figure 6.21a, the excellent electrochemical performance of MnO@HCNb electrode can be attributed to the such unique structure. The cavities not only provide adequate space for volume change, but also shorten the ionic diffusion path in solid phase. Moreover, new Mn/Na_2O

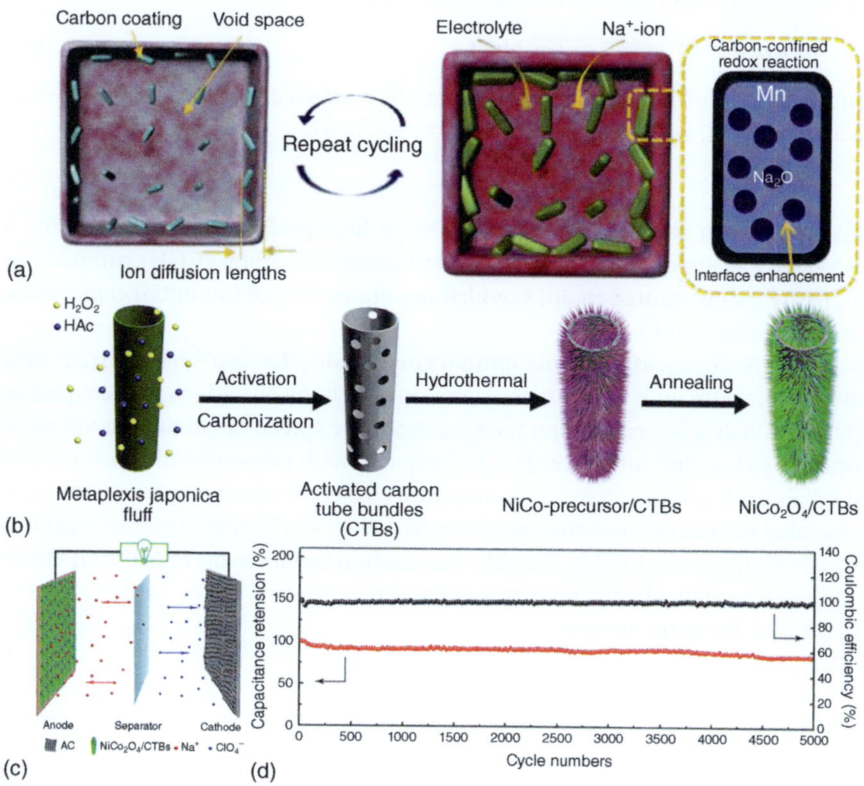

Figure 6.21 (a) Illustration of structural merits in the energy-storage mechanism for the MnO@HCNb electrode. Source: Qin et al. [153]/Reproduced with permission from ELSEVIER. (b) Illustration of preparation process of the $NiCo_2O_4$/CTBs composite. (c) Illustration of energy-storage mechanism and (d) cyclic performance at 5 A g^{-1} of $NiCo_2O_4$/CTBs//AC SIC. Source: Reproduced with permission Zhao et al. [154]. Copyright 2020, Wiley.

interfaces generated from conversion reaction are more conducive to reducing the ion diffusion barrier. Tested in Na half-cell, MnO@HCNb electrode exhibits an outstanding rate performance and an ultralong lifespan (more than 10 000 cycles with 88.6% of capacity retention). When paired with AC cathode, the as-built SIC can achieve a maximum energy density of 116 Wh kg^{-1} and a maximum power density of 4.2 kW kg^{-1}, respectively.

MoO_2: It is notable that the conversion reaction of metal oxides is mainly dependent a heterogeneous charge transfer at the interfaces, Li^+ and O^{2-} in the solid state and M—O bond cleavage. For instance, the bond dissociation energy of Mo–O unit in MoO_2 is about 678 kJ mol^{-1}, which is much higher than 368 kJ mol^{-1} for Co–O moiety of CoO. Therefore, the insertion reaction could appear in the electrochemical behavior of MoO_2 without bond cleavage. Nowadays, according to the analysis of in situ characterization measurements, the sodium-storage mechanism of MoO_2 is the combination of insertion and conversion reactions [155]. It has been found that the insertion process of Na^+ ions for MoO_2 is initially occurred the relatively high potential region. This reaction can be described as Eq. (6.5):

$$MoO_2 + xNa^+ + xe^- \rightarrow Na_xMoO_2 \tag{6.5}$$

Subsequently, orthorhombic Na_xMoO_2 can be converted into metal Mo and Na_2O at a low potential, which can be expressed by Eq. (6.6):

$$Na_xMoO_2 \rightarrow Mo + x/2\,Na_2O \tag{6.6}$$

Inspired by the above-mentioned features, a high-performance SIC by using a MoO_2@rGO as anode and goat hair derived activated carbon (CGH) cathode has been firstly manufactured in 2018, which maintains 78% of the initial capacitance after 1000 cycles [156].

$NiCo_2O_4$: $NiCo_2O_4$, as a kind of multinary metal oxide, haves a ferrimagnetic cubic spinel architecture with mixed redox states, in which the Ni cations are occupied in merely octahedral interstices and the Co cations are spread about among the tetrahedral and octahedral interstices [157]. Compared with monometallic nickel oxides and cobalt oxides, $NiCo_2O_4$ exhibits an enhanced electronic conductivity because of the blended valences of the same positive ions, which can be regarded as prospective anode in energy-storage fields. Notably, the reaction mechanisms of $NiCo_2O_4$ anode for SICs can be stated by the following equations:

In the first discharge process:

$$NiCo_2O_4 + 8Na^+ + 8e^- \rightarrow Ni + 2Co + 4Na_2O \tag{6.7}$$

In the first charge process:

$$Ni + 2Co + 3Na_2O \rightarrow NiO + 2CoO + 6Na^+ + 6e^- \tag{6.8}$$

$$CoO + 1/3\,Na_2O \rightarrow 1/3\,Co_3O_4 + 2/3\,Na^+ + 2/3\,e^- \tag{6.9}$$

The original SIC prototype based on $NiCo_2O_4$ anode has been designed in 2013, which exhibits desirable energy/power performance and a good cycling stability [158]. Later, $NiCo_2O_4$ hollow spheres with chestnut shell morphology have been prepared, which can deliver a reversible capacity of 313 mAh g^{-1} at 1 A g^{-1} in sodium half-cell. After the pre-sodiation treatment of $NiCo_2O_4$ anode, the SIC paired with AC cathode manifests the improved energy density of 120.3 Wh kg^{-1} at the power density of 200 W kg^{-1}, which is almost five to nine times larger than a pristine SIC without pre-sodiation [159]. Moreover, a composite composed of $NiCo_2O_4$ nanowires with carbon tubular bundles (CTBs) has been explored as a novel anode for SICs (Figure 6.21b) [154]. Such the unique hollow tubular structure with the large space can ensure the good structural stability of electrode. As expected, the $NiCo_2O_4$/CTBs electrode delivers an enhanced specific capacity of 298 mAh g^{-1} at 1 A g^{-1} in Na half-cell. Moreover, a $NiCo_2O_4$/CTBs//AC SIC shows a high energy density of 99 Wh kg^{-1} at a power density of 600 W kg^{-1} and retains a desired energy density of 48 Wh kg^{-1} even at a high-power density of 15 000 W kg^{-1} in a wide potential range in 0.5–4.0 V (Figure 6.21c). In addition, 85% of initial specific capacitance can be maintained at 5 A g^{-1} after 5000 cycles (Figure 6.21d).

$CoWO_4$: $CoWO_4$ material has been firstly proposed as a novel conversion-type anode of SIC in 2019, which can be transformed into Co, W, and amorphous Na_2O during the reduction process. Then, a SIC prototype by using the rGO-modified $CoWO_4$ nanoparticles as anode and AC as cathode has been fabricated, displaying a high specific energy of 87.2 Wh kg^{-1} at the power density of 224.6 W kg^{-1} [160].

6.3.2.2 Metal Sulfides

MoS_2: Featured with the unique layer structure, MoS_2 material has been regarded as a promising anode for accommodating of metal ions. Notably, the MoS_2 layer is consisted of two S layers and one Mo layer (S–Mo–S) (Figure 6.22a) [161]. Meanwhile, the adjacent MoS_2 layers are reciprocally stacked by weak van der Waals force. Typically, MoS_2 species can be classified into three main crystal structure (1T-, 2H-, and 3R-MoS_2), which is resulted from the different atom arrangements (Figure 6.22b). It is noticed that 2H and 3R types of MoS_2 have trigonal prismatic coordination (the stacking sequences are ABA and ABC, respectively). Furthermore, both of them are existent in nature, whereas 1T-MoS_2 is an artificial structure. Due to the arrangement of the Mo atom, 1T-MoS_2 is metastable and metallic, and 2H-MoS_2 and 3R-MoS_2 show the semiconducting properties [164]. According to the reported results originated from the in situ measurements, the Na^+-ion storage mechanism of MoS_2 is insertion-conversion reaction, which can be described as follows:

$$MoS_2 + xNa^+ + xe^- \rightarrow Na_xMoS_2\ (> 0.4\ \text{V vs.Na/Na}^+) \tag{6.10}$$

$$Na_xMoS_2 + (4-x)Na^+ + (4-x)e^- \rightarrow Mo + 2Na_2S\ (< 0.4\ \text{V vs.Na/Na}^+) \tag{6.11}$$

In 2017, a new SIC configuration based on MoS_2 anode has been proposed by Wang et al. In this system, a novel MoS_2–carbon (MoS_2-C) monolayer interoverlapped superstructure prominently bringing out the fast pseudocapacitive characteristics and robust structural stability, further delivering an enhanced rate capability and

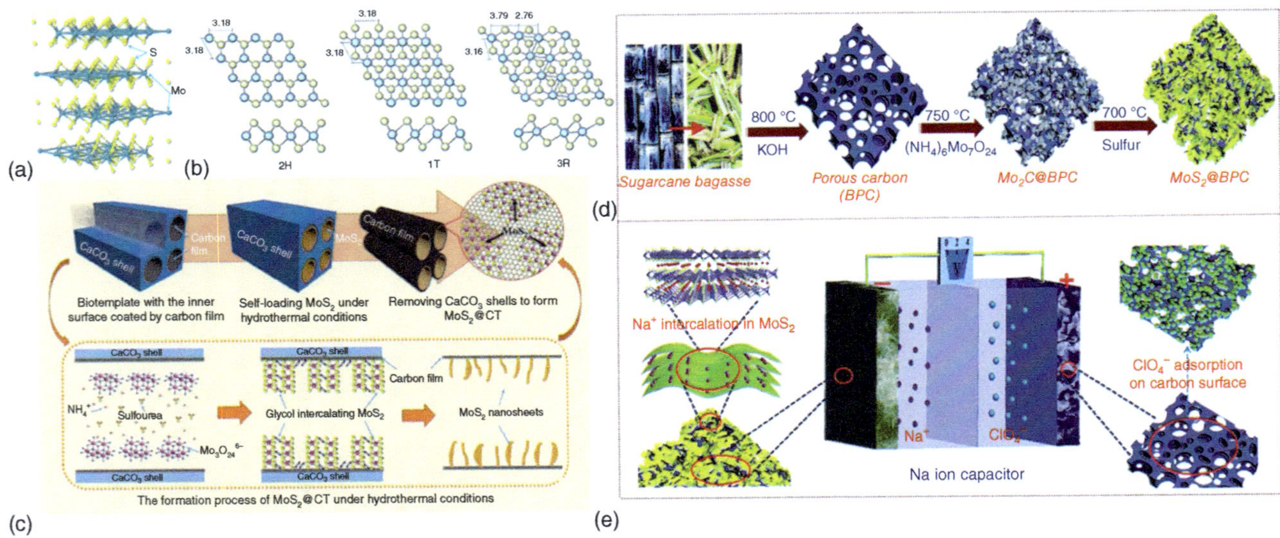

Figure 6.22 (a) Schematic of MoS_2 molecular structure. (b) Three main types of MoS_2 structures. Source: Reproduced with permission Fan et al. [161]. Copyright 2014, Royal Society of Chemistry. (c) Illustration of the preparation produces of MoS_2@CT composites. Source: Reproduced with permission Zhang et al. [162]. Copyright 2020, Wiley. (d) Schematic of synthesis of MoS_2@BPC material. (e) Illustration of energy-storage mechanism of MoS_2@BPC//BPC SIC. Source: Li et al. [163]/Reproduced with permission from Royal Society of Chemistry.

an improved cyclability. When paired with AC cathode, the as-built SIC can be operated up to 3.8 V, which displays a maximum energy density of 111.4 Wh kg^{-1} and an extremely high-power density of 12 000 W kg^{-1} [165]. It is well known that expanding the interlayer spacing of MoS_2 is an effective strategy to boost its electrochemical performance, which is attributed to the decreased ion diffusion pathways and increased active areas. Hence, a 3D network-structured interlayer-expanded MoS_2/rGO nanocomposite (3D-IEMoS_2@G) has been employed as the anode of SICs, which achieves a considerable specific capacity of 320 mAh g^{-1} at a high current density of 10 A g^{-1} [166]. When paired with nitrogen-doped hierarchically porous 3D graphene (N-3DG), the obtained SIC can deliver a high energy density of 140 Wh kg^{-1} at 630 W kg^{-1} and remain 99% capacitance retention after 10 000 cycles, which surpasses LIC based on the same material system. In addition, the MoS_2 nanoflowers with super wide interlayer spacing (larger than 1.34 nm) supported on carbon fibers (E-MoS_2/carbon fibers) has been prepared as flexible and binder-free anode for SICs, which possesses nearly twice as large as that of the original MoS_2 sample (theoretical value, 0.62 nm) [167]. Such a super wide interlayer spacing can not only decrease the ion diffusion pathways and resistance, but also increase the available and accessible active surface area, which provides the rapid mass transport, thus ensuring the high energy/power features of the E-MoS_2/carbon fibers//AC SIC. Interestingly, a novel MoS_2@carbon tubes (MoS_2@CT) with semienclosed architecture has been designed by space-confined strategy [162]. As shown in Figure 6.22c, the $CaCO_3$-based biostructures containing an impressively regulate arrangement of nanotubes is employed as a good temple to construct the semienclosed architectures. Furthermore, thiourea and ammonium molybdate ingredients are reacted to form MoS_2@nanotube composite in ethylene glycol solution under the hydrothermal condition. Notably, such a unique architecture not only affords adequate accommodation for ameliorating volume expansion, but also boosts the Na-ion kinetics of MoS_2, guaranteeing the superb electrochemical performance for SICs. Therefore, the as-assembled SIC successfully achieves a maximum energy density of 82 Wh kg^{-1} and an ultrahigh power density of 22 000 W kg^{-1} as well as an excellent cycling capability of an 89% retention after 5000 cycles. Interestingly, Mo_2C-induced ultrathin MoS_2 nanosheet arrays on bagasse-derived porous carbon frameworks (MoS_2@BPC) through a facile solid-phase reaction strategy have been proposed as an appropriate anode of SIC (Figure 6.22d) [163]. It is noteworthy that Mo_2C intermediate serves a vital role in the growth of powder-type MoS_2 nanosheet arrays on BPC substance. Notably, the tight coupling between MoS_2 and BPC could enable the fast reaction kinetics of the composite along with enhanced pseudocapacitive feature. The as-fabricated MoS_2@BPC//BPC SIC with the optimal mass ratio of 1/4 can display a maximum energy density of 112.2 Wh kg^{-1} at a power density of 55 W kg^{-1} (Figure 6.22e).

WS_2: Inspired by the similar layered structure of 2D MoS_2, WS_2 as a potential anode has been widely investigated in the energy-storage filed. Compared with MoS_2, WS_2 has relatively weaker interactions between WS_2 layers and a higher intrinsic conductivity, which are beneficial for accommodating Na^+ ions [168]. Unfortunately, WS_2 is also suffered from the notorious volume change triggered by the conversion reaction. In order to overcome this encountered problem, a special layer-structured WS_2 nanoflake has been designed. Benefited from the special layered nanostructure with the adequate interlayer spacing, the structural damage can be remarkably relieved, which ensures the better stability during the sodiation/desodiation processes [169]. Moreover, the as-built WS_2//AC SIC manifests the outstanding electrochemical properties (maximum energy/power densities of 132.7 Wh kg^{-1}/1134.4 W kg^{-1}). In addition, WS_2 nanosheets with the outstanding pseudocapacitive contribution have been coupled with mesoporous hollow carbon (m-HC) cathode for fabricating SICs. Due to the fast kinetics of Na-ion storage for WS_2 nanosheets, the as-prepared SIC system presents an energy density of 63 Wh kg^{-1} at a power density of 350 W kg^{-1} and a remarkable cycling stability [170].

TiS_2: It is well known that TiS_2 as the pioneering electrode has been employed to construct the rechargeable TiS_2/Li cell configuration. However, this system has been replaced by other evolutive anode/cathode materials due to potential security issues of the Li dendrite growth on the anode part. A SIC configuration composed of TiS_2 anode and AC cathode is firstly constructed in 2019, which can deliver an energy density of ~38 Wh kg^{-1} and sustain up to 3600 cycles [171]. Moreover, carbon coated TiS_2 nanosheets (TiS_2@C_{pvp}) have been prepared by in situ sulfidation of PVP modified $Ti_3C_2T_x$ MXene precursor [172]. The unique 2D structure with the enlarged large interlayer spaces enables the ultrafast rate capability and high reversibility. As a result, the TiS_2@C_{pvp} anode delivers a superior rate capability with 387 mAh g^{-1} even at 10 A g^{-1} and an excellent cycling stability with 92.5% capacity retention after 5000 cycles. When paired with AC cathode, the as-constructed SIC delivers a maximum energy density of 101.7 Wh kg^{-1} and an ultrahigh power density of 15 kW kg^{-1}.

Iron sulfides: Due to the various merits of abundance, low cost, and nontoxicity, iron sulfides have been developed as a promising anode for SIBs. Generally, the crystalline phases of iron sulfides are FeS, FeS_2, Fe_3S_4, and Fe_7S_8. Nowadays, depended on the crystal structure, FeS and FeS_2 are extensively investigated in energy-storage filed, while Fe_3S_4 and Fe_7S_8 are rarely studied. Notably, totally eight polymorphs of FeS have been discovered, which are the cubic, monoclinic, orthorhombic, tetrahedral, and hexagonal space groups, respectively [173]. Moreover, the structures can be classified into pyrrhotite, troilite, mackinawite, and their modifications. Meanwhile, FeS_2 have two kinds of polymorphs (pyrite and marcasite). For example, 3D inverse opal (IO)-structured nanohybrid of carbon-confined FeS quantum dots (3D-IO FeS-QDs@NC) has been designed as a superior host anode for SICs (Figure 6.23a) [174]. It has been noticed that the robust IO architecture can provide adequate active areas for alleviating the volumetric expansion and enhancing the utilization efficiency of electrode material (Figure 6.23b). As a result, the 3D-IO FeS-QDs@NC anode delivers a superior rate capacity and ultralong cyclic stability over 2500 cycles as well as increased surface pseudocapacitive contribution. Coupled with AC cathode, the constructed SIC device can achieve a maximum energy output and power density of 151.8 Wh kg^{-1} and 9280 W kg^{-1}, respectively. Interestingly, a novel SIC prototype using the $Fe_{1-x}S$ anode has been fabricated in an ether-based electrolyte (Figure 6.23c). Specially, in $NaPF_6$-diglyme (DGM)

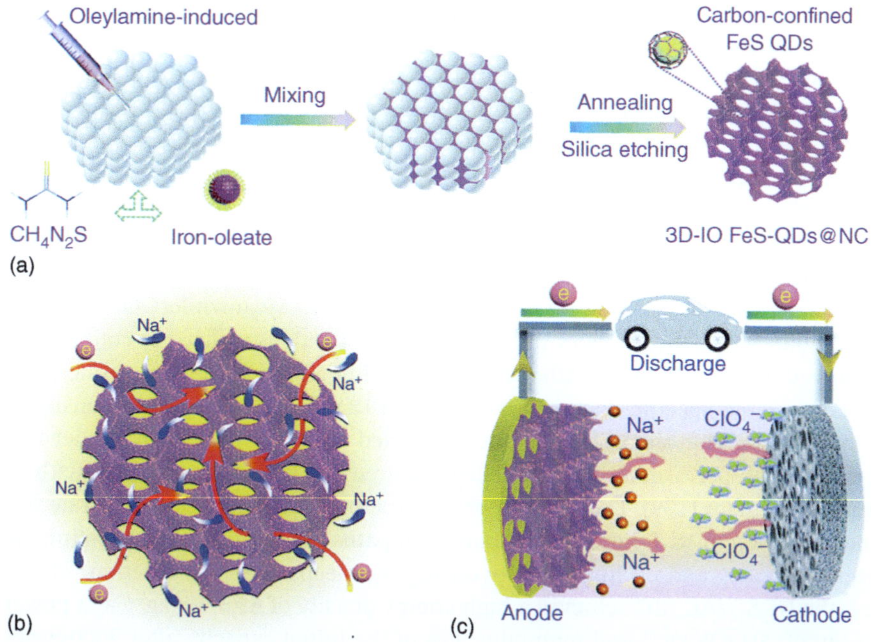

Figure 6.23 (a) Illustration of the preparation procedures of the 3D-IO FeS-QDs@NC composite. (b) Schematic of the Na^+/electron transportation in the 3D-IO FeS-QDs@NC anode. (c) Illustration of the 3D-IO FeS-QDs@NC//AC SIC configuration. Source: Reproduced with permission Hu et al. [174]. Copyright 2019, Royal Society of Chemistry.

electrolyte, the $Fe_{1-x}S$ nanoplate anode delivers a remarkable rate performance (179 mAh g^{-1} at 5 A g^{-1}) and outstanding cycling stability (95% capacity retention after 7000 cycles at 10 A g^{-1}). Importantly, due to the high ICE of $Fe_{1-x}S$ anode, the pre-sodiation process is unnecessary in the preparation process of SICs [175]. The as-assembled nonpresodiated SIC manifests an excellent high energy density of 88 Wh kg^{-1} and superior power output of 11 500 W kg^{-1} as well as an ultralong lifetime of 9000 cycles with over 93% capacity retention. A unique anion-modified strategy has been employed to synthesize $FeS_{2-x}Se_x$ [176]. And, the $FeS_{2-x}Se_x$ exhibits the superior electrochemical features than those of FeS_2 and $FeSe_2$. When paired with AC cathode, the constructed SIC displays the remarkable cycle life (up to 1000 cycles with slight energy loss).

Nickel sulfides: Nickel sulfides including Ni_3S_2, NiS, and NiS_2 as versatile anodes from SICs have been investigated. Among them, NiS_2 material owns a relatively high theoretical capacity of 873 mAh g^{-1} and the detailed multistep reaction mechanism is listed:

$$NiS_2 + xNa^+ + xe^- \rightarrow Na_xNiS_2 \tag{6.12}$$

$$Na_xNiS_2 + (4-x)Na^+ + (4-x)e^- \rightarrow Ni + 2Na_2S \tag{6.13}$$

For instance, well-oriented NiS_2 nanosheets on porous carbon microtubes (NiS_2/pCMTs) have been exploited as a novel anode material for SICs [177].

The as-obtained pCMT substrate with unique hollow tubular structure not only provides abundant active sites for the heterogeneous growth of uniform NiS_2 nanosheets, but also enables the robust structural ability during the Na^+ insertion/extraction processes. Moreover, it is confirmed that the NiS_2 (100) surface affords the high affinity to Na^+ ions, further leading to enhanced redox reaction kinetics on/nearby the surface. As a result, the NiS_2/pCMT anode delivers an impressive rate performance (a specific capacity of 431 mAh g^{-1} at a high current density of 8 A g^{-1}) and a long lifespan. Paired with AC cathode, the as-built SIC device achieves an excellent energy density of 136 Wh kg^{-1} and a high capacitance retention of 83% at 1 A g^{-1} after 5000 cycles. MeanwhilerGO decorated amorphous NiS_2 nanosheets as high-performance anode is utilized to couple with biomass-derived carbon as cathode for fabricating SICs [178]. The resulted SIC displays a large specific energy density of 163.18 Wh kg^{-1} at a power density of 224.99 W kg^{-1}. The hierarchically interconnected Ni_3S_2 nanofibers as binder-free anode for high-performance SIC has been acquired by a hydrothermal route with Ni foam as precursor [179]. The unique porous structure can effectively ameliorate volume expansion of Ni_3S_2 and shorten diffusion paths of Na^+ ions, thus boosting the specific capacity and cycling stability during the electrochemical reaction. Therefore, the Ni_3S_2//AC SIC achieves a high energy density of 65.6 Wh kg^{-1} at a power density of 204.6 W kg^{-1} and maintains 98% of the initial capacity after 2000 cycles with a high CE of nearly 100%. On the other hand, NiS_x nanoparticles decorated on 3D porous carbon matrix (NiS_x@PCM) has been fabricated by a one-step hydrothermal process, which is utilized as an appropriate anode material for SICs [180]. Notably, the 3D carbon network effectively prevents the agglomeration of discharge products or the pulverization of active materials. Such the unique feature affords the remarkable energy and power densities of the as-assembled NiS_x@PCM//AC SIC.

MnS: MnS as a promising anode for sodium storage has attracted tremendous attentions due to its natural abundance, environment friendliness, and the theoretical capacity up to 616 mAh g^{-1}. For example, MnS nanocrystals embedded into the N, S-co-doped carbon matrix (MnS@NSC) composite has been designed [181]. It has been noticed that the downsized MnS can expose more active sites for charge storage, which can promote pseudocapacitive kinetics behavior. Moreover, the N, S-co-doped carbon substance can effectively buffer the volume changes of MnS nanocrystals. Due to the above-mentioned advantages, the MnS@NSC electrode shows superb rate capability (205.6 mAh g^{-1} at 10 A g^{-1}) and outstanding cycling stability without obvious capacity degradation after 2000 cycles. A SIC prototype by employing MnS@NSC as anode and carbon cathode can achieve high energy and power features (139.8 Wh kg^{-1} and 11 500 W kg^{-1}, respectively).

6.3.2.3 Metal Selenides

$MoSe_2$: Molybdenum selenide ($MoSe_2$) as a typical layered transition metal selenide possesses the interlayer distance 0.646 nm, which advantageously allows the insertion of Na^+ into the host material [182]. 3D honeycomb $MoSe_2$/rGO nanocomposites have been investigated as a high-performance anode of SICs [183]. The sandwich-like $MoSe_2$ structure provides adequate pathways and active sites for Na^+

storage, while the rGO alleviates the structural damage. Featured with the combined merits of $MoSe_2$ and rGO, the $MoSe_2$/rGO composite anode can achieve a high capacity of 214.7 mAh g^{-1} at 1 A g^{-1}. When paired with AC cathode, the as-built SIC system successfully delivers an energy density of 87.7 Wh kg^{-1} at a power density of 208.8 W kg^{-1}. By virtue of the existence of Mo—C chemical bonding, the strong interfacial coupling between graphene and $MoSe_2$ has been acquired by the assistance of hexadecyltrimethyl ammonium bromide (CTAB), which facilitates the Na-ion/electron transfer of composite anode (Figure 6.24a) [184]. Moreover, exposed (001) facets and expanded interlayer distance of oriented $MoSe_2$ nanosheets enables the fast Na-ions intercalation behavior. As a result, the as-assembled SIC composed of $MoSe_2$/G anode and AC cathode could achieve high energy and power density (82 Wh kg^{-1} and 10 752 W kg^{-1}, respectively), convincingly bridging the gap between batteries and supercapacitors. Furthermore, 81% capacitance retention at 5 A g^{-1} after 5000 cycles is obtained, which illustrates that this SIC device shows an outstanding cycling durability. Interestingly, the rationally designed hollow bowl-like carbon (HBC) material possessed abundant mesochannels and high accessible surface area, which serves as an appropriate 3D host to anchor the $MoSe_2$ nanocrystals (Figure 6.24b) [185]. The resulted $MoSe_2$@HBC composite demonstrates the fast kinetics enhanced by surface pseudocapacitive effect, thus resulted in the high-rate capability (298.8 mAh g^{-1} even at 20 A g^{-1}). Analyzed by the electrochemical behavior, it has been found that the Na-ion storage of $MoSe_2$ is similar to that of MoS_2 material. Firstly, the intercalation/de-intercalation of Na^+ into the interlayer of $MoSe_2$ has been appeared accompanied by the formation of Na_xMoSe_2 ($MoSe_2 + xNa^+ + xe^- \leftrightarrow Na_xMoSe_2$). Subsequently, the conversion reaction (from Na_xMoSe_2 to Mo/Na_2Se) has been realized in the relatively low potential range. When paired with HBC as capacitive-type cathode, the as-assembled $MoSe_2$@HBC//HBC SIC can display a maximum energy density of 118 Wh kg^{-1} and an ultrahigh power density of 50 kW kg^{-1} (Figure 6.24c).

WSe_2: Notably, WSe_2 presents a similar 2D layered structure to $MoSe_2$, exhibiting the interlayer distance of 0.65 nm, which can be regarded as a potential anode for sodium storage. Unfortunately, the related studies of WSe_2 are rarely reported. Onion-like WSe_2 nanoparticles synthesized by a homemadeCVD method have been evaluated as a SIC anode. Notably, the sodiation/de-sodiation mechanism in this study has been carefully analyzed (Figure 6.24d) [186]. It is proposed that the proceeded electrochemical reactions can be divided into two procedures (namely, intercalation and conversion reactions), which can be expressed by the following equations:

$$WSe_2 + 4Na^+ + 4e^- \rightarrow W + 2Na_2Se \tag{6.14}$$

$$Na_2Se \leftrightarrow Se + 2Na^+ + 2e^- \tag{6.15}$$

Coupling with AC cathode, the resulted SIC based on WSe_2 anode can achieve a superior energy/power features (a maximum energy density of 123.1 Wh kg^{-1} and a maximum power density of 14.1 kW kg^{-1}, respectively). Moreover, rGO as a highly conductive and flexible framework has been proposed to modify WSe_2 anode,

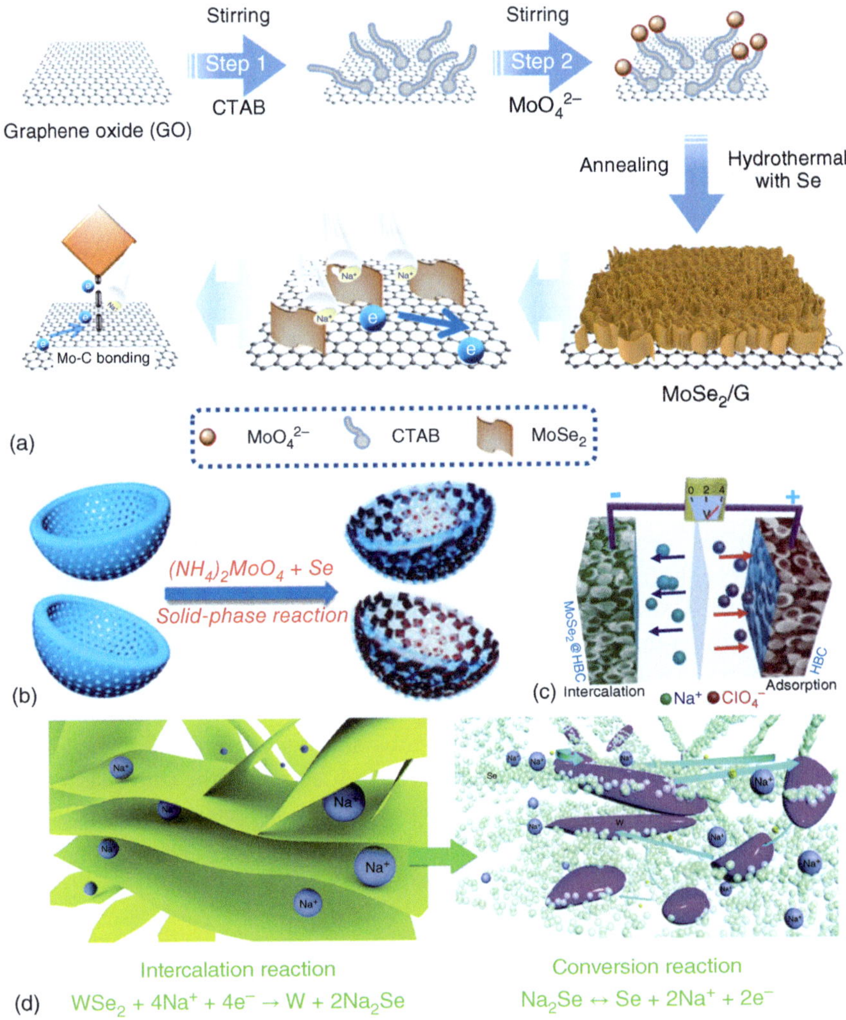

Figure 6.24 (a) Illustration of ordered 2D $MoSe_2$/graphene nanocomposite through the CTAB assisted hydrothermal method. Source: Reproduced with permission Zhao et al. [184]. Copyright 2018, Elsevier. (b) Schematic of the preparation procedure of $MoSe_2$@HBC. (c) Illustration of the $MoSe_2$@HBC//HBC SIC configuration. Source: Reproduced with permission Fei et al. [185]. Copyright 2020, Wiley. (d) Illustration of sodiation/desodiation mechanism of WSe_2 anode. Source: Reproduced with permission Wang et al. [186]. Copyright 2018, Royal Society of Chemistry.

which provides the more active interface and facilitates the electron transport [187]. As a result, WSe_2/rGO anode can deliver outstanding rate ability up to 60 A g^{-1}. Moreover, the WSe_2/rGO//AC SIC achieves a maximum energy density of 42.8 Wh kg^{-1} and a maximum power output of 14.4 kW kg^{-1} as well as the superior cyclic stability.

$CoSe_2$: $CoSe_2$ as a one of cobalt selenide materials has been widely investigated in sodium-storage field. It has been found that the $CoSe_2$ can undergo the multistep reaction mechanism consisted of insertion and conversion processes within voltage window of 0.5–3 V. Typically, $CoSe_2$@vertically oriented graphene (VG) on carbon fiber cloth (CC) has been firstly reported as a high-performance anode of SIC [188]. Benefiting from the enhanced pseudocapacitive Na storage originating from hierarchical architecture, the resulted $CoSe_2$@VG/CC//AC SIC demonstrates a high energy and power density of 116 Wh kg^{-1} and 7298 W kg^{-1}. Furthermore, 88% of its initial energy density can be well maintained without obvious attenuation at 0.5 A g^{-1} after 1800 cycles.

6.3.3 Alloying Materials

6.3.3.1 Sn-Based Anode

Sn: Metallic Sn is a promising anode material for sodium-ion storage, due to its high theoretical specific capacity (847 mAh g^{-1}, equivalently storing 3.75 Na^+), low potential plateau and high electrical conductivity [189]. It has been noticed that such a large theoretical specific capacity is resulted from the alloying reaction with Na, and the detailed mechanism is described as follows:

$$4Sn + 15Na^+ + 15e^- \rightarrow Na_{15}Sn_4 \tag{6.16}$$

However, due to the formation of Na-rich phase of $Na_{15}Sn_4$, the volume expansion of 420% during the sodiation/de-sodiation can give rise to the drastic structural damage, further causing a rapid fading of capacity [190]. It is well known that the phase transformation of Sn anode in SIBs has been firstly explored by an operando transmission electron microscopy (TEM) in 2012. It has been found that the sodiation process of Sn nanoparticles is divided into two steps (Figure 6.25a) [191]. Firstly, the crystalline Sn substance is sodiated via a two-phase mechanism to form a Na-poor, amorphous Na_xSn alloy ($x \sim 0.5$). Subsequently, the intermediate phase is further sodiated via a single-phase mechanism to amorphous Na_9Sn_4, Na_3Sn, and crystalline $Na_{15}Sn_4$ phases ($x = 3.75$).

In order to overcome the low reaction kinetics of Na^+ and the volume expansion, the composition optimization strategy has been utilized for boosting the sodium storage performance of Sn anode. Thus, the Sn-graphite composite (SntGraphite) has been prepared by a simple annealing of graphite with $SnCl_2$ (Figure 6.25b) [192]. When evaluated in Na half-cell, SntGraphite anode can present a very high ICE of 90% with a large Na^+-storage capacity of 223 mAh g^{-1} at 50 mA g^{-1}. Moreover, the SntGraphite//AC SIC device shows a maximum energy density of ~93 Wh kg^{-1} at a power density of 495 W kg^{-1} and maintains an energy density of 33 Wh kg^{-1} at a high-power density of 7.8 kW kg^{-1}. Meanwhile, high-level N/P co-doped Sn–carbon nanofibers (Sn-NP-CNFs) have been designed for ameliorating the kinetics mismatch between the battery-type anode and capacitive-type cathode [195]. In this system, the increased defects and enlarged interlayer distance of N/P-doped carbon can enable the remarkable rate capability of Sn-NP-CNF. Significantly, high-level N/P-doping greatly enhances the Na-adsorption and brings out the high pseudocapacitance. When paired with AC cathode, the as-assembled

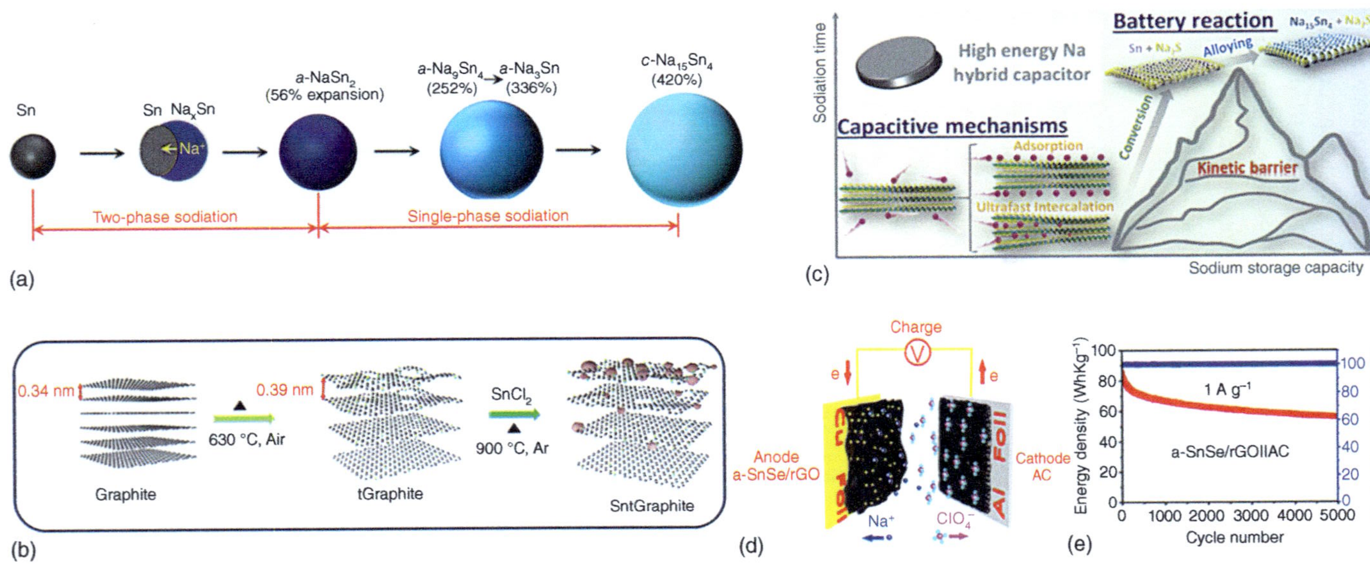

Figure 6.25 (a) Schematic illustration of the structural evolution mechanism of Sn during the sodiation. Source: Reproduced with permission Wang et al. [191]. Copyright 2012, American Chemical Society. (b) Illustration of the preparation of SntGraphite. Source: Palaniselvam et al. [192]/John Wiley & Sons/CC BY 4.0. (c) Schematic illustration of the sodiation mechanism of SnS_2/GCA. Source: Reproduced with permission Cui et al. [193]. Copyright 2018, Wiley. (d) Illustration of energy-storage behavior and (e) long-term cycling performance at a 1 A g^{-1} of a-SnSe/rGO//AC SIC. Source: Reproduced with permission Wang et al. [194]. Copyright 2020, Elsevier.

SIC achieves an ultrahigh energy density of 134.6 Wh kg^{-1} and a superior energy retention of 79.8% after 10 000 cycles at 5 A g^{-1}.

SnO_2: Tin dioxide (SnO_2) has also been considered as promising anode material, which can react with sodium by two reaction procedures. The conversion reaction ($SnO_2 + 4Na^+ + 4e^- \rightarrow Sn + 2Na_2O$, 711 mAh g^{-1}) is firstly occurred during the sodiation process. Consequently, the second sodiation behavior is the alloying reaction of formed Sn species ($4Sn + 15Na^+ + 15e^- \rightarrow Na_{15}Sn_4$, 847 mAh g^{-1}). However, the experimental specific capacity of SnO_2 is substantially lower than the impressive theoretical specific capacity of 1398 mAh g^{-1}, resulted from the sluggish kinetics. Typically, the nanoconfinement strategy originated from the micropores of N, O-doped porous carbon nanosheets (PCNSs) is capable of the formation of amorphous SnO_2 (A-SnO_2) clusters with a subnano size [196]. Moreover, due to the existence of the strong interfacial Sn—O and Sn—Sn bonds, the structural stability of A-SnO_2 subnanoclusters can be effectively increased. Thus, the as-prepared A-SnO_2@PCNS composite manifests excellent rate capability and cycling performance as anode for Na-ion storage. The optimized SIC device based on A-SnO_2@PCNS anode can display a large energy density of 196.4 Wh kg^{-1} with a power density of 100 W kg^{-1}. Moreover, a high energy density of 59.3 Wh kg^{-1} could still be remained even at an ultrahigh power density of 28.1 kW kg^{-1}.

SnS_x: Similar to Sn-based oxides, Sn-based sulfides presents higher theoretical capacity than that of the pristine Sn, due to the additional capacity resulted from the initial conversion reaction. Particularly, hexagonal SnS_2 (space group P3m1) and orthorhombic SnS (space group Pnma) show the unique layered structure with interlayered spaces of 0.589 and 0.433 nm, respectively. Moreover, the corresponding conversion reactions of Sn-based sulfides with Na are $SnS_2 + 4Na^+ + 4e^- \leftrightarrow Sn + 2Na_2S$ and $SnS + 2Na^+ + 2e^- \leftrightarrow Sn + Na_2S$. Subsequently, both of them undergo the same alloying reaction ($4Sn + 15Na^+ + 15e^- \rightarrow Na_{15}Sn_4$). Their theoretical specific capacities are 1137 and 1022 mAh g^{-1}, respectively. Generally, carbonaceous materials are always employed to combine with Sn-based sulfides to alleviate the large volume variation for superb electrochemical performance. For instance, the SnS_2 nanoplatelets anchored on graphene–CNT aerogels (SnS_2/GCA) have been designed as a novel anode for SIC [193]. Significantly, the in situ TEM results and DFT calculations reveal that the excellent pseudocapacitance is originated from the combination of Na^+ adsorption on the surface/Sn edge of SnS_2 nanoplatelets and ultrafast Na^+ ion intercalation into the SnS_2 layers. Moreover, the slow conversion reaction served as a "kinetic barrier" can impede the sodiation at high current densities, and thus Na^+ ion intercalation becomes the dominant electrochemical reaction (Figure 6.25c). A SIC prototype consisted of a SnS_2/GCA anode and an AC cathode displays the excellent energy densities of 108.3 and 26.9 Wh kg^{-1} at power densities of 130 and 6053 W kg^{-1}, respectively. On the other hand, ultrasmall-sized SnS nanosheets vertically aligned on carbon microtubes (SnS/aCMTs) have also been explored as a promising anode for SIC [197]. The incorporation of the conductive carbon framework and porous SnS nanosheets with abundant ultrasmall nanocrystals (5 nm) could provide more active sites and shorten electron/ion transport pathways, greatly endowing the improved structural stability, rate performance, and sodium storage capability. When studied as a SIB anode, the SnS/aCMT delivers a high capacity of 517 mAh g^{-1} at 0.1 A g^{-1} and long-term cycling stability (88.7% capacity retention after 500 cycles). Moreover, the as-assembled SIC device based on the SnS/aCMT anode can demonstrate a large energy density of 115 Wh kg^{-1} in the voltage range of 0.5–4.0 V. Meanwhile, oriented SnS nanosheets confined in S and N co-doped graphene by electrostatic self-assembly method have been acquired [198]. This structure is beneficial for intercalation of Na^+ ions along the *Z*-axis with open edges. Due to the presence of Sn^{4+}-O, the active open edges in the nanostructure can provide rich active sites for Na^+ pseudocapacitance storage on the surface of SnS. The voltage window of SnS/rGO//AC SIC device is optimized to 1–4 V, which can effectively avoid the side reactions. Furthermore, the energy densities of 113 Wh kg^{-1} and 54 Wh kg^{-1} are obtained at the power densities of 101 W kg^{-1} at 11 100 W kg^{-1}, respectively.

SnSe: By virtue of the lower band-gap energy (band gap of 0.9 eV), the SnSe owns a higher electrical conductivity than those of tin oxide and tin sulfide, which is regarded as a potential anode in SIBs. However, suffered from the large volume expansion issue like other Sn-based anode materials, the SnSe inevitably encounters the unsatisfactory electrochemical stability during sodiation/desodiation process. Therefore, amorphous SnSe quantum dots (2 nm) anchoring on the nitrogen doped graphene (a-SnSe/rGO) have been designed, in which the ultrasmall SnSe quantum dot structure not only can prominently suppress the particle pulverization resulted from the tin alloying reaction process, but also affords more active sites for Na^+ pseudocapacitance storage [194]. In addition, strong Sn—C, Sn—O—C, and Se—C chemical bonds between a-SnSe and graphene can ensure the rapid electrical transport channels. As a result, the a-SnSe/rGO composites exhibit good electrochemical performance as SIB anode. Moreover, the a-SnSe/rGO//AC SIC with an optimal cathode/anode mass ratio is operated in the voltage window of 1–4 V, which can achieve high energy density of 110 Wh kg^{-1} at 0.05 A g^{-1} (Figure 6.25d). Furthermore, a-SnSe/rGO//AC SIC can still maintain a high energy density of 58 Wh kg^{-1} at 1 A g^{-1} after 5000 cycles with a high energy retention rate of about 70% (Figure 6.25e).

Sn_4P_3: Sn_4P_3 with rhombohedral structure is considered as another promising conversion plus alloying material, which combines the merits of P (high capacity) and Sn (high conductivity). Notably, the high theoretical capacity of 1132 mAh g^{-1} is mainly attributed to the sodiation behavior of P and Sn to the terminal Na_3P and $Na_{15}Sn_4$, respectively. The total reaction can be proposed to be the following equation: $Sn_4P_3 + 24Na^+ + 24e^- \rightarrow Na_{15}Sn_4 + 3Na_3P$ during the initial sodiation. In 2019, a novel SIC prototype using the Sn_4P_3 anode and AC cathode has been proposed by Chojnacka et al. The as-built SIC displays a stable performance in the voltage range from 2.2 to 3.8 V (94% capacitance retention after 6500 cycles at 0.2 A g^{-1}).

6.3.3.2 Sb Anode

Sb as one of the popular alloy-type anode materials possesses a theoretical sodium-storage capacity of 660 mAh g^{-1} (corresponding to the maximum stoichiometry Na–Sb alloy system-Na_3Sb), suitable working voltage and excellent electronic conductivity [199]. Unfortunately, Sb also confronts the troublesome challenges of drastic volume expansion about 390% and slow Na^+ reaction kinetics. In order to improve the cycling stability of Sb anode, the rational engineering of M-Sb alloys (M represents other metal elements) has been explored. For example, the $TiSb_2$ anode with excellent rate capability has been coupled with AC cathode to construct an innovative SIC configuration [200]. This SIC device can obtain the high energy densities of 132 Wh kg^{-1} at 114 W kg^{-1} and 65 Wh kg^{-1} at 11 kW kg^{-1}, respectively.

6.3.3.3 Bi Anode

Bi has gained tremendous attentions in sodium-ion storage due to its large crystal lattice along *c*-axis ($d(003) = 0.395$ nm) for ion insertion, low melting point, low thermal conductivity, and high theoretical gravimetric capacity of 384 mAh g^{-1} [201]. However, the volume change and instable chemical interface can limit the development of Bi anode in electrochemical energy storage field. Hence, rational material design serves a vital role in improvement of sodium storage performance. A porous NaBi anode has been engineered for pairing with AC cathode to fabricate a novel SIC device with high power and energy density [202]. Notably, Na^+ is intercalated into NaBi to form Na_3Bi, and the PF_6^- anions in the electrolyte are adsorbed into the AC cathode in the charging process of SIC. On the contrary, the Na_3Bi is desodiated to NaBi and eventually Bi, and the adsorbed PF_6^- anions are released into the electrolyte during the discharging behavior. Benefited from 3D porous architecture of anode, this SIC achieves a maximum energy density of 106.5 Wh kg^{-1} and a maximum power density of 11.1 kW kg^{-1} for proving an ingenious combination of batteries and supercapacitors.

6.4 Other Novel Materials

It is well known that the organic compound is regarded as a novel electrode material in electrochemical energy storage systems, due to the advantages of the lightweight elements (such as H, C, N, and O), low-cost, abundance, environmental friendliness, and high sustainability [203]. Therefore, utilization of organic electrode materials could effectively avoid the presence of metal element, greatly decreasing the expenses and guaranteeing the security [204]. Most importantly, inspired by the structural diversity, the organic principal framework can be purposefully designed via molecular engineering, which is beneficial for improving the electrochemcial performances [205]. Until now, in terms of Na^+ storage mechanism of organics, there are five kinds of reaction mechanisms, including the carbonyl (C=O) reaction, imine (C=N) reaction, azo (N=N) reaction, thioketone/thioester (C=S) reaction, and free radical (N–O) reaction [206]. Typically, Figure 6.26 explicitly exhibits the refered five reaction mechanisms by taking the representative organic compounds as examples.

Notably, the electroactive carbonyl compound is a promising candidate for energy storage [207]. Nowadays, most carbonyl compounds are n-type organics, which can transfer at least two electrons during the charge/discharge process [208].

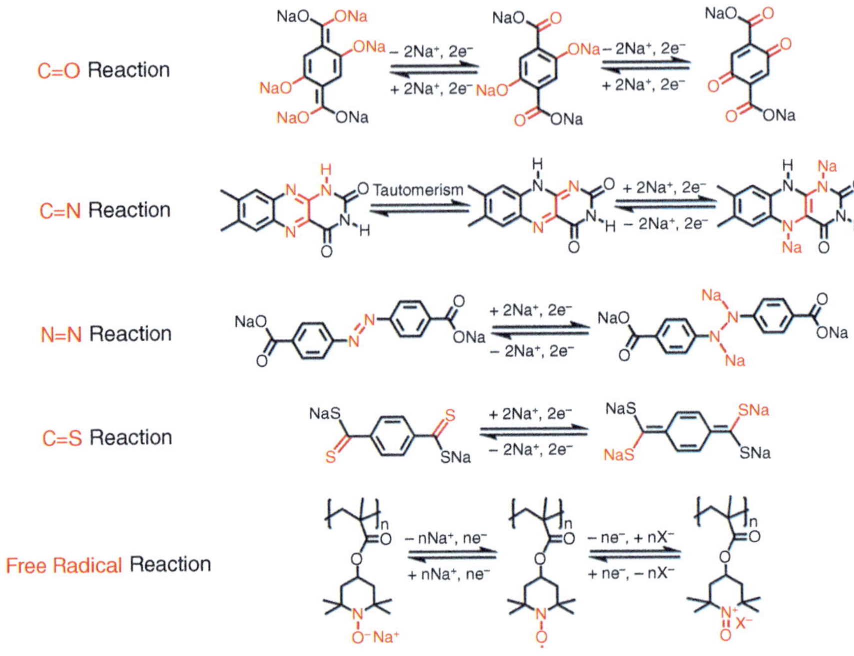

Figure 6.26 Illustration of five types of Na^+ reaction mechanisms of organic electrode materials. Source: Reproduced with permission Holguin et al. [206]. Copyright 2018, Royal Society of Chemistry.

Moreover, it is confirmed that n-type electrodes are inclined to undergo reduction reaction, which brings about the negatively charged anions, thus acquiring a salt structure balanced by a suitable electrolyte counterion [209]. According to the structural differences, carbonyl compound materials are mainly divided into four categories: quinones, conjugated carboxylates, imides, and anhydrides [210]. In views of quinones, the carbonyl motifs are linked to the conjugated aromatic ring in the *ortho*-position or *para*-position of the rings [211]. Due to the small molecular mass, quinones generally possess the high theoretical specific capacities, which could ensure the output of large energy density. For conjugated carboxylates, the carboxyl groups are connected to the aromatic ring, which can generate an enlarged conjugated system [212]. Nevertheless, owing to the electron donor feature of –ONa linked with the carbonyl group, the conjugated carboxylates commonly deliver the relatively low potential, which could merely be empoloyed as anode materials. On the other hand, the aromatic imides and anhydrides both own the large π-conjugated skeleton accompanied with four carbonyl groups [213].

In 2017, Thangavel et al. have successfully fabricated an all-organic sodium hybrid capacitor (OHC), which effectively prevents the usage of toxic inorganic complexes with high cost [214]. In this system, perylene-3,4,9,10-tetracarboxylic acid dianhydride (PTCD) and PANI have been employed as the Na^+ host and the anionic host, respectively. PTCD with good structural reversibility can interact with two sodium ions through an enolation process. According to the charge–discharge profile of

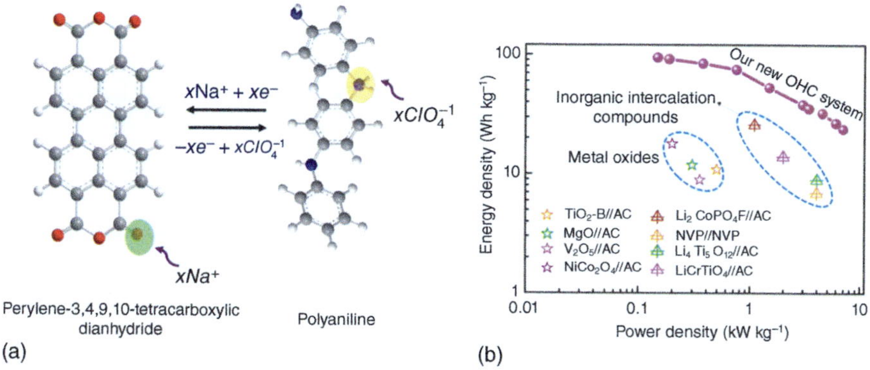

Figure 6.27 (a) Illustration of the working principle of the PTCD//PANI OHC. (b) Ragone plots of the OHC and other reported inorganic hybrid capacitor systems. Source: Reproduced with permission Thangavel et al. [214]. Copyright 2011, American Chemical Society.

PTCD at various current densities, it is observed that the sloppy plateaus in the potential of 1–3 V illustrates that the sequential storage and extraction of two Na^+ ions in the structure. Significantly, PTCD electrode exhibits a reversible discharge capacity of 110 mAh g^{-1} at a current density of 0.25 A g^{-1}, which is very closed to theoretical specific capacity based on the two-electron transfer mechanism. Moreover, PTCD presents an excellent rate capability, delivering a relatively large discharge capacity of 80 mAh g^{-1}. When assembled by PANI cathode with pseudocapacitive reaction of ClO_4^- anion, the constracuted OHC exhibits the splendid electrochemical performances, delivering a maximum energy density of 95 Wh kg^{-1} and an ultrahigh power density of 7 kW kg^{-1} (based on both electrodes), which is far superior to conventional inorganic hybrid capacitor systems (Figure 6.27a,b).

Unfortunately, the organic compounds suffer from the troublesome issue of high solubility in the electrolyte, which seriously reduces the utilization of active carbonyl species, thus leading to rapid capacity fading [215]. It is well known that the salinization method is a feasible and convenient strategy to solve the solubility problem of organic molecule [216]. Endowed by the polar ionic bond (like O–Na pair), the derived organic salts present the insoluble characteristic of an aprotic electrolyte. Therefore, the organic-based salts could deliver the better long-term cycling stabilities. For example, disodium rhodizonate (DSRH, $Na_2C_6O_6$) featured with active C=O center is an appropriate material for hosting Na^+ ions [217]. As shown in Figure 6.28a, DSRH is a keto-carbonyl compound, which is consisted of a six-membered carbon ring bearing with one C=C bond and four C=O functionalities [218]. More importantly, DSRH molecule crystallizes in space group *Fddd* and possesses the alternating layers of hexagonally packed Na^+ ions and rhodizonate anions ($C_6O_6^{2-}$), which is beneficial for the sodium-ion storage (Figure 6.28b). Furthermore, DSRH as anode could exhibit a splendid rate performance with a high discharge capacity of 112 mAh g^{-1} even at a large current density of 5 A g^{-1}. The prepaed biomass-derived carbon exhibits the abundant porosity characteristic, which could present the strong ClO_4^- adsorption/desorption behavior.

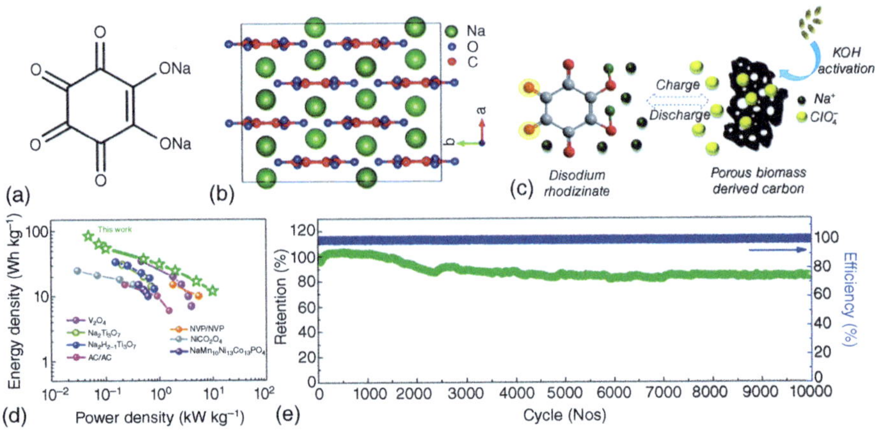

Figure 6.28 (a) Chemical structure and (b) crystal structure of DSRH molecule. (c) Scheme of the working mechanism of the SIC. (d) Ragone plots of the SIC and other reported hybrid capacitor systems. (e) Cycling stability of the SIC at 4 A g^{-1}. Source: Reproduced with permission Thangavel et al. [218]. Copyright 2018, Royal Society of Chemistry.

Therefore, the fabricated SIC could shows the two working mechanisms integrating with the functions of both a battery and a capacitor, potentially realizing the high energy/power output (Figure 6.28c). Notably, this SIC based on DSRH anode successfully achieves a maximum energy density of 87 Wh kg^{-1} and a maximum power density of 10 kW kg^{-1}, which is superior to other reported SIC systems, such as V_2O_5, $NaTi_3O_7$, and $NiCo_2O_4$ anode hosts (Figure 6.28d). As shown in Figure 6.28e, SIC demonstrates the excellent long-term stability during the 10 000 cycles, showing an energy loss of 0.15% per 100 cycles with 100% coulombic efficiency.

In recent years, metal–organic frameworks (MOFs) have attracted great interest in the field of electrochemical energy storage and conversion, due to the various advantages of high porosity, versatile functionality, and structural tunability [219]. MOFs is elaborately designed by the metal nodes and organic linkers, in which the desirable metal element and modifed group not only can combine the respective mertis of organic and inorganic components, but also surprisingly bring out the novel properties [220]. Typically, IRMOF series have been regarded as the first models to regulate the pore size of MOF by using linker extension strategy [221]. It is noticed that the extension of bridging linker length would directly leads to the large pores in MOFs, which is expected for acquiring the mesoporosity. The MOF-5 exhibits cubic topology, which is assembled by the octahedral $Zn_4O(CO_2)_6$ clusters as nodes and ditopic linear carboxylate ligands as linkers. When the ligand BDC (1,4-benzenedicarboxylate) has been replaced with the extender linkers, a series of IRMOFs with larger pore sizes can be obtained, respectively. As a result, the pore size of IRMOF-16 could be enlarged up to 2.88 nm. However, the targeted MOFs with extended linear linkers are mostly thermodynamically instable, which would be inclined to generate the interpenetrated structure, which in turn reduces the pore sizes. On the other hand, increasing the size of the metal nodes could also straightway induce the formatioin of large pore spaces, which has been supposed to be a

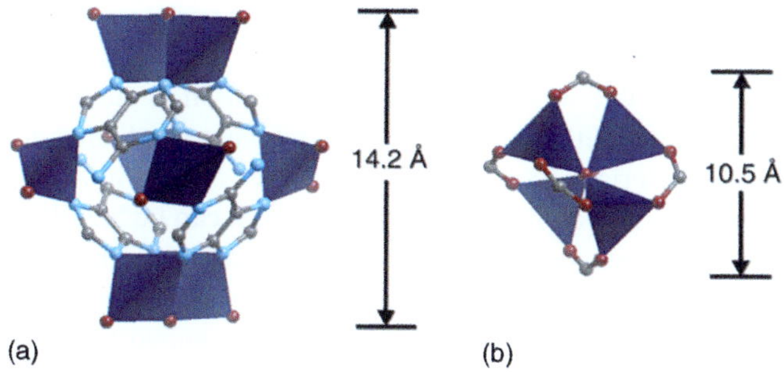

Figure 6.29 Comparison of the structure and size of (a) zinc-adeninate octahedral and (b) basic zinc-carboxylate building units. Source: Reproduced with permission An et al. [222]. Copyright 2012, Springer Nature.

promising alternative strategy for controlling the porosity of MOFs. For example, Bio-MOF-100 equipped with the 2.8 nm mesoporous channel has been successfully prepared by using the huge metal-biomolecule clusters (zinc-adeninate octahedral building units [ZABUs]) as the metal node [222]. As shown in Figure 6.29a,b, the size of ZABUs is 14.2 Å, which is observably larger thant that of the basic zinc-carboxylate cluster (10.5 Å). Notably, the ~2.5 time's larger volume of building cluster unit is beneficial for constructing highly porous MOFs.

In view of the structural diversity and unique physicochemical properties of MOFs, the MOF-related materials have been widely studied in rechargeable battery system (Figure 6.30) [223]. Typically, the intrinsically porous peculiarity of MOFs

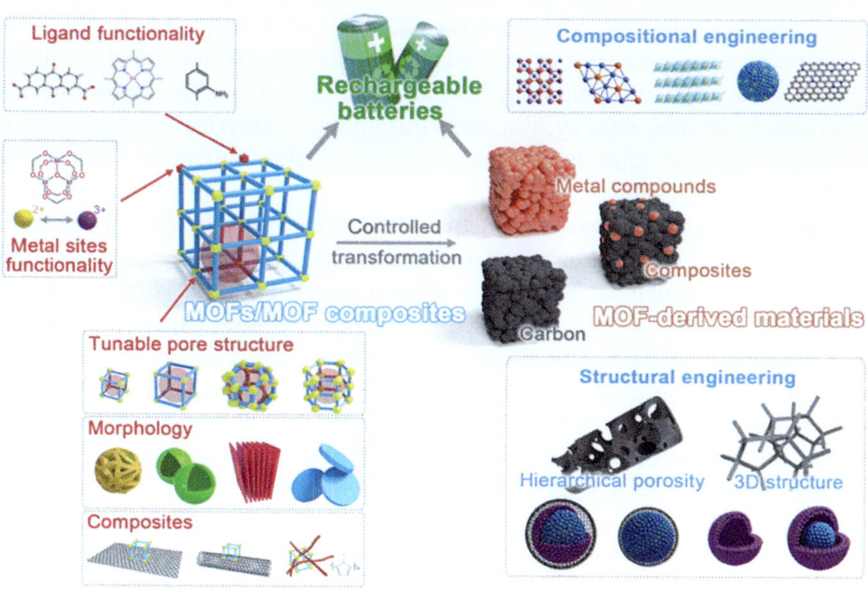

Figure 6.30 Schematic of MOF-related electrode materials for rechargeable batteries. Source: Reproduced with permission Zhao et al. [223]. Copyright 2018, Elsevier.

can provide a favorable electrolyte penetration and fast ion transportation, which could ensure the sufficient reaction degree for electroactive sites of electrode materials. Moreover, the controllable architecture and composition at molecular level could enable the more possibilities for exploiting promising candidates for high-performance electrode materials. Moreover, besides the direct utilization, MOFs are composed of metal sources and organic ligands, which can be identified as the self-sacrificial templates to prepare various nanostructured electrode materials, which profitably expands the practical applications of MOF-based materials [224]. It is confirmed that the in situ generated carbon matrix resulted from the decompositions of organic components can effectively prevent the aggregation of metal/metallic compound particles, which ensures the good integrated morphology of pristine MOF precursor, leading to the prominent electrochemical behaviors [225].

In recent years, MOFs as electrode materials have been investigated in SIC applications [226]. For instance, a conductive MOF, Ni_3(hexaaminobenzene)$_2$, has been designed as a high-performance anode electrode for SICs by Dong et al. [227]. Ni-MOF possesses the high electrical conductivity of framework accompanied with the big channel structure, which is beneficial for the accelerated sodium-ion diffusion and electron transfer. In the potential of 0.5–3.0 V, the Ni-MOF electrode could exhibit a high specific capacity of 297 mAh g^{-1} at 50 mA g^{-1}. Moreover, Ni-MOF presents an excellent rate capability, remaining a reversible capacity of about 107 mAh g^{-1}. Importantly, a SIC prototype has been successfully constructed by employing Ni-MOF as the anode and $Na_3V_2O_2(PO_4)_2F$/activated carbon (NVOPF/AC) as the cathode. The assembled SIC could achieve the good energy/power characteristics, showing a large energy density of 127 Wh kg^{-1} at a power density of 190 W kg^{-1} (based on the total mass of Ni-MOF and NVOPF/AC). Moreover, an energy density of 38 Wh kg^{-1} can be still maintained at an ultra-high power density of 17 309 W kg^{-1}. It has been found that the hybrid device demonstrates the long-term cycle stability during the 5000 cycles at 1 A g^{-1}, which obtains the splendid capacity retention of 85% with the Coulombic efficiency of near 100% per cycle.

Meanwhile, a copper-based conductive MOF with a metal catechol structure (Cu-CAT) has been synthesizing on the compressed nickel foam by a one-step solution process, which has been firstly utilized as anode of SICs [228]. As shown in Figure 6.31a, Cu ions are coordinated to 2,3,6,7,10,11-hexahydroxytriphenylene (HHTP) ligands in the *ab* plane to construct a 2D hexagonal lattice. Subsequently, the 2D layer could be arranged along the *c*-axis with a slipped-parallel AB stacking model to form a honeycomb-like porous structure. As seem from the SEM image, the Cu-CAT nanowires with approximately 25 nm in diameter have been uniformly grown on the surface of nickel foam substrate (Figure 6.31b). Moreover, the Cu-CAT anode demonstrates the outstanding electrochemical performances, delivering the reversible capacities of 260, 146, and 76 mAh g^{-1} at current densities of 0.1, 0.5, and 2 A g^{-1}, respectively, which illustrates that the framework structure could be maintained well at the high rates (Figure 6.31c). According to the XPS results, it is confirmed that The Cu-CAT nanowire anode can reversibly store sodium

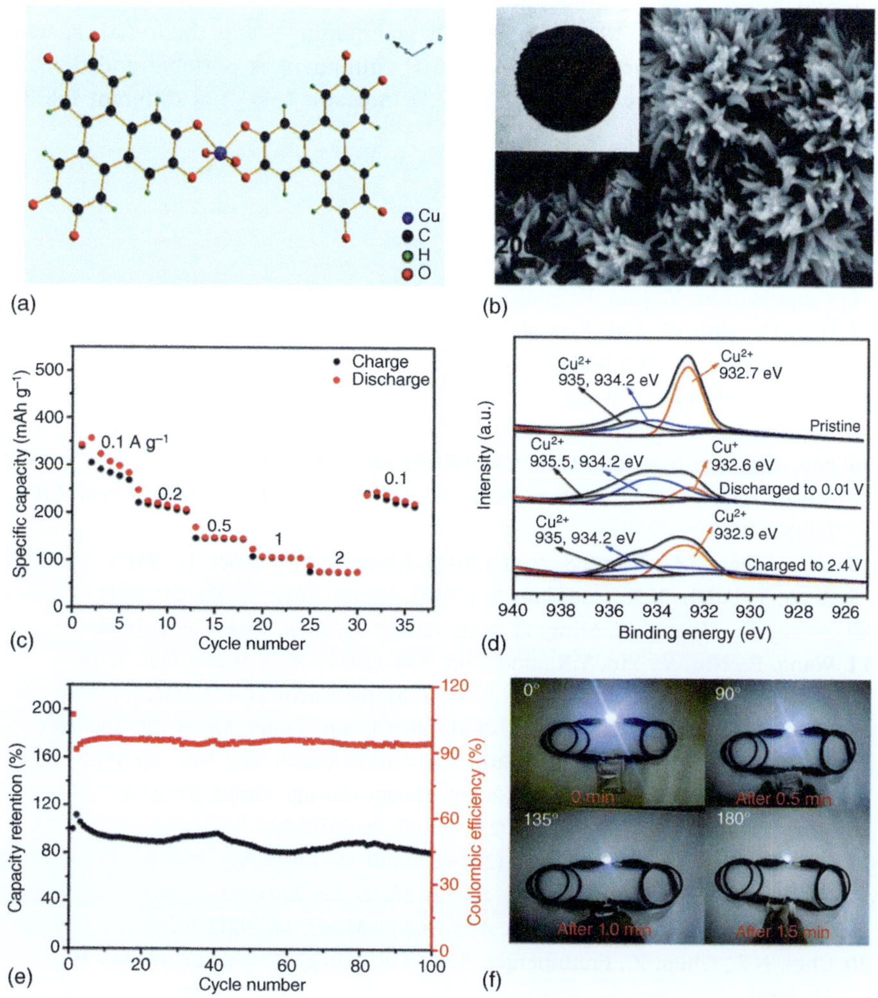

Figure 6.31 (a) Coordinated structure of the *ab* plane of Cu-CAT. (b) SEM image of Cu-CAT nanowires on nickel foam. (c) Rate performance and (d) Cu 2p XPS spectra at different states of Cu-CAT electrode. (e) Cycling stability of Cu-CAT//AC SIC device. (f) Optical photographs of a white LED lightened by this SIC device. Source: Reproduced with permission Zhou et al. [228]. Copyright 2019, Royal Society of Chemistry.

ions through the coordination effect with O atom of C—O bonds accompanied by the reduction of Cu^{2+} to Cu^{+} instead of zero valence Cu (Figure 6.31d). In order to evaluate the practical application of Cu-CAT in SIC, the Cu-CAT anode has been pre-sodiated for pairing with the AC cathode. And the the mass ratio of the cathode to anode is 4.6 : 1, which could accomplish the charge balance of SIC device. Notably, the assembled Cu-CAT//AC device demonstrates a maximum gravimetric energy density of 132.85 Wh kg^{-1}. Moreover, the maximum volumetric power and energy densities of 74.5 mW cm^{-3} and 3.32 mWh cm^{-3} have been also achieved, respectively. As shown in Figure 6.31e, this device presents a remarkable cycling stability over the 100 cycles, feasibly maintaining 84% of the initial capacity. Interestingly, the prepared soft-packing SIC equipment is portable and flexible, which can continuously light a white LED indicator (3.0 V) at different folding states (Figure 6.31f).

References

1 Kang, H., Liu, Y., Cao, K. et al. (2015). *J. Mater. Chem. A* 3: 17899.
2 Deng, X., Zou, K., Cai, P. et al. (2020). *Small Methods* 4: 2000401.
3 Fleischmann, S., Mitchell, J.B., Wang, R. et al. (2020). *Chem. Rev.* 120: 6738.
4 Zhao, X., Zhang, Y., Wang, Y., and Wei, H. (2019). *Batteries & Supercaps* 2: 899.
5 Ding, J., Hu, W., Paek, E., and Mitlin, D. (2018). *Chem. Rev.* 118: 6457.
6 Niu, Y.B., Yin, Y.X., and Guo, Y.G. (2019). *Small* 15: 1900233.
7 Nayak, P.K., Yang, L., Brehm, W., and Adelhelm, P. (2018). *Angew. Chem. Int. Ed.* 57: 102.
8 Li, L., Zheng, Y., Zhang, S. et al. (2018). *Energy Environ. Sci.* 11: 2310.
9 Zhao, Y., Adair, K.R., and Sun, X. (2018). *Energy Environ. Sci.* 11: 2673.
10 Zhang, H., Huang, Y., Ming, H. et al. (2020). *J. Mater. Chem. A* 8: 1604.
11 Wang, E., Niu, Y., Yin, Y.X., and Guo, Y.G. (2021). *ACS Mater. Lett.* 3: 18.
12 Mei, Y., Huang, Y., and Hu, X. (2016). *J. Mater. Chem. A* 4: 12001.
13 Fang, L., Wang, C., Huangfu, L. et al. (2019). *Adv. Funct. Mater.* 29: 1906680.
14 Hwang, J.Y., Myung, S.T., and Sun, Y.K. (2017). *Chem. Soc. Rev.* 46: 3529.
15 He, H., Sun, D., Tang, Y. et al. (2019). *Energy Storage Mater.* 23: 233.
16 Sun, L., Liu, Y., Wu, J. et al. (2022). *Small* 18: 2102894.
17 Sun, Y., Shi, P., Xiang, H. et al. (2019). *Small* 15: 1805479.
18 Xu, Z. and Wang, J. (2022). *Adv. Energy Mater.* 12: 2201692.
19 Zhao, J. and Burke, A.F. (2021). *Adv. Energy Mater.* 11: 2002192.
20 Choi, N.S., Chen, Z., Freunberger, S.A. et al. (2012). *Angew. Chem. Int. Ed.* 51: 9994.
21 Gan, Z., Yin, J., Xu, X. et al. (2022). *ACS Nano* 16: 5131.
22 Liu, Z.G., Du, R., He, X.X. et al. (2021). *ChemSusChem* 14: 3724.
23 Rosa Palacín, M. (2009). *Chem. Soc. Rev.* 38: 2565.
24 Natarajan, S., Subramanyan, K., and Aravindan, V. (2019). *Small* 15: 1904484.
25 Wang, X., Tang, S., Guo, W. et al. (2021). *Mater. Today* 50: 259.
26 Lu, Y., Yu, L., and Lou, X.W. (2018). *Chem* 4: 972.
27 Xue, Y., Li, Y., Luo, G. et al. (2020). *Adv. Energy Mater.* 10: 2002644.
28 Hosaka, T., Kubota, K., Hameed, A.S., and Komaba, S. (2020). *Chem. Rev.* 120: 6358.
29 Liu, S., Kang, L., Henzie, J. et al. (2021). *ACS Nano* 15: 18931.
30 Zhao, L.F., Hu, Z., Lai, W.H. et al. (2021). *Adv. Energy Mater.* 11: 2002704.
31 Wahid, M., Puthusseri, D., Gawli, Y. et al. (2018). *ChemSusChem* 11: 506.
32 Sun, N., Qiu, J., and Xu, B. (2022). *Adv. Energy Mater.* 12: 2200715.
33 Cao, Y., Xiao, L., Sushko, M.L. et al. (2012). *Nano Lett.* 12: 3783.

34 Luo, P., Zheng, C., He, J. et al. (2022). *Adv. Funct. Mater.* 32: 2107277.
35 Zhang, H., Hasa, I., and Passerini, S. (2018). *Adv. Energy Mater.* 8: 1702582.
36 Guo, S., Feng, Y., Wang, L. et al. (2021). *Small* 17: 2005248.
37 Fang, L., Bahlawane, N., Sun, W. et al. (2021). *Small* 17: 2101137.
38 Wang, L., Światowska, J., Dai, S. et al. (2019). *Mater. Today Energy* 11: 46.
39 Shao, H., Wu, Y.C., Lin, Z. et al. (2020). *Chem. Soc. Rev.* 49: 3005.
40 Wang, Y., Song, Y., and Xia, Y. (2016). *Chem. Soc. Rev.* 45: 5925.
41 Wang, E., Chen, M., Guo, X. et al. (2019). *Small Methods* 3: 1900163.
42 Park, M.S., Veerasubramani, G.K., Thangavel, R. et al. (2019). *ChemElectroChem* 6: 653.
43 Kaliyappan, K. and Chen, Z. (2018). *Nano Energy* 48: 107.
44 Kim, H.J., Ramasamy, H.V., Jeong, G.H. et al. (2020). *ACS Appl. Mater. Interfaces* 12: 10268.
45 Qu, Q., Yun, J., Wan, Z. et al. (2014). *RSC Adv.* 4: 64692.
46 Gu, H., Kong, L., Cui, H. et al. (2019). *J. Energy Chem.* 28: 79.
47 Wu, L., Dong, S., Pang, G. et al. (2019). *J. Mater. Chem. A* 7: 1030.
48 Naguib, M., Mochalin, V.N., Barsoum, M.W., and Gogotsi, Y. (2014). *Adv. Mater.* 26: 992.
49 Wang, X., Kajiyama, S., Iinuma, H. et al. (2015). *Nat. Commun.* 6: 6544.
50 Xie, X., Zhao, M.Q., Anasori, B. et al. (2016). *Nano Energy* 26: 513.
51 Kurra, N., Alhabeb, M., Maleski, K. et al. (2018). *ACS Energy Lett.* 3: 2094.
52 Luo, J., Fang, C., Jin, C. et al. (2018). *J. Mater. Chem. A* 6: 7794.
53 Liu, S., Hu, F., Shao, W. et al. (2020). *Nano-Micro Lett.* 12: 135.
54 Luo, J., Zheng, J., Nai, J. et al. (2019). *Adv. Funct. Mater.* 29: 1808107.
55 Brady, A., Liang, K., Vuong, V.Q. et al. (2021). *ACS Nano* 15: 2994.
56 Fan, Z., Wei, C., Yu, L. et al. (2020). *ACS Nano* 14: 867.
57 Or, T., Kaliyappan, K., Li, G. et al. (2020). *Electrochim. Acta* 342: 136024.
58 Wei, Q., Li, Q., Jiang, Y. et al. (2021). *Nano-Micro Lett.* 13: 55.
59 Luo, W., Shen, F., Bommier, C. et al. (2016). *Acc. Chem. Res.* 49: 231.
60 Wang, H., Zhu, C., Chao, D. et al. (2017). *Adv. Mater.* 29: 1702093.
61 Li, Y., Lu, Y., Adelhelm, P. et al. (2019). *Chem. Soc. Rev.* 48: 4655.
62 Pan, H., Hu, Y.S., and Chen, L. (2013). *Energy Environ. Sci.* 6: 2338.
63 Yoon, G., Kim, H., Park, I., and Kang, K. (2017). *Adv. Energy Mater.* 7: 1601519.
64 Divya, M.L., Natarajan, S., Lee, Y.S., and Aravindan, V. (2020). *ChemSusChem* 13: 5654.
65 Han, P., Han, X., Yao, J. et al. (2015). *J. Power Sources* 297: 457.
66 Huang, T., Liu, Z., Yu, F. et al. (2020). *ACS Appl. Mater. Interfaces* 12: 52635.
67 Saurel, D., Orayech, B., Xiao, B. et al. (2018). *Adv. Energy Mater.* 8: 1703268.
68 Chen, D., Zhang, W., Luo, K. et al. (2021). *Energy Environ. Sci.* 14: 2244.
69 Zhang, M., Li, Y., Wu, F. et al. (2021). *Nano Energy* 82: 105738.
70 Kuratani, K., Yao, M., Senoh, H. et al. (2012). *Electrochim. Acta* 76: 320.
71 Zhu, J., Roscow, J., Chandrasekaran, S. et al. (2020). *ChemSusChem* 13: 1275.
72 Ajuria, J., Redondo, E., Arnaiz, M. et al. (2017). *J. Power Sources* 359: 17.
73 Xu, Z., Wu, M., Chen, Z. et al. (2019). *Adv. Sci.* 6: 1802272.
74 Ding, Y., Yang, B., Chen, J. et al. (2018). *Sci. China Mater.* 61: 285.

75 Liu, H., Liu, X., Wang, H. et al. (2019). *ACS Sustainable Chem. Eng.* 7: 12188.

76 Kang, H.J., Huh, Y.S., Im, W.B., and Jun, Y.S. (2019). *ACS Nano* 13: 11935.

77 Cai, P., Momen, R., Li, M. et al. (2021). *Chem. Eng. J.* 420: 129647.

78 Li, X. and Zhi, L. (2018). *Chem. Soc. Rev.* 47: 3189.

79 Thangavel, R., Kannan, A.G., Ponraj, R. et al. (2020). *Energy Storage Mater.* 25: 702.

80 Li, G., Li, Y., Liu, H. et al. (2010). *Chem. Commun.* 46: 3256.

81 Wang, K., Wang, N., He, J. et al. (2017). *ACS Appl. Mater. Interfaces* 9: 40604.

82 Shen, X., Yang, Z., Wang, K. et al. (2018). *ChemElectroChem* 5: 1435.

83 Shi, C., Owusu, K.A., Xu, X. et al. (2019). *Small* 15: 1902348.

84 Li, D., Ye, C., Chen, X. et al. (2018). *J. Power Sources* 382: 116.

85 Liao, K., Wang, H., Wang, L. et al. (2019). *Nanoscale Adv.* 1: 746.

86 Yu, Q., Dong, T., Qiu, R., and Wang, H. (2021). *Mater. Res. Bull.* 138: 111211.

87 Yang, C., Zhang, M., Kong, N. et al. (2019). *ACS Sustainable Chem. Eng.* 7: 9291.

88 Ding, Y., Li, Y., Li, J., and Yan, X. (2020). *Chin. Chem. Lett.* 31: 2219.

89 Yuan, G., Liu, D., Feng, X., and Zhang, Y. (2021). *ChemPlusChem* 86: 1135.

90 Liu, M., Niu, J., Zhang, Z. et al. (2019). *J. Power Sources* 414: 68.

91 Yang, B., Chen, J., Lei, S. et al. (2018). *Adv. Energy Mater.* 8: 1702409.

92 Li, J., Wang, B., Hu, T. et al. (2021). *J. Mater. Chem. A* 9: 3360.

93 Zou, K., Cai, P., Liu, C. et al. (2019). *J. Mater. Chem. A* 7: 13540.

94 Wang, P., Yang, B., Zhang, G. et al. (2018). *Chem. Eng. J.* 353: 453.

95 Chen, L., Duan, W., Yang, B. et al. (2020). *ChemistrySelect* 5: 5824.

96 Hu, X., Wang, G., Li, J. et al. (2021). *Energy Environ. Sci.* 14: 4564.

97 Kan, J., Wang, H., Zhang, H. et al. (2019). *Electrochim. Acta* 304: 192.

98 Qiu, D., Gao, A., Xie, Z. et al. (2018). *ACS Appl. Mater. Interfaces* 10: 44483.

99 Liu, L., Sun, X., Dong, Y. et al. (2021). *J. Power Sources* 506: 230170.

100 Xiao, Z., Song, Q., Guo, R. et al. (2018). *Small* 14: 1703569.

101 Zhu, S., Dong, X., Huang, H., and Qi, M. (2020). *J. Power Sources* 459: 228104.

102 Yan, R., Leus, K., Hofmann, J.P. et al. (2020). *Nano Energy* 67: 104240.

103 Xu, J., Liu, Z., Zhang, F. et al. (2020). *RSC Adv.* 10: 7780.

104 Yan, R., Josef, E., Huang, H. et al. (2019). *Adv. Funct. Mater.* 29: 1902858.

105 Lou, S., Zhao, Y., Wang, J. et al. (2019). *Small* 15: 1904740.

106 Aravindan, V., Lee, Y.S., Yazami, R., and Madhavi, S. (2015). *Mater. Today* 18: 345.

107 Lee, G.H. and Kang, J.K. (2020). *Adv. Sci.* 7: 1902986.

108 Feng, W., Maça, R.R., and Etacheri, V. (2020). *ACS Appl. Mater. Interfaces* 12: 4443.

109 Li, H., Lang, J., Lei, S. et al. (2018). *Adv. Funct. Mater.* 28: 1800757.

110 Wang, R., Wang, S., Zhang, Y. et al. (2018). *J. Mater. Chem. A* 6: 1017.

111 Luo, S., Yuan, T., Soule, L. et al. (2020). *Adv. Funct. Mater.* 30: 1908309.

112 Liu, S., Cai, Z., Zhou, J. et al. (2016). *J. Mater. Chem. A* 4: 18278.

113 Bauer, D., Roberts, A.J., Patnaik, S.G. et al. (2018). *J. Electrochem. Soc.* 165: A1662.

114 Le, Z., Liu, F., Nie, P. et al. (2017). *ACS Nano* 11: 2952.

115 Zhu, Y.E., Yang, L., Sheng, J. et al. (2017). *Adv. Energy Mater.* 7: 1701222.
116 Fang, Y., Zhang, Y., Miao, C. et al. (2020). *Nano-Micro Lett.* 12: 128.
117 e Silva, F.L.R., Filho, A.A.A., da Silva, M.B. et al. (2018). *J. Raman Spectrosc.* 49: 538.
118 Chen, H., Wu, Y., Duan, J. et al. (2019). *ACS Appl. Mater. Interfaces* 11: 42197.
119 Dong, S., Shen, L., Li, H. et al. (2015). *J. Mater. Chem. A* 3: 21277.
120 Yin, J., Qi, L., and Wang, H. (2012). *ACS Appl. Mater. Interfaces* 4: 2762.
121 Liu, Z., Zhang, X., Huang, D. et al. (2020). *Chem. Eng. J.* 379: 122418.
122 Babu, B. and Shaijumon, M.M. (2017). *J. Power Sources* 353: 85.
123 Kim, M.S., Roh, H.K., Jeong, J.H. et al. (2019). *J. Power Sources* 409: 48.
124 Wu, M., Ni, W., Hu, J., and Ma, J. (2019). *Nano-Micro Lett.* 11: 44.
125 Wei, T., Yang, G., and Wang, C. (2017). *ACS Appl. Mater. Interfaces* 9: 31861.
126 Xu, G.B., Chen, Z., Liu, X. et al. (2019). *Electrochim. Acta* 309: 177.
127 Qu, D., Chen, Z., Xu, G. et al. (2020). *J. Mater. Sci.: Mater. Electron* 31: 17550.
128 Chen, M., Zhou, Q., Iqbal, A. et al. (2020). *ACS Appl. Mater. Interfaces* 12: 50388.
129 Wei, P., Liu, Y., Wang, Z. et al. (2018). *ACS Appl. Mater. Interfaces* 10: 27039.
130 Roh, H.K., Kim, M.S., Chung, K.Y. et al. (2017). *J. Mater. Chem. A* 5: 17506.
131 Deng, Q., Fu, Y., Zhu, C., and Yu, Y. (2019). *Small* 15: 1804884.
132 Chen, D., Wang, J.H., Chou, T.F. et al. (2017). *J. Am. Chem. Soc.* 139: 7071.
133 Li, H., Zhu, Y., Dong, S. et al. (2016). *Chem. Mater.* 28: 5753.
134 Han, X., Russo, P.A., Goubard-Bretesché, N. et al. (2019). *Adv. Energy Mater.* 9: 1902813.
135 Lim, E., Jo, C., Kim, M.S. et al. (2016). *Adv. Funct. Mater.* 26: 3711.
136 Wu, Y., Fan, X., Gaddam, R.R. et al. (2018). *J. Power Sources* 408: 82.
137 Lian, Y., Zheng, Y., Bai, Y. et al. (2021). *J. Power Sources* 507: 230267.
138 She, L., Li, Q., Zhang, F. et al. (2020). *J. Power Sources* 451: 227744.
139 Wu, Y., Fan, X., Chen, Y. et al. (2019). *J. Mater. Chem. A* 7: 20813.
140 Tong, Z., Liu, S., Zhou, Y. et al. (2018). *Energy Storage Mater.* 13: 223.
141 Subramanian, Y., Veerasubramani, G.K., Park, M.S., and Kim, D.W. (2019). *J. Electrochem. Soc.* 166: A598.
142 Liu, M., Su, B., Tang, Y. et al. (2017). *Adv. Energy Mater.* 7: 1700885.
143 Chen, Z., Augustyn, V., Jia, X. et al. (2012). *ACS Nano* 6: 4319.
144 Kiruthiga, R., Nithya, C., Karvembu, R., and Reddy, B.V.R. (2017). *Electrochim. Acta* 256: 221.
145 Yuan, J., Hu, X., Li, J. et al. (2021). *ACS Appl. Mater. Interfaces* 13: 10001.
146 Huang, H., Kundu, D., Yan, R. et al. (2018). *Adv. Energy Mater.* 8: 1802800.
147 Shinde, P.A. and Jun, S.C. (2020). *ChemSusChem* 13: 11.
148 Kim, M.S., Lim, E., Kim, S. et al. (2017). *Adv. Funct. Mater.* 27: 1603921.
149 Kaliyappan, K., Jauhar, M.A., Yang, L. et al. (2019). *Electrochim. Acta* 327: 134959.
150 Cheng, W., Wan, B., Xu, S. et al. (2020). *ACS Appl.Mater. Interfaces* 12: 29218.
151 Wei, X., Wang, X., Tan, X. et al. (2018). *Adv. Funct. Mater.* 28: 1804458.
152 Feng, W., Avvaru, V.S., Maça, R.R. et al. (2021). *ACS Appl. Mater. Interfaces* 13: 27999.

153 Qin, J., Sari, H.M.K., Wang, X. et al. (2020). *Nano Energy* 71: 104594.
154 Zhao, J., Zhou, C., Li, Y. et al. (2020). *Int. J. Energy Res.* 44: 3883.
155 Hu, X., Zhang, W., Liu, X. et al. (2015). *Chem. Soc. Rev.* 44: 2376.
156 Ramakrishnan, K., Nithya, C., and Karvembu, R. (2018). *ACS Appl. Energy Mater.* 1: 841.
157 Han, X., Gui, X., Yi, T.F. et al. (2018). *Curr. Opin. Solid St. M.* 22: 109.
158 Ding, R., Qi, L., and Wang, H. (2013). *Electrochim. Acta* 114: 726.
159 Yang, D., Sun, X., Lim, K. et al. (2017). *J. Power Sources* 362: 358.
160 Zhang, H., Bai, R.J., Lu, C. et al. (2019). *Ionics* 25: 533.
161 Fan, X.L., Yang, Y., Xiao, P., and Lau, W.M. (2014). *J. Mater. Chem. A* 2: 20545.
162 Zhang, Y., Liu, W., Wang, T. et al. (2020). *Adv. Mater. Interfaces* 7: 2000953.
163 Li, Y., Wang, H., Huang, B. et al. (2018). *J. Mater. Chem. A* 6: 14742.
164 Wang, T., Chen, S., Pang, H. et al. (2017). *Adv. Sci.* 4: 1600289.
165 Wang, R., Wang, S., Peng, X. et al. (2017). *ACS Appl. Mater. Interfaces* 9: 32745.
166 Zhan, C., Liu, W., Hu, M. et al. (2018). *NPG Asia Mater.* 10: 775.
167 Zhao, C., Yua, C., Zhang, M. et al. (2017). *Nano Energy* 41: 66.
168 Huang, J., Wei, Z., Liao, J. et al. (2019). *J. Energy Chem.* 33: 100.
169 Liu, M.C., Zhang, H., Hu, Y.X. et al. (2019). *Sustainable Energy Fuels* 3: 1239.
170 Ding, C., Huang, T., Tao, Y. et al. (2018). *J. Mater. Chem. A* 6: 21010.
171 Chaturvedi, A., Hu, P., Long, Y. et al. (2019). *Scripta Materialia* 161: 54.
172 Tang, J., Huang, X., Lin, T. et al. (2020). *Energy Storage Mater.* 26: 550.
173 Xu, Q.T., Li, J.C., Xue, H.G., and Guo, S.P. (2018). *J. Power Sources* 379: 41.
174 Hu, X., Liu, Y., Chen, J. et al. (2019). *J. Mater. Chem. A* 7: 1138.
175 Li, S., Chen, J., Gong, X. et al. (2018). *Small* 14: 1804035.
176 Long, Y., Yang, J., Gao, X. et al. (2018). *ACS Appl. Mater. Interfaces* 10: 10945.
177 Zhao, J., Wang, G., Cheng, K. et al. (2020). *J. Power Sources* 451: 227737.
178 Xu, Y.G., Liu, J., and Kong, L.B. (2021). *Ionics* 27: 3315.
179 Liu, M.C., Li, J., Yang, Q.Q. et al. (2019). *ACS Appl. Nano Mater.* 2: 2634.
180 Li, S., He, W., Liu, B. et al. (2020). *Energy Storage Mater.* 25: 636.
181 Li, S., Chen, J., Xiong, J. et al. (2020). *Nano-Micro Lett.* 12: 34.
182 Ge, P., Hou, H., Banks, C.E. et al. (2018). *Energy Storage Mater.* 12: 310.
183 Zhang, B.M., Zhang, C.B., Zhang, H. et al. (2020). *Applied Surface Science* 513: 145826.
184 Zhao, X., Cai, W., Yang, Y. et al. (2018). *Nano Energy* 47: 224.
185 Fei, R., Wang, H., Wang, Q. et al. (2020). *Adv. Energy Mater.* 10: 2002741.
186 Wang, Y., Zhang, X., Xiong, P. et al. (2018). *J. Mater. Chem. A* 6: 21605.
187 Wang, Q., Wang, Y., Zhang, X. et al. (2018). *J. Electrochem. Soc.* 165: A3642.
188 Xia, Z., Sun, H., He, X. et al. (2019). *Nano Energy* 60: 385.
189 Li, Z., Ding, J., and Mitlin, D. (2015). *Acc. Chem. Res.* 48: 1657.
190 Zhu, J. and Deng, D. (2015). *J. Phys. Chem. C* 119: 21323.
191 Wang, J.W., Liu, X.H., Mao, S.X., and Huang, J.Y. (2012). *Nano Lett.* 12: 5897.
192 Palaniselvam, T., Babu, B., Moon, H. et al. (2021). *Batteries & Supercaps* 4: 173.
193 Cui, J., Yao, S., Lu, Z. et al. (2018). *Adv. Energy Mater.* 8: 1702488.
194 Wang, M., Peng, A., Xu, H. et al. (2020). *J. Power Sources* 469: 228414.
195 Yang, C., Ren, J., Zheng, M. et al. (2020). *Electrochim. Acta* 359: 136898.

196 Niu, J., Liang, J., Gao, A. et al. (2019). *J. Mater. Chem. A* 7: 21711.
197 Zhao, J., Wang, G., Hu, R. et al. (2019). *J. Mater. Chem. A* 7: 404.
198 Wang, M., Xu, H., Yang, Z. et al. (2019). *ACS Appl. Mater. Interfaces* 11: 41363.
199 Ruiz, O., Cochrane, M., Li, M. et al. (2018). *Adv. Energy Mater.* 8: 1801781.
200 Arnaiz, M., Gómez-Cámer, J.L., Ajuria, J. et al. (2018). *Chem. Mater.* 30: 8155.
201 Wang, A., Hong, W., Yang, L. et al. (2020). *Small* 16: 2004022.
202 Yuan, Y., Wang, C., Lei, K. et al. (2018). *Sci.* 4: 1261.
203 Lee, S., Kwon, G., Ku, K. et al. (2018). *Adv. Mater.* 30: 1704682.
204 Huang, J., Dong, X., Guo, Z., and Wang, Y. (2020). *Angew. Chem. Int. Ed.* 59: 18322.
205 Song, Z. and Zhou, H. (2013). *Energy Environ. Sci.* 6: 2280.
206 Holguin, K., Mohammadiroudbari, M., Qina, K., and Luo, C. (2021). *J. Mater. Chem. A* 9: 19083.
207 Häupler, B., Wild, A., and Schubert, U.S. (2015). *Adv. Energy Mater.* 5: 1402034.
208 Tie, Z. and Niu, Z. (2020). *Angew. Chem. Int. Ed.* 59: 21293.
209 Lee, S., Hong, J., and Kang, K. (2020). *Adv. Energy Mater.* 10: 2001445.
210 Zhu, L., Ding, G., Xie, L. et al. (2019). *Chem. Mater.* 31: 8582.
211 Han, C., Li, H., Shi, R. et al. (2019). *J. Mater. Chem. A* 7: 23378.
212 Fédèle, L., Ouari, O., Sauvage, F. et al. (2020). *ChemSusChem* 13: 2321.
213 Li, L., Yin, Y.J., Hei, J.P. et al. (2021). *Small* 17: 2005752.
214 Thangavel, R., Kaliyappan, K., Kim, D.U. et al. (2017). *Chem. Mater.* 29: 7122.
215 Zhao, Q., Zhu, Z., and Chen, J. (2017). *Adv. Mater.* 29: 1607007.
216 Xiao, F., Liu, P., Li, J. et al. (2022). *Chem. Commun.* 58: 697.
217 Wang, C., Fang, Y., Xu, Y. et al. (2016). *Adv. Funct. Mater.* 26: 1777.
218 Thangavel, R., Ponraj, R., Kannan, A.G. et al. (2018). *Green Chem.* 20: 4920.
219 Cai, G., Yan, P., Zhang, L. et al. (2021). *Chem. Rev.* 121: 12278.
220 Qiu, T., Liang, Z., Guo, W. et al. (2020). *ACS Energy Lett.* 5: 520.
221 Eddaoudi, M., Kim, J., Rosi, N. et al. (2002). *Science* 295: 469.
222 An, J., Farha, O.K., Hupp, J.T. et al. (2012). *Nat. Commun.* 3: 604.
223 Zhao, R., Liang, Z., Zou, R., and Xu, Q. (2018). *Joule* 2: 2235.
224 Zou, K.Y. and Li, Z.X. (2018). *Chem. Eur. J.* 24: 6506.
225 Wang, L., Zhu, Y., Du, C. et al. (2020). *J. Mater. Chem. A* 8: 24895.
226 Yan, W., Su, J., Yang, Z.M. et al. (2021). *Small* 17: 2005209.
227 Dong, S., Wu, L., Xue, M. et al. (2021). *ACS Appl.Energy Mater.* 4: 1568.
228 Zhou, W., Lv, S., Liu, X. et al. (2019). *Chem. Commun.* 55: 11207.

7

Flexible Sodium-Ion Capacitor Devices

Taoqiu Zhang and Huanwen Wang

China University of Geosciences, Faculty of Material and Chemistry, No. 68, Jincheng Street, Zuoling Street, Hongshan District, Wuhan 430074, China

7.1 Flexible SICs Devices

Growing market demands for portable electronic devices are driving the development of advanced electrochemical energy storage technologies. Future research of such portable devices, represented by flexible electronic devices, foldable displays, and wearable products, is directed toward the development of smaller, lighter, and thinner [1, 2]. The advent of flexible electronics has been silently changing every aspect of our daily lifestyle, which also requires corresponding advances in energy supply devices. However, the bulky nature of current energy storage devices makes it difficult to meet the fundamental demand for flexible energy storage. Therefore, great efforts must be devoted to optimizing the structure, reducing weight, and increase flexibility to meet the requirements of energy storage for FLEXIBLE electronic devices [3, 4]. Lithium-ion batteries (LIBs), as one of the mature energy storage technologies that have been commercialized and nearly reached the ideal performance after a long period of exploration. In the current mature market for flexible electronic devices, lithium-based energy storage dominates with absolute advantage, however, limited reserves of lithium leading to its rising price, have become an insurmountable obstacle to its development. In hence, it is crucial to explore and develop new energy storage technologies [5, 6].

Over the past few years, sodium-based energy storage technologies have been widely considered to complement or replace lithium-based energy storage systems due to the abundance of sodium resources (2.75 wt% of the earth's crust, ≈1850 times the abundance of lithium) and their wide availability, as well as the advantage of the close physicochemical properties of sodium and lithium [7, 8]. Therefore, the development of flexible sodium-based energy storage devices for applications in portable electronics is extremely promising. Unfortunately, sodium-based energy storage is still in its infancy compared to commercialized lithium-based energy storage systems. So far, although sodium-ion batteries (SIBs) have been extensively and intensively researched and Chinese companies are leading efforts to promote

Sodium-Ion Capacitors: Mechanisms, Materials, and Technologies, First Edition.
Edited by Guoqiang Zou, Xiaobo Ji, and Hongshuai Hou.

their industrialization, the large radius of sodium ions, slow kinetics, and safety issues have been widely recognized as a great challenge [9, 10].

Moreover, to the best of our knowledge, SIBs, similar to LIBs, can provide high energy density (150–200 Wh kg^{-1}), but their power density is limited by the Faradic reaction inside the cell, which has been at a low level (below 1 kW kg^{-1}) [11, 12]. Electric-double-layer capacitors (EDLCs, referred to as the most common supercapacitors [SCs]) can exhibit the highest known power capability (100 kW kg^{-1}) and long-term electrochemical stability (up to 10^5 cycles), while the limited charge accumulation results in significantly low gravimetric and volumetric energy densities for EDLCs [13]. Therefore, combination of these two technologies according to their own merits and drawbacks is an efficient strategy to develop a suitable power source for satisfying the requirements of state-of-the-art equipment. In this regard, a new hybrid energy storage device (called as "mental-ion capacitor" or "mental-ion hybrid SCs") is proposed to provide both high energy and power density [14]. The first sodium-ion capacitor (SIC) was reported in 2012, in which V_2O_5/carbon nanotube (CNT) composites were used as anode and active carbon (AC) as cathode. The assembled SIC, with a cell voltage of up to 2.8 V, can provide a maximum energy density of about 40 Wh kg^{-1} [15]. Since then, research on SICs has been rapidly spreading, and impressive results have been achieved so far.

Despite the potential advantages of low cost, building high-performance flexible Na^+ ion energy storage devices still faces many great challenges such as poor electrochemical performance, and safety issues. Accordingly, the development of intrinsically flexible electrode materials and materials with special structural designs seems to be the best solution for now [16]. However, various technical requirements need to be met to develop suitable flexible materials. As for the manufacturing process of electrodes, conventional electrodes based on the slurry coating method not only led to bulky and stiff energy storage devices, but also exhibited low weight/volume energy density. Poorly insulated and electrochemically active binders also have a detrimental effect on conductivity and cycling stability. In addition, the potential safety of liquid organic electrolytes in most flexible hybrid devices is of great concern in most cases [17]. Thus, conventional slurry coating strategies and organic liquid electrolytes in general must be modified or even replaced in order to improve electrochemical performance, flexibility, safety, and simplicity of the fabrication process.

While several works on flexible SIBs and LIBs have been systematically summarized, advances in flexible SICs, including the development of flexible solid-state electrolytes and flexible electrodes based on metal substrates, graphene, carbon cloth, carbon nanofiber (CNF), CNT, and other conductive materials, have not attracted much attention. However, it is worth being explored further. In this chapter, we will sort out and summarize the work related to flexible SICs [18, 19]. The focus of this chapter is on intrinsically flexible materials for flexible SICs, including flexible electrode materials based on metal substrates, graphene, commercial carbon cloths, CNFs, CNTs, and their composites with inorganic substances and polymers also including flexible solid-state electrolytes. By summarizing the review in this chapter, we hope to gain a better understanding of flexible electrode

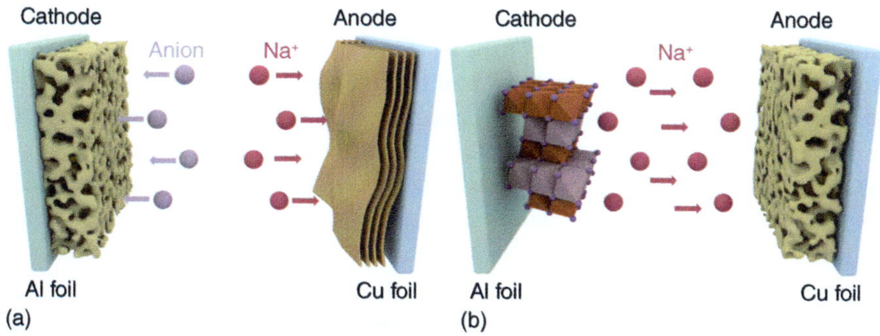

Figure 7.1 Schematic diagram of two types of sodium-ion capacitors (a) conventional SIC. (b) Rocking-chair SIC.

materials, which will lead to a deeper understanding and innovative design of future flexible energy storage devices.

In general, SICs normally consist of a battery-type anode and a capacitor-type cathode, which work by charging or discharging with the help of Faraday reaction or absorption and desorption of two ions (such as Na^+ ion and PF_6^- ion) in the electrode material, so they are also called "electrolyte-consuming capacitor." As shown in Figure 7.1a, during charging, Na^+ ions and the corresponding anion (e.g. PF_6^- ions) enter the anode and cathode, respectively; while during discharging, Na^+ ions and PF_6^- ions are removed from both electrodes and returned to the electrolyte. There is also a less common "rocking chair capacitor," which usually uses a capacitive material as the anode and a battery-type material as the cathode (Figure 7.1b). The operating principle of these rocking chair SICs differs from that of conventional SICs in that during charging, Na^+ ions are de-inserted from the lattice of the cathode material, a process similar to that of the cathode of LIBs, and then the de-inserted Na^+ ions are adsorbed on the surface of the capacitive cathode material through the electrolyte [20]. Despite the difference in the working mechanism, these rocking chair SICs only need to be consistent with the conventional SICs electrolyte. Based on the known excellent work, both types of flexible SICs will be elaborated in this chapter.

7.1.1 Flexible Battery-Type Anode and Capacitive Cathode SICs Cell Configurations

As mentioned above, SICs integrate the complementary advantages of energy-dense rechargeable SIBs and high-power SCs to become another promising energy supply device in the field of flexible electronics. Flexible SICs can provide a large amount of power in a short period of time, which is their most prominent advantage over flexible SIBs. However, so far, not much has been reported about flexible SICs due to the lack of suitable electrode materials to combine carriers with excellent ion/electron conductivity and good mechanical flexibility. Moreover, flexible SICs are complex devices compared to conventionally packaged SICs, and the effective energy density and cycling stability they can provide depend on many design factors, such as the choice of electrode materials, the current collector, and the electrolyte [21]. Recently, great efforts have been made to match the fast kinetics of capacitive cathodes, mainly in terms of increased power capability and rational anode structure design. To the best of our knowledge, flexible SICs are constructed by coating or growing electroactive materials on flexible substrates such as titanium foil, carbon fabric, grapheme, and CNF. This section is centered on flexible cell-based anode and capacitive cathode SICs cell configurations, which will be expanded below based on different flexible substrate materials.

7.1.1.1 Flexible Electrodes Based on Carbon Nanofiber

CNTs are one-dimensional (1D) fibers with large aspect ratios and high specific surface areas; thanks to their structure, they can provide efficient electron transport rates along the longitudinal direction while shortening the ion transport distances [22, 23]. So far, the level of industrial manufacturing and production technology for CNTs is quite mature and ready for mass production on a large scale, and their cost depends mainly on the purity of the precursor. Based on the structural characteristics of CNTs, they can be easily assembled into different flexible/stretchable structures, including 1D CNT fibers, 2D CNT films, and 3D CNT foam, and almost all of the above have the conditions that make them ideal flexible substrates to fabricate flexible electrodes for various flexible energy storage devices [24–26]. Flexible electrodes based on CNTs whose active materials can be easily combined with them into integral units by simple methods such as one-pot method and in situ growth [27, 28]. What's more, the means to enhance the electrochemical performance of flexible electrode materials based on CNTs will become abundant, for example, through heteroatom doping treatment, where the introduction of heterogeneous atoms makes a beneficial structural transformation of the material to enhance the electrochemical performance [29].

Transition metal disulfides (TMDs) with a unique 2D layered structure allowing Na^+ ion intercalation are considered as promising candidates for sodium storage [30]. As a typical 2D material, MoS_2 exhibits structurally large interlayer spacing (6.2 Å between adjacent S–Mo–S layers) with a theoretical capacity of up to 670 mAh g^{-1}. The Na^+ ion storage mechanism of MoS_2 consists of a reversible intercalation process ($MoS_2 + 2Na^+ + 2e^- \rightarrow 1T\ Na_2MoS_2$; 0.4–3 V vs. Na/Na^+) and a low potential conversion reaction ($Na_2MoS_2 + 2Na^+ + 2e^- \rightarrow 2Na_2S + Mo$; 0.01–0.4 V vs. Na/Na^+). In general, the conversion process is accompanied by a large volume change (~200%), leading to a large capacity decay, especially when cycling in the full potential range of 0.01–3 V (compared to Na/Na^+). When the cut-off potential is set to 0.4 V, the transition reaction can be avoided, but a large part of the capacity will be sacrificed [31, 32]. Therefore, maintaining the high capacity of MoS_2 active materials by, for example, growing 3D array structures is a feasible solution.

Wang's group developed a novel anode of MoS_2@PEDOT core/shell nanosheet arrays grown on a flexible porous lightweight carbon nanofiber network (denoted as PCNF@MoS_2@PEDOT) (Figure 7.2a–c) [33]. In this design, the PCNF film can withstand the mechanical bending of the integrated electrode. This design also

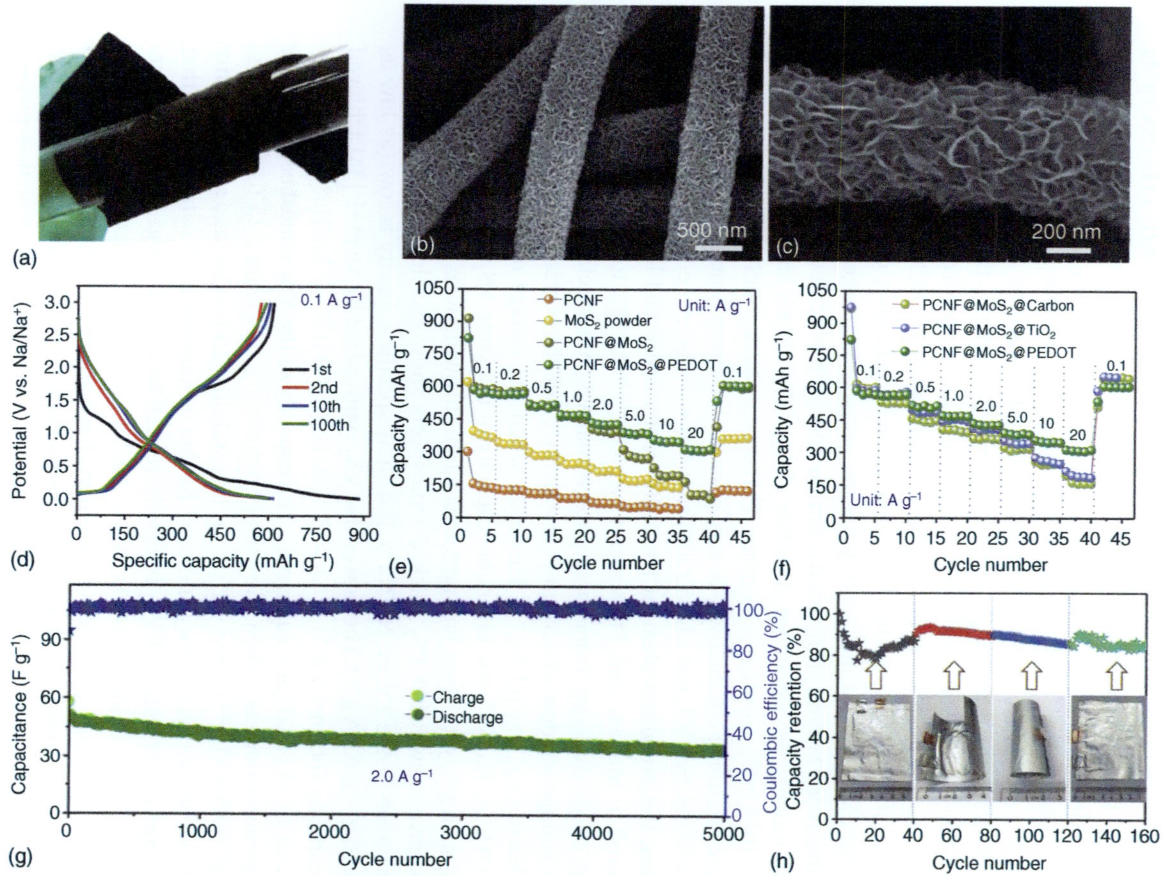

Figure 7.2 Characterization and Na^+ ion storage property in half-cells of PCNF@MoS_2@PEDOT fibers network. (a) Photograph. (b, c) SEM images. Na^+ ion storage property in half-cells. (d) GCD curves of PCNF@MoS_2@PEDOT at 0.1 A g^{-1}. (e, f) Rate performance for different electrodes. Electrochemical performance of sodium-ion hybrid capacitor (g) cycling performance. (h) Capacity retention at 0.1 A g^{-1} at different bending states. Source: Wang et al. [33]/Reproduced with permission from Elsevier.

eliminates the "dead volume" of conventional collectors, as well as polymer binders and conductive components. In particular, the poly(3,4-ethylenedioxythiophene) (PEDOT) coating proved to be more effective in maintaining the stability of MoS_2 as a protective layer compared to conventional atomic layer deposition (ALD) TiO_2 films and glucose-derived carbon layers. Based on the specific capacity of PCNF@MoS_2@PEDOT (~461 mAh g^{-1}) and PCNF@PEDOT (115 mAh g^{-1}) at 1 A g^{-1}, the capacity contribution of the MoS_2 component in the composite was determined to be 661 mAh g^{-1}, with a high MoS_2 utilization of 98%, which is very close to the theoretical capacity (670 mAh g^{-1}). Moreover, the reversible capacity of PCNF@MoS_2@PEDOT at a current density of 0.1 A g^{-1} is 558 mAh g^{-1} Even at a high current density of 20 A g^{-1}, the reversible capacity of PCNF@MoS_2@PEDOT remains as high as 313 mAh g^{-1} (Figure 7.2d–f). By coupling the PCNF@MoS_2@PEDOT anode and the PCNF@AC cathode to GPE. The flexible SIC exhibits excellent capacity retention during high-current charge/discharge cycles after 5000 cycles and low self-discharge rate (Figure 7.2g,h).

Furthermore, Wang's group used a similar approach to achieve high performance sodium storage for vanadium disulfide (VS_2). VS_2 shows metallic conductivity (5.0×10^2 S m^{-1}) compared to other TMDs (e.g. MoS_2, SnS_2, and WS_2) whose intrinsic electronic conductivity is very poor [34]. Density functional theory (DFT) also indicates that VS_2 has a lower diffusion potential barrier compared to MoS_2. In addition, the vanadium element has a different valence and lower atomic weight compared to Mo, Sn, and Ti elements, which enables VS_2 to be a high-capacity storage anode material. It has been shown that VS_2 has excellent cycling stability in the voltage window of 0.3–3.0 V vs. Na/Na^+, but the corresponding capacity is low (386 mAh g^{-1} at 0.1 A g^{-1}). However, when the voltage range is extended to 0.01–3.0 V vs. Na/Na^+, the reversible capacity of the same VS_2 structure is as high as 657 mAh g^{-1}, yet the cycling performance in this window is rather poor. Several works have shown that VS_2 stacked nanosheets show excellent rate and cycling performance (0.4–2.2 V vs. Na/Na^+), but their specific capacity is low (as low as 252 mAh g^{-1} at 0.1 A g^{-1}). The flower-like VS_2 synthesized by Xu et al. [34] achieved high capacity (600 mAh g^{-1}) at cut-off voltage (0.3–3.0 V vs. Na/Na^+) in an electrolyte of 1.0 M $NaCF_3SO_3$ (diglyme). The capacity of the VS_2 nanosheet assembly was further enhanced when the voltage window was widened to 0.01–3.0 V vs. Na/Na^+ (680 mAh g^{-1}); however, the cycling stability (200 cycles) and rate capability (2 A g^{-1}) were not satisfactory and need further enhancement. It can be found that the storage performance of VS_2 is closely related to the discharge cut-off voltage. The poor cycling ability below 0.3 V vs. Na/Na^+ is due to the fact that deep Na^+ intercalation induces a conversion reaction that leads to a large volume expansion of the electrode, resulting in a rapid capacity decay. This Na^+ storage mechanism of VS_2 is similar to that of MoS_2; both involve a reversible intercalation process and a subsequent conversion reaction, which may be a common feature of the 2D layered TMDs during the insertion of alkali metal ions. Therefore, it is crucial to achieve high capacity and long-term cycling capability of VS_2 materials by rational integration or optimization of the microstructure. Thus, his group successfully grew ultrathin bowl-shaped VS_2 nanosheet arrays on highly conductive CNFs (denoted as CNF@VS_2) by a one-step solvothermal process in order to achieve high sodium storage performance of VS_2. The VS_2 arrays exhibited large area, high uniformity and vertical distribution.

The synthesis process of CNF@VS2 involves electrospinning into a homogeneous nanofiber network and carbonization to obtain flexible carbon nanofiber membranes, O_2 plasma treatment, and growth of VS2 arrays on CNF substrates (Figure 7.3a–d). The storage performance of CNF@VS2 in EC: DMC: EMC with 1M $NaClO_4$ as the electrolyte for the half-cell is shown in Figure 7.3e–f.

Na^+ ion storage performances of CNF@VS_2 in half-cell.

The storage performance of CNF@VS_2 in EC: DMC: EMC with 1 M $NaClO_4$ as the electrolyte for the half-cell is shown in Figure 7.3. A potential window of 0.01–3 V was chosen to maximize the specific capacity. In most reported VS_2 electrodes, a potential range of 0.3–3 V is generally utilized to achieve a stable storage capacity, but the corresponding capacity is low (usually below 300 $mAh\,g^{-1}$). An electrode had an ICE of 56.7% (0.2 $A\,g^{-1}$) and the galvanostatic charge/discharge (GCD) curves of CNF@VS_2 maintained good overlap over the subsequent 250 cycles, indicating satisfactory stability, while CNF@VS_2 had an initial coulombic efficiency of 83.2% in a diethylene glycol electrolyte with 1 M $NaCF_3SO_3$, but was not as capacitive at 0.2 $A\,g^{-1}$ as in the ester electrolyte (403 $mAh\,g^{-1}$). The rate capability of CNF@VS_2 exhibited high reproducibility in all three parallel electrodes. Not only that, after 250 cycles at 0.2 $A\,g^{-1}$, the CNF@VS_2 electrode still maintained a high capacity of 659 $mAh\,g^{-1}$ indicating the excellent cycling stability of the CNF@VS_2 electrode, while the VS_2 electrode showed a rapid capacity loss, retaining only 117 $mAh\,g^{-1}$ at the 250th cycle. The current density of 2 $A\,g^{-1}$ measurements were performed for 6000 cycles and the CNF@VS_2 electrode exhibited excellent cycling performance, retaining a high reversible capacity of 345 $mAh\,g^{-1}$. At the same time, the coulombic efficiency can be maintained at almost 100% throughout the cycling process (VS_2 electrode exhibits severe capacity decay). Thus, the results suggest that the coupling of the bowl-shaped VS_2 array with the CNF matrix can effectively suppress the destruction of VS_2 nanocrystals during repeated Na^+ ion insertion/de-insertion.

To match the flexible CNF@VS_2 anode, CNF@GS with flexibility was also synthesized (Figure 7.3a), when assembled into a full cell, the CNF@VS_2//CNF@GS SIC exhibited excellent electrochemical performance, and not only that, the SIC energy density and power density were as high as 100 $Wh\,kg^{-1}$ and 40 $kW\,kg^{-1}$, respectively, in addition, to demonstrate the fabricated device's flexibility, the SIC was repeatedly bent to approximately 90° (Figure 7.3g,h) the GCD curve is almost distortion free, and the capacity loss is negligible, which demonstrates its excellent flexibility.

In addition to in situ growth of active materials on CNF to obtain flexible high-performance sodium storage materials, the flexible electrodes can also be prepared by a simple one-pot method of embedding the active materials into CNF. So far, various promising active materials for sodium storage candidates, including monolayer/bilayer graphene, titanium dioxide (TiO_2), Sb nanoparticles, Sn nanodots, Mo_2C quantum dots, and MoS_2 nanosheets [35–40], have been embedded in flexible CNFs. For example, Wang's group successfully prepared sulfur-covalently linked TiO_2/C doped with sulfur heteroatoms with a 3D network

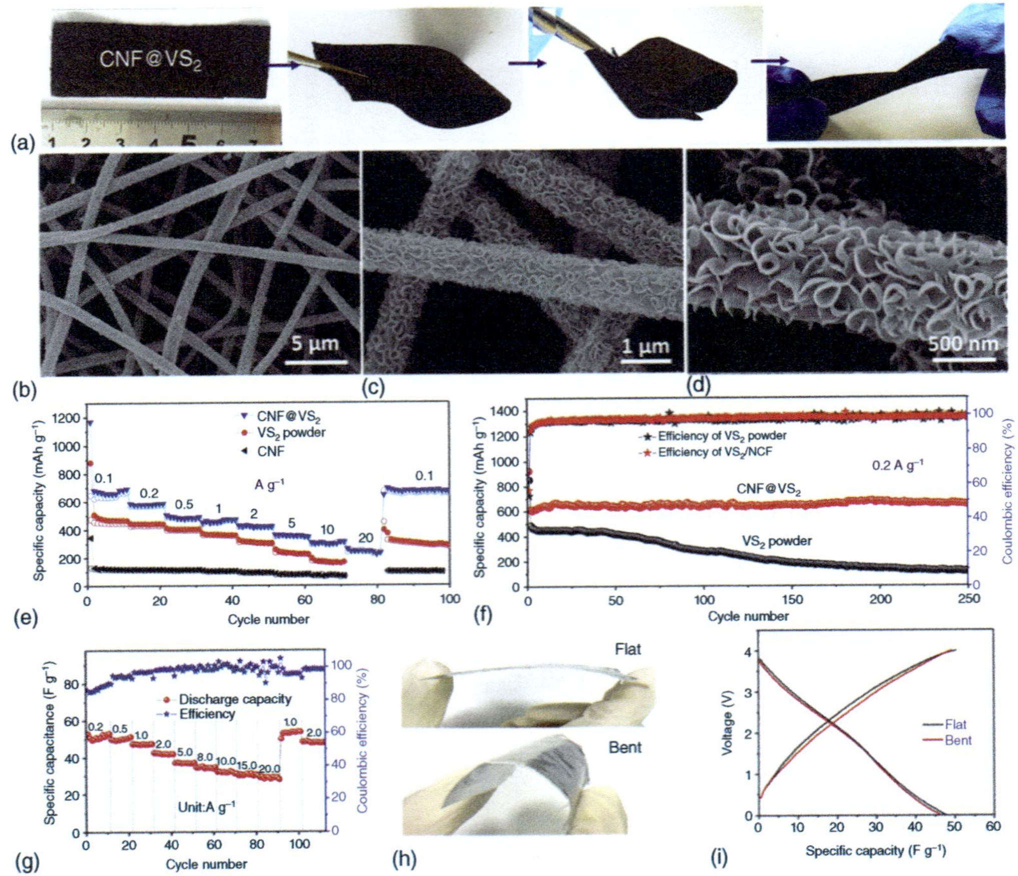

Figure 7.3 (a) Schematic synthetic process of bowl-like VS_2 arrays on flexible CNFs. Photographs of the CNF@VS_2 membrane under various bending states. (b–d) SEM images of CNF@VS_2. Na^+ ion storage performances of CNF@VS_2 in half-cell. (e) Rate capability at different current densities for three parallel electrodes. (f) Rate capability for CNF@VS_2, VS_2 powder and CNF. Electrochemical performance of SIC in the voltage range of 0–4 V. (g) Specific capacitance values at 0.2–20 A g^{-1}. (h) The photograph of flexible SIC pouch cell. (i) GCD curves of the SIC under flat and bent states. Source: Xu et al. [34]/ Reproduced with permission from Elsevier/CC BY-NC-ND 4.0.

nanofiber structure and abundant defects by electrospinning and simple heat treatment techniques [41]. The sulfur-doped titanium dioxide (S-TiO_2)/C electrode has excellent sodium storage performance, and it can provide 358 mAh g^{-1} after 500 cycles at a current density of 0.2 A g^{-1} in a half-cell with a high reversible specific capacity (353 mAh g^{-1} at 0.2 A g^{-1}) and remarkable long-cycle stability (no capacity loss after 10 000 cycles at 5 A g^{-1}). When assembled with AC in a SIC device, the S-TiO_2/C//AC device shows a high energy density of 158 Wh kg^{-1} at a power density of 200 W kg^{-1} and maintains an energy density of 42 Wh kg^{-1} at an ultrahigh-power density of 30 kW kg^{-1}. In this work, TiO_2 with uniform distribution is an important factor for achieving excellent performance. In addition, the S-doped carbon not only induces many defects to exhibit more favorable binding to Na^+ ions, but also improves the electronic conductivity.

Wang's group reports a flexible freestanding T-Nb_2O_5/CNF film with a 3D interconnected fiber network structure fabricated by an electrostatic spinning and subsequent SiO_2 etching strategy (Figure 7.4a–g) [42]. This T-Nb_2O_5/CNF film can provide high excellent rate capability (178 mAh g^{-1} at an ultrahigh current density of 30 A g^{-1}, completing the fast-charging process in 24 seconds) and remarkable cycling stability (high reversible capacity of 175.8 mAh g^{-1} at 200 mA g^{-1} even after 10 000 cycles) (Figure 7.4e,f). As shown in Figure 7.4j, T-Nb_2O_5/CNF//GF/CNF shows excellent cycling stability when assembled with a capacitive graphene fiber (GF)/CNF cathode (Figure 7.4g–i) to form a SIC. A stable discharge capacity of 28.1 mAh g^{-1} can be achieved at a large rate of 30 A g^{-1}, and the SIC also provides a high-capacity retention rate of 80% at 2 A g^{-1} after 3000 cycles. In addition, no obvious capacity loss is observed for m-Nb2O5/CNF//GF/mCNF NIC under continuous bending conditions and go back to the initial status (Figure 7.4k,l), indicating its excellent mechanical flexibility and great potential for application in portable electronic devices.

In addition, Liu's team successfully fabricated flexible free-standing sulfur-doped Nb_2O_5 quantum dot/nitrogen–sulfur co-doped microporous carbon nanofiber (S-Nb_2O_5@NS-PCNF) films by electrospinning PAN/PMMA/$NbCl_5$ solution followed by sulfidation treatment (Figure 7.5a) [43]. The obtained S-Nb2O5@NS-PCNF films have a three-dimensional interconnected fiber network structure with abundant microporosity, structural stability and high elasticity(Figure 7.5 b–d). S-Nb_2O_5@NS-PCNF thin film electrodes show excellent cycling stability (173 mAh g^{-1} after 10 000 cycles at 2 A g^{-1}) and excellent rate capability (even at 4 A g^{-1} with 124 mAh g^{-1}) (Figure 7.5 e–h). It was demonstrated that the homogeneous Nb_2O_5 quantum dots and micropores in CNFs could suppress the aggregation of particles and provide short ion diffusion paths. More importantly, S doping in Nb_2O_5 and porous carbon nanofibers (PCNFs) can synergistically provide more abundant active sites for highly conductive Na^+ storage. S-Nb_2O_5@NS-PCNF thin film electrodes show excellent cycling stability (173 mAh g^{-1} after 10 000 cycles at 2 A g^{-1}) and excellent rate capability (even at 4 A g^{-1} with 124 mAh g^{-1}) (Figure 7.5e,h). By using S-Nb_2O_5@NS-PCNF as the anode and commercial activated carbon as the cathode, the assembled SIC system exhibited high energy and power densities of 112 Wh kg^{-1} and 7949 W kg^{-1}, respectively (even with an energy

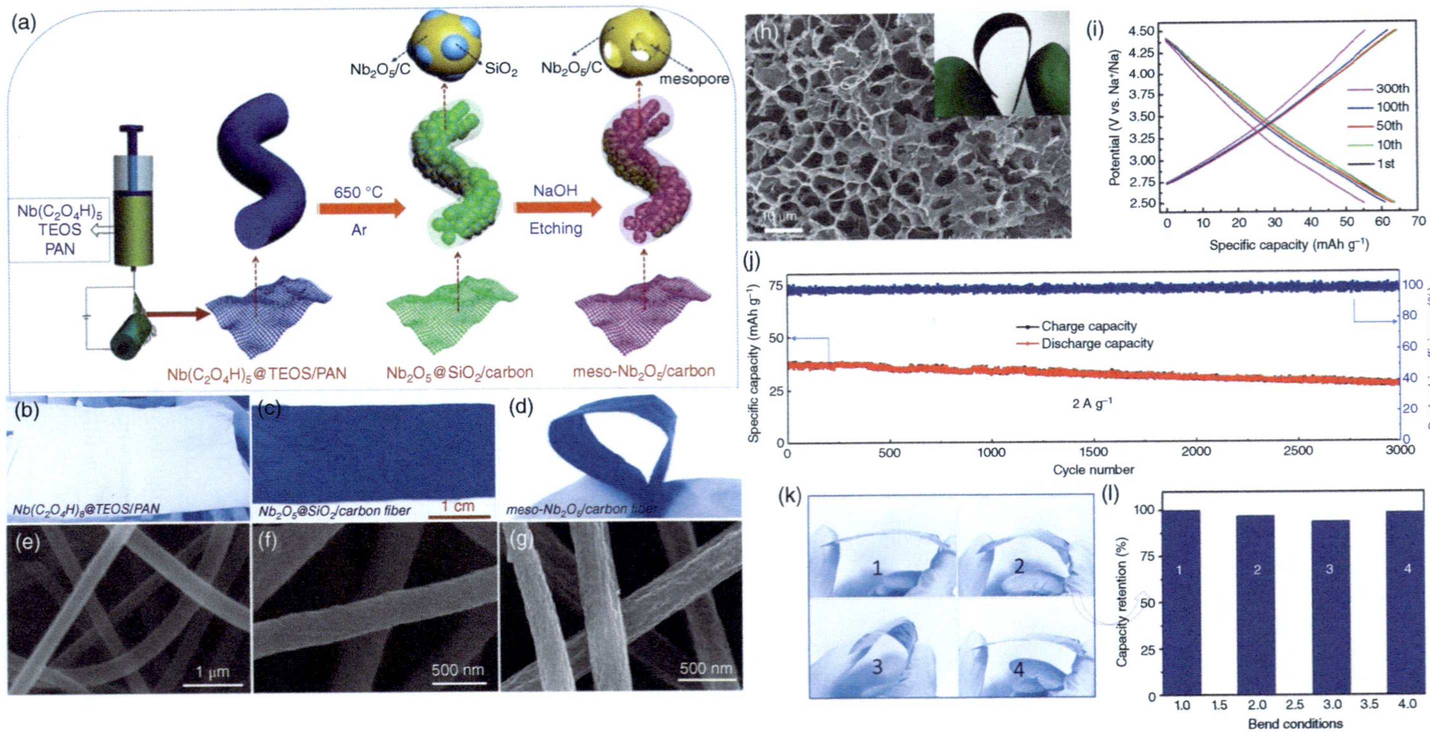

Figure 7.4 (a) Schematic illustration of the fabrication of the flexible m-Nb_2O_5/CNF films. (b–d) Photograph. (e–g) SEM images of $Nb(C_2O_4H)_5$/TEOS/PAN precursor Nb_2O_5/SiO_2/CNF and m-Nb_2O_5/CNF. Graphene framework/mesoporous carbon nanofiber (GF/mCNF) as cathode of SIC. (h) SEM images. (i) GCD curves at $0.2\ A\ g^{-1}$. The cycling performance at $2\ A\ g^{-1}$ of the SIC. (k) Photographs of the SIC device in different bent conditions. (l) Capacity retentions of flexible SIC measured at $0.2\ A\ g^{-1}$ under different bent conditions. Source: Li et al. [42]/Reproduced with permission from John Wiley & Sons.

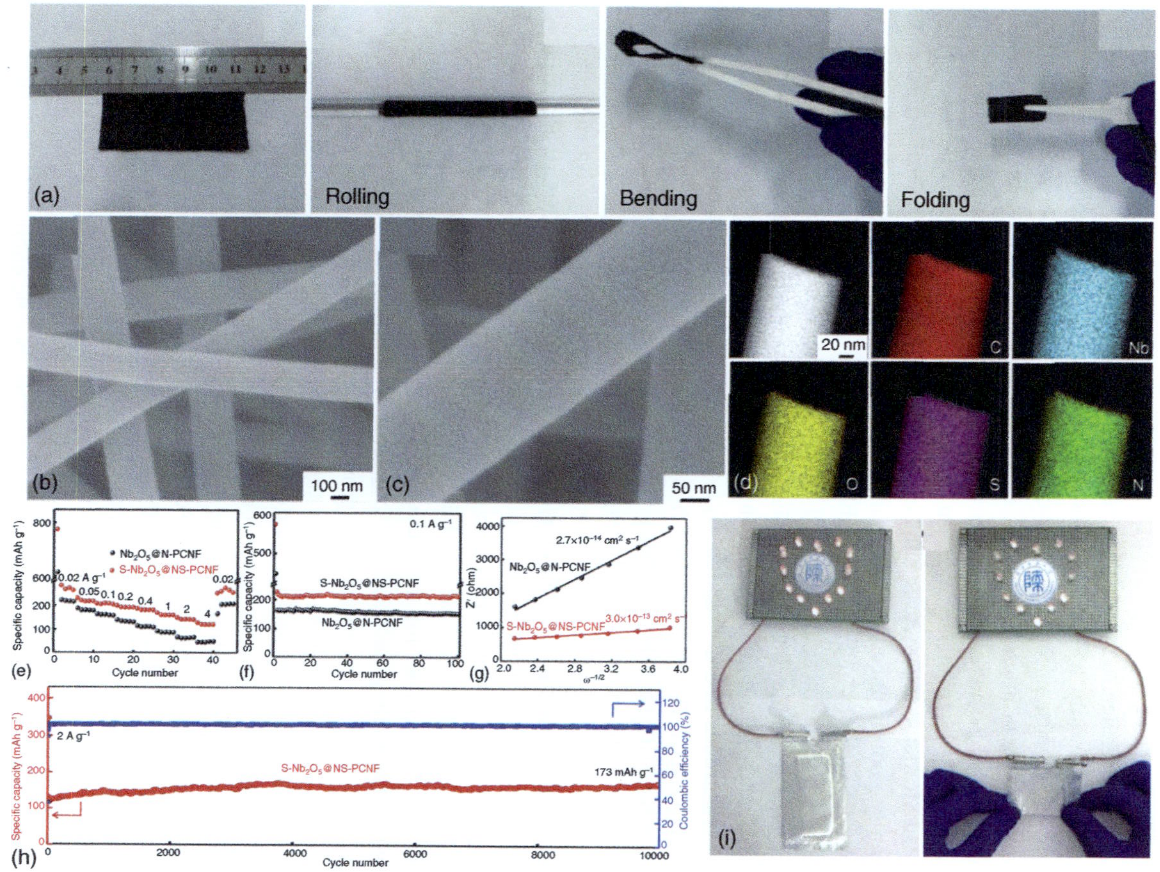

Figure 7.5 (a) Digital photographs for the appearance and flexibility (rolling, bending, and folding) of S–Nb_2O_5@NS-PCNF. FE-SEM images (b, c), (d) EDS elemental mappings. (e) Rate capability at different current densities (f), cycling performances at 0.1 A g^{-1}. (g) The Nyquist plots. (h) Long cycling performance at 2 A g^{-1}. (i) Digital photographs lightening a LED under flat and bent state (d) the flexible S–Nb_2O_5@NS-PCNF//AC SIC. Source: She et al. [43]/Reproduced with permission from Elsevier.

density of 55 Wh kg^{-1}), as well as ultralong-term cycling stability (81% capacity retention after 1 A g^{-1} cycling). These excellent electrochemical properties make the designed heteroatom-doped Nb_2O_5-based composite a promising anode material for SICs, and to further demonstrate the potential of this flexible anode material for practical applications, they assembled pocket SICs to successfully light several LEDs with little effect even in the bent state (Figure 7.5i).

The outstanding electrochemical properties of lithium titanate ($Li_4Ti_5O_{12}$) make it the most commonly used anode material in LICs [44], and sodium titanate ($Na_2Ti_3O_7$) is likewise used as an ideal candidate for SICs due to its low cost and environmental friendliness. $Na_2Ti_3O_7$ exhibits a herringbone-layered structure consisting of TiO_6 octahedra and has a low Na^+ insertion potential (0.3 V vs. Na^+/Na), which can effectively increase the operating voltage and specific energy of practical SIC devices. However, the large band gap of $Na_2Ti_3O_7$ (3.73 eV) makes it inherently poor in electronic conductivity. In addition, the surface reaction, electrolyte decomposition, and irreversible trapping of Na^+ ions make the $Na_2Ti_3O_7$ electrode susceptible to capacity decay [45]. In addition, it has been demonstrated that the sodium storage behavior of $Na_2Ti_3O_7$ is strongly influenced by its specific microstructure. So far, strategies to improve the sodium storage kinetics of $Na_2Ti_3O_7$ electrodes mainly include. (i) Construction of carbon-based composites (e.g. $Na_2Ti_3O_7$@CNTs, $Na_2Ti_3O_7$@C hollow spheres, $Na_2Ti_3O_7$@graphene networks, and $Na_2Ti_3O_7$@CNFs). (ii) Design of 1D or 2D nanostructures (e.g. $Na_2Ti_3O_7$ nanowires, nanosheets, and nanotubes). (iii) Surface engineering (e.g. S-doping, F-doping, and self-integration of abundant Ti^{3+}). Despite recent advances, the high-rate capability of these $Na_2Ti_3O_7$ electrodes is typically below 20 C, which still cannot compete with $Li_4Ti_5O_{12}$ in lithium storage tests [46]. It has been widely demonstrated that 3D nanoarrays grown on highly conductive substrates are an effective way to achieve high-rate capability and long-term cyclability, as the ordered array configuration allows for mechanical stress release during ion insertion/exit. In Wang's group, they successfully fabricated self-supported $Na_2Ti_3O_7$ arrays through the conversion reaction of pre-grown TiO_2 arrays on CNF substrates [47]. In addition to investigating the effect of the microscopic morphology of the arrays on the $Na_2Ti_3O_7$ sodium storage performance, they synthesized two different $Na_2Ti_3O_7$ nanostructures (i.e. nanoribbons and nanowires) to investigate their favorable properties for sodium storage. Notably, the mass loading of $Na_2Ti_3O_7$ on CNF is high, reaching 6.5 mg cm^{-2}. The comparison results show that the $Na_2Ti_3O_7$ nanoparticles have a high-rate capability (123 mAh g^{-1} at 113 C, corresponding to 20 A g^{-1}) and a stable cycling capability of maintaining 130 mAh g^{-1} at 56 C (corresponding to 10 A g^{-1}), which surpasses the many reported nanowire electrodes and other $Na_2Ti_3O_7$ materials. For practical applications, they further assembled a SIC device based on $Na_2Ti_3O_7$ nanowire electrodes. The specific energy (126 Wh kg^{-1}) and power (38 kW kg^{-1}) of this SIC full cell are higher than those reported so far for sodium-ion hybrid devices (Figure 7.6). This result emphasizes the importance of electrode structure to improve the performance of sodium-ion hybrid capacitors.

So far, various intercalation-type anode materials for SICs have been intensively investigated, such as $NaTiO_2$, $NaTi_2(PO_4)_3$, $Ti_3C_2T_x$ MXene, TiO_2, $Na_2Ti_3O_7$, and

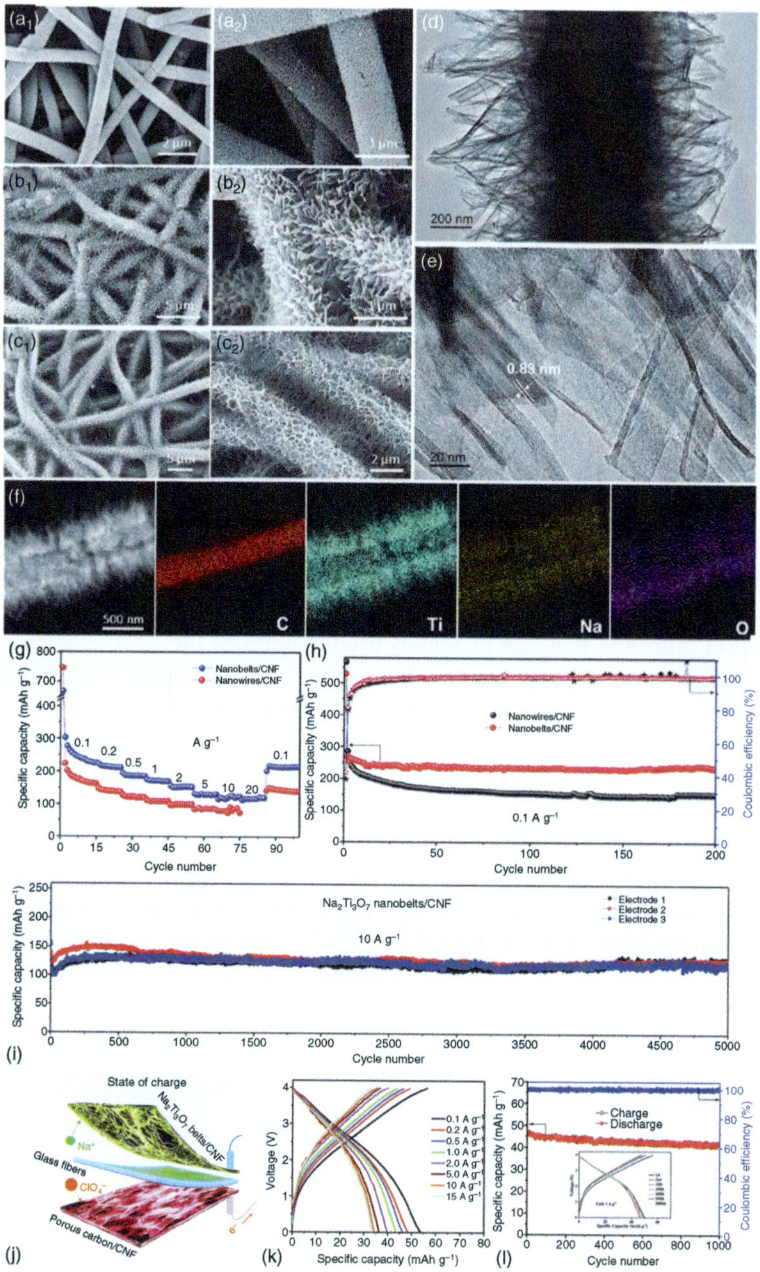

Figure 7.6 SEM and TEM images of (a) TiO_2 nanosheets/CNF, (b, d) $Na_2Ti_3O_7$ nanobelts/CNF, and (c, e) $Na_2Ti_3O_7$. (f) Energy dispersive spectrum mapping images nanowires/CNF. Electrochemical performance of $Na_2Ti_3O_7$ nanobelts/CNF and $Na_2Ti_3O_7$ nanowires/CNF, (g) rate capability, (h, i) cycling stability at 0.1 and 10 A g^{-1}. (j) Schematic of the SIC device using the $Na_2Ti_3O_7$ nanobelts/CNF anode and the biomass-derived porous carbon (BPC)/CNF cathode. (k) Charging/discharging curves. (l) Cycling performance at 1 A g^{-1}. Source: Wang et al. [47]/Reproduced with permission from John Wiley & Sons.

Nb_2O_5. Unfortunately, these materials usually have low specific capacities (below 200 mAh g^{-1}), especially at high rates, which limits their use in high energy and high power SIC applications. Therefore, the exploration of novel intercalation anode materials with high capacity is crucial for building high-performance SICs. Recently, hybrid Ti–Nb oxides with multiple redox mates ($Ti^{+4/+3}$, $Nb^{+5/+4}$, and $Nb^{+4/+3}$) have been reported as promising anodes for SIBs, and especially $Ti_2Nb_2O_9$ has been extensively investigated for Na^+ storage because it possesses not only a good theoretical capacity (252 mAh g^{-1}) and a high working potential (0.7 V vs. Na/Na^+), but also shows a pseudocapacitive Na^+ intercalation behavior [48]. However, the unsatisfactory rate capability at high current densities and poor cycling stability of the $Ti_2Nb_2O_9$ electrode due to its low electronic conductivity and unavoidable nanosheet overlap caused by strong van der Waals interactions between neighboring nanosheets greatly hinder its application in SICs. Therefore, highly conductive and stack-free $Ti_2Nb_2O_9$-based composite anode materials are expected to be realized. Inspired by the modification of other metal oxide electrode materials with highly conductive carbonaceous nanomaterials, the electron conductivity of the anode with electrochemical sodium-ion storage can also be improved by hybridization of carbon-based materials. 1D continuous CNFs have attracted much attention due to their high electrical conductivity, good flexibility, and high aspect ratio, and they can be widely used to improve the electrochemical sodium-ion storage properties of anodes. Using the same approach, Liu's group fabricated self-supporting films composed of $Ti_2Nb_2O_9$ nanosheets and CNFs (denoted as $Ti_2Nb_2O_9$/CNF) by electrospinning and subsequent carbonization treatment [49]. The recombination process of $Ti_2Nb_2O_9$ nanosheets was effectively prevented because the $Ti_2Nb_2O_9$ nanosheets were embedded within the CNFs. The prepared $Ti_2Nb_2O_9$/CNF films exhibit fast Na^+ ion intercalation kinetics and high conductivity during Na^+ ion storage and show excellent rate capability in half-cell tests with a capacity of 204 mAh g^{-1} at 4 A g^{-1} and good cycling stability with 97% capacity retention in 2000 cycles at 1 A g^{-1} (Figure 7.7). When the $Ti_2Nb_2O_9$/CNF anode and the AC cathode are combined together to form a SIC, it has a high energy density (129 Wh kg^{-1} at 75 W kg^{-1}), a high power density (7560 W kg^{-1} with 63 Wh kg^{-1}), and an excellent lifetime (85% capacity retention after 10 000 cycles at 1 A g^{-1}), which is very competitive with the reported SICs compared to those reported in the literature.

As mentioned above, transition metal oxides (TMOs) (e.g. Nb_2O_5, TiO_2, and MoO_2) have attracted great interest as sodium storage hosts due to their high theoretical capacity, ease of fabrication, and structural diversity. However, due to their inherent low electron conductivity, bulk expansion, and slow Na^+ diffusion rate, their bulk structures have always experienced low-rate capability and frustrating cycling performance. To solve this dilemma, TMO-based nanostructures have been reported and exhibit better electrochemical properties than bulk structures, benefiting from "nano-effects" such as large specific surface areas and short ion/electron transport paths. Sub-nanomaterials are reported to bridge the gap between nanomaterials and molecules, and it has atomic-scale dimensions compared to its nano counterparts with larger dimensions and thus can exhibit

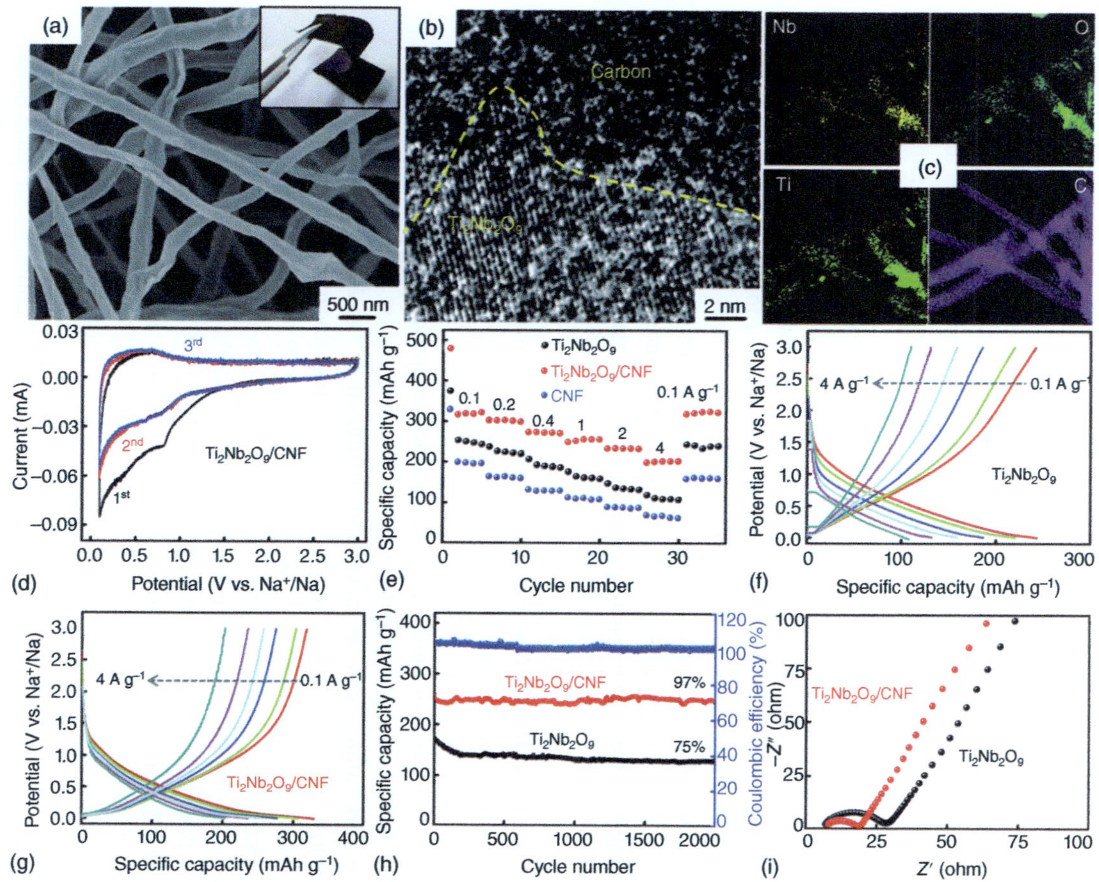

Figure 7.7 FE-SEM image (inserted for the digital image) (a, b) TEM images, high resolution transmission electron microscope (HRTEM), (c) and the corresponding EDS elemental mappings. (d) CV curves for the first three cycles at 0.1 mV s^{-1}, (e) rate capabilities, (f) the charge and discharge curves for $Ti_2Nb_2O_9$, (g) $Ti_2Nb_2O_9$/CNF, (h) long cycling performance at 1 A g^{-1}, and (i) the Nyquist plots. Source: She et al. [49]/Reproduced with permission from Royal Society of Chemistry.

unique properties. As the only 0D sub-nanomaterials (SCs) can be considered as a combination of atoms. Due to "sub-nanoscale effects," including ultrasmall size, highly exposed active atoms, and molecular-like chemistry, sub-nanoclusters show promising applications in many fields, such as catalysis, energy storage, and conversion. Carbon nanomaterials, especially nitrogen-doped carbon (NC) nanomaterials, have been shown to be suitable substrates for loading SCs with metals or metal compounds because of the strong interaction between carbon and metal atoms, which greatly improves the thermal and chemical stability of the subnanostructures, and in addition, carbon substrates facilitate the fabrication of flexible electrodes with strong mechanical strength and high electrical conductivity, adapting to the requirements of flexible electronics. However, obvious drawbacks such as complex synthesis process and low SC loading limit their further application in alkali ion storage systems.

Here, Yuan's group first developed a novel method to construct highly flexible composite films consisting of MoO_2 SCs encapsulated in nitrogen-doped CNFs (MoO_2 SCs@N-CNFs) (Figure 7.8a) [50]. High-quality fractions of MoO_2 SCs formed in situ were tightly bound to N-CNFs. The in situ-formed sub-nanohybrids are capable of high pseudocapacitive charge storage, fast electron/ion diffusion rates, and tiny volume expansion during repeated intercalation/deintercalation of Na^+. Thus, the unique MoO_2 SCs@N-CNFs act as stand-alone anodes for SICs, exhibiting high weight ratio/actual capacity, excellent rate capability, and long cycle life in both liquid and quasi-solid electrolytes. In addition, MoO_2 SCs@N-CNFs exhibit better sodium storage behavior by forming more pores through polymethylmethacrylate (PMMA). The thin film electrode exhibited significant reversible weight and area capacities of ≈219.0–65.8 mAh g^{-1} and ≈0.48–0.66 mAh cm^{-2} with an area mass loading of 2.2–10.1 mg cm^{-2} over 700 cycles at 0.5 A g^{-1}. In addition, the thin film anode exhibited attractive rate capabilities (0.1–2.0 A g^{-1} at ≈231.3–82.5 mAh g^{-1}) and long-term cycle life (Figure 7.8). The superb Na^+ storage performance can be reasonably attributed to the MoO_2 SCs with amorphous structures, which not only confer ideal pseudocapacitance contribution to the electrodes, but also provide sufficient active sites and pathways for rapid Na^+ intercalation and diffusion. Attractively, the reversible capacity of MoO_2 SCs@N-CNFs is further enhanced with more pore channels constructed by the PMMA pore generator. Moreover, the SICs of MoO_2 SCs@N-CNFs have remarkable energy density and stable cycling performance, and the paper electrodes of P(VDF-HFP)-based quasi-solid-state electrolytes (QSSE) achieve a reversible capacity of ≈192.4–45.5 mAh g^{-1} (0.1–2.0 A g^{-1}). The flexible MoO_2 SCs@N CNFs-based SICs show an encouraging energy density of ≈71.2 Wh kg^{-1} at a power density of 75.0 W kg^{-1} with good mechanical flexibility and exceptional cycling stability (Figure 7.9). More promisingly, this sub-nanocluster engineering strategy will further facilitate the development of sub-nanoscale materials for other electrochemical energy storage/conversion applications.

In addition to TMO, which is considered as a promising anode material for sodium storage, transition metal carbides (TMCs) with superb electrical conductivity are also considered as ideal anode materials for sodium storage. Mo_2C, a new

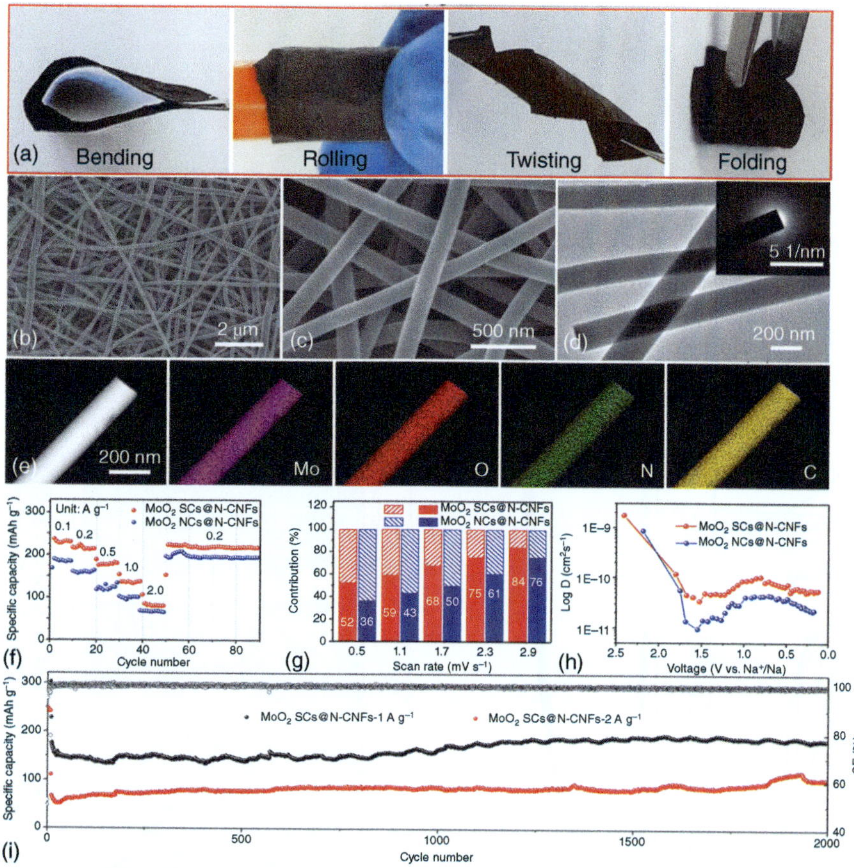

Figure 7.8 (a) Digital images for the flexibility of MoO_2 SCs@N-CNFs film with various states including bending, rolling, twisting, and even folding twice. Morphology characterizations of MoO_2 SCs@N-CNFs. (b, c) FESEM images, (d) TEM images, and (e) corresponding EELS map. (f) Rate capabilities. (g) Capacitive contributions. The striped columns represent the diffusion-controlled part. (h) The Na^+ diffusion coefficient as a function of discharge states. (i) Long-term cycling performances. Source: Liu et al. [50]/ Reproduced with permission from John Wiley & Sons.

star in the TMC family, has attracted new research on efficient sodium storage due to its excellent electrical conductivity ($1.02 \times 10^2\,S\,cm^{-1}$) and remarkable mechanical/chemical stability. Unfortunately, the sodium storage mechanism of active Mo_2C remains unclear, for which there are different views, including the coexistence of de-insertion mechanisms, conversion reactions, and even hybrid processes of both. Moreover, the poor cycle life and low capacity of Mo_2C-based anodes at high current densities remain the main obstacles for their application in sodium storage. To alleviate these problems, great efforts have been made to construct Mo_2C nanoparticles by reducing the particle size, which reduces the volume expansion of Mo_2C materials, provides abundant electrochemical active sites, and shortens the diffusion path of Na^+, while the increased exposed surface

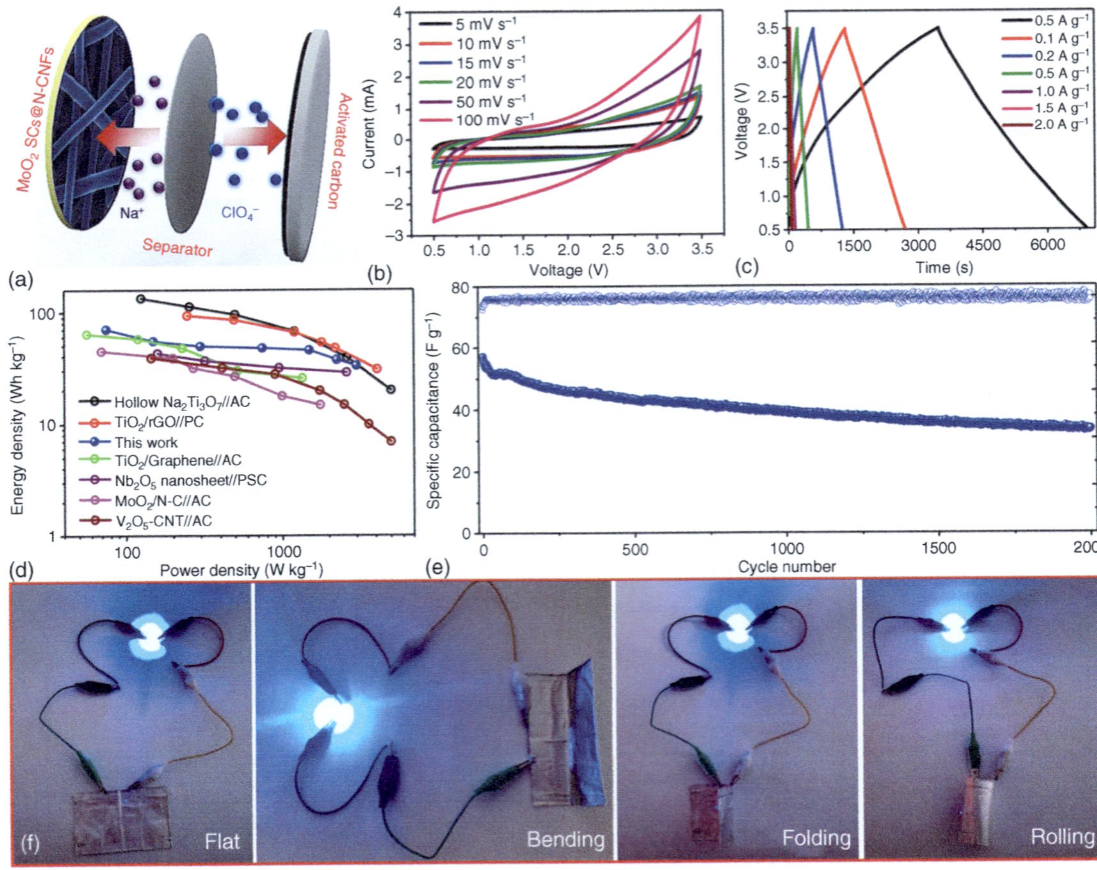

Figure 7.9 Electrochemical performance of the hybrid SICs. (a) Schematic diagram of SICs. (b) CVs at different scan rates. (c) GCD profiles. (d) Ragone plot of as assembled SICs compared with others. (e) Long-term cycling performance at 0.5 A g^{-1}. (f) Digital photographs of the flexible pouch SICs lightening a LED under various states. Source: Liu et al. [50]/Reproduced with permission from John Wiley & Sons.

and high surface energy of nanostructures accelerate the severe aggregation of Mo_2C particles and more solid electrolyte interphase (SEI) layers growth. For this reason, further combining buffer substrates (e.g. conductive carbon) with Mo_2C nanoparticles would bring several benefits that would favor enhanced cycling/rate characteristics. First, the conductive carbon backbone, especially when doped with N, can uniformly disperse and immobilize the nano-Mo_2C. Second, the directly exposed nano-Mo_2C/electrolyte surface is suppressed. Finally, the exfoliation and aggregation of nano-Mo_2C into larger grains during electrochemical sodium storage are largely prevented. However, further insight should be gained into the intrinsic charge migration at the heterogeneous Mo_2C-carbon matrix interface, which plays an important role in enhancing the electrochemical behavior. In addition to this, an intelligent design at the electrode level, i.e. a flexible self-supporting Mo_2C-based electrode, would be better than a conventional slurry-coated electrode, which gets rid of the insulating polymer binder and efficiently adapts to the stress changes during the sodization/de-sodization process.

Considering all the above factors, Yuan's team first explored a heterogeneous interface engineering strategy to construct advanced three-dimensional (3D) flexible hybrid thin-film electrodes in which metallic Mo_2C quantum dots are embedded in functionalized nitrogen-doped carbon nanofibers (denoted as Mo_2C QDs@N-CNFs) films toward SICs as self-supported anodes (Figure 7.10a–d) [51]. The electroactive Mo_2C QDs (41.1 wt%) are uniformly distributed in the elastic carbon framework, which effectively relieves the mechanical stress, strengthens the structural integrity, and alleviates the volume expansion, while bringing many phase boundaries/nano heterogeneous surfaces, which favor the properties of Na^+ storage and optimize the formation of SEI layers. More particularly, the unique interfacial coupling, i.e. Mo–N–C bonding at the phase interface of heterogeneous Mo_2C/N-CNFs, stabilizes/immobilizes the ultrafine Mo_2C in N-CNFs, promotes ion/electron transport, and accelerates the reaction kinetics. It is due to the synergistic effect of both that the Mo_2C QDs@N-CNFs films exhibit high reversible capacity and long-term electrochemical stability, especially for efficient sodium storage at high current densities. It exhibited a high-rate capacity of 341.6/160.9 $mAh\,g^{-1}$ at 0.1/5.0 $A\,g^{-1}$ and maintained a large capacity of 252 $mAh\,g^{-1}$ after 2000 uninterrupted cycles at 2.0 $A\,g^{-1}$, as well as a high CE value of 99.7% during cycling (Figure 7.10e–g). When assembled with an AC cathode to form SICs, the constructed SICs show good promise for commercial applications with their large material-level energy density (87.0 $Wh\,kg^{-1}$ at 100 $W\,kg^{-1}$), ideal capacitance retention (0.015% SC degradation per cycle at 1.0 $A\,g^{-1}$), small leakage current (8 μA), moderate self-discharge capability (0.5% loss of open-circuit potential). This work also provides important guidance for the future design of advanced TMCs in energy-related fields and beyond.

7.1.1.2 Flexible Electrodes Based on Graphene Substrates

Although several advances have been made in the preparation of flexible electrodes using electrospinning, poor adhesion between the active material and the substrate through in situ growth, not only hydrothermal reactions carry the

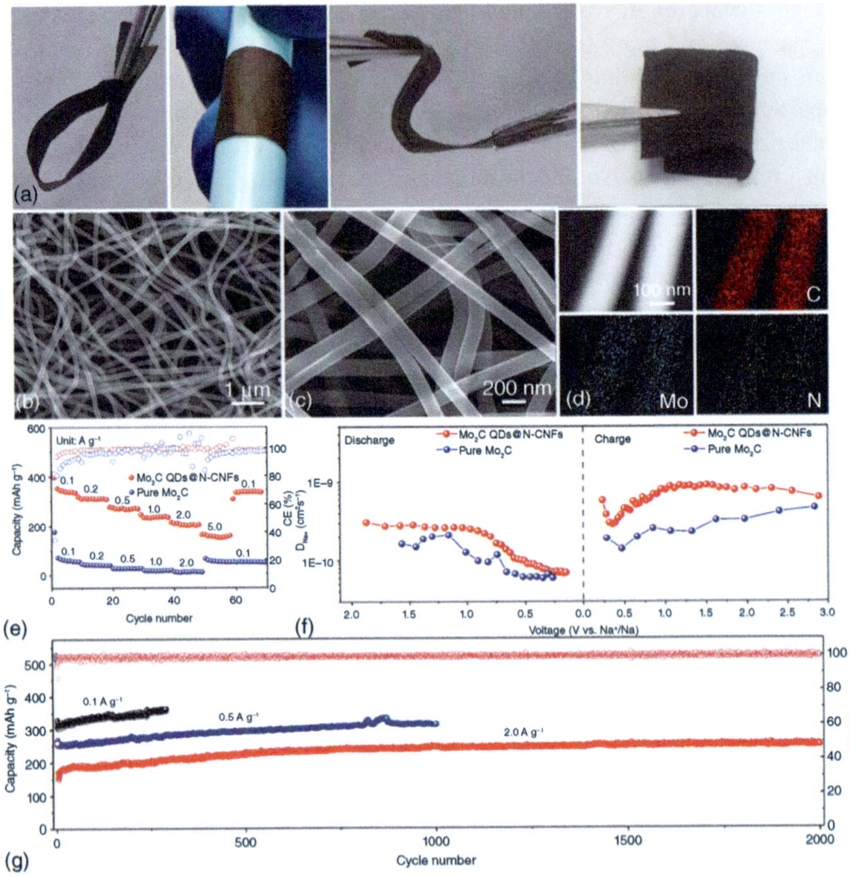

Figure 7.10 (a) Digital images for the flexibility of the resultant Mo_2C QDs@N-CNFs films with various states. (b, c) FESEM images, (d) corresponding elemental mapping. (e) Rate behaviors. (f) Variation of DNa^+ values at different discharge/charge states. (g) Cycling properties at various current densities. Source: Wang et al. [51]/Reproduced with permission from American Chemical Society.

risk of deteriorating electrochemical stability, but also layer-by-layer stacking of 2D materials has limited ion diffusion and transport behavior, especially in the vertical direction. Furthermore, the one-pot method for the preparation of flexible electrodes requires high temperature (>500 °C) sintering in order to convert electrospun polymer nanoparticles into conductive carbon, however, which limits the choice of active materials [52–54]. Based on this, the preparation of nonwoven GF fabrics by using wet spinning technique has been recently reported. The nonwoven fabric exhibits high electrical conductivity in the 3D direction while demonstrating high active material compatibility. Multidimensional active materials, including 1D $Na_2Ti_7O_{15}$ (NTO) nanorods, 2D $MoSe_2$ nanosheets, 3D bismuth (Bi), and AC powders, have been readily incorporated into the fabrics. The introduced active materials are surrounded by graphene sheets, and their comminution is strongly hindered during the fast charging and discharging process, therefore, the electrochemical reactivity of the active materials can be well maintained. The naturally porous nature of the nonwoven fabric allows a high ionic diffusivity, while the ionic permeability of the individual GFs is further optimized by pore etching on the graphene sheets. These simultaneously facilitate the diffusion and transport of electrolyte ions and contribute to the overall capacity, as well as the excellent electrochemical performance. In addition, the whole process is carried out at low temperatures, expanding the compatibility and application range of the active materials.

Nonwoven GF fabric-reinforced ferroconcrete electrodes were prepared and a pore etching process was applied to improve the ion accessibility by Liu and coworkers [55]. And successfully incorporated 1D NTO nanorods, 2D $MoSe_2$ nanosheets, and 3D Bi and AC into the fabric. The study showed that in the structure of the fabric, GF interwoven to form a complete nonwoven fabric without breakpoints, meanwhile, the interleaved structure reduced the contact resistance, improved the electron transport in 3D direction throughout the nonwoven fabric and improved the mechanical robustness. The balance between electron transport and ion permeability of the GF is achieved by pore etching, which facilitates the diffusion and transport of electrolyte ions and helps to improve the rate capability as well as the overall capacity. The SIC based on Bi and AC fabric electrodes can complete the entire charge/discharge process in 15 seconds and store energy up to 44 Wh kg^{-1} (excellent weight and volume energy/power density; 107 Wh kg^{-1} at 611 W kg^{-1}, 44 Wh kg^{-1} at 21.5 kW kg^{-1}, 12 mWh cm^3 at 37 mW cm^3, and 6 mWh cm^3 at 1.9 W cm^3). In addition, the electrochemical properties of the fabricated flexible SIC are almost unaffected after bending and can light up a red light-emitting diode (LED) and drive a digital watch (3 V), proving their practicality (Figure 7.11).

Similar to this work, Gao et al. [57] reported a flexible graphene fiber fabric (GFF) with good dispersion of bio-nanoparticles by a simple and effective method of microwave (MW) irradiation. By adjusting the irradiation time and the π–π conjugate structure of the GFF, the original micro-sized Bi particles can be uniformly redispersed on the GFF within five second with an average particle size of 20 nm after dispersion. The Bi/GFF composite exhibits good electrical conductivity and flexibility and can provide 113 mAh g^{-1} at a high current density of 30 A g^{-1} with excellent cycling stability. When assembled with AC/GFF and gel polymer electrolyte (GPE) the quasi-solid state SIC exhibits great mechanical–electrochemical stability, integration, long-term stability, and high energy/power density (up to 16.2 mWh cm^{-3} at 12.2 W cm^{-3} and 81.8 W cm^{-3} at 5 mWh cm^{-3}). Self-supported electrodes were prepared and flexible SIC were constructed using GFs by Dong et al. [58] and Bi et al. [59]. In their work, both exemplified further improvement of the electrochemical properties of the materials through rational structural design. Another flexible electrode was also prepared using graphene as a substrate by Zhao et al. [60] They used different graphene substrates to prepare a flexible and lightweight polyimide-graphene nanocomposite (PI/rGO–rGO) using a two-step self-assembly method, which exhibited a capacity of 225 mAh g^{-1} after 200 cycles in a half-cell, much higher than that of pure polyimide (50 mAh g^{-1}). This preparation method synthesizes a very simple and green process without

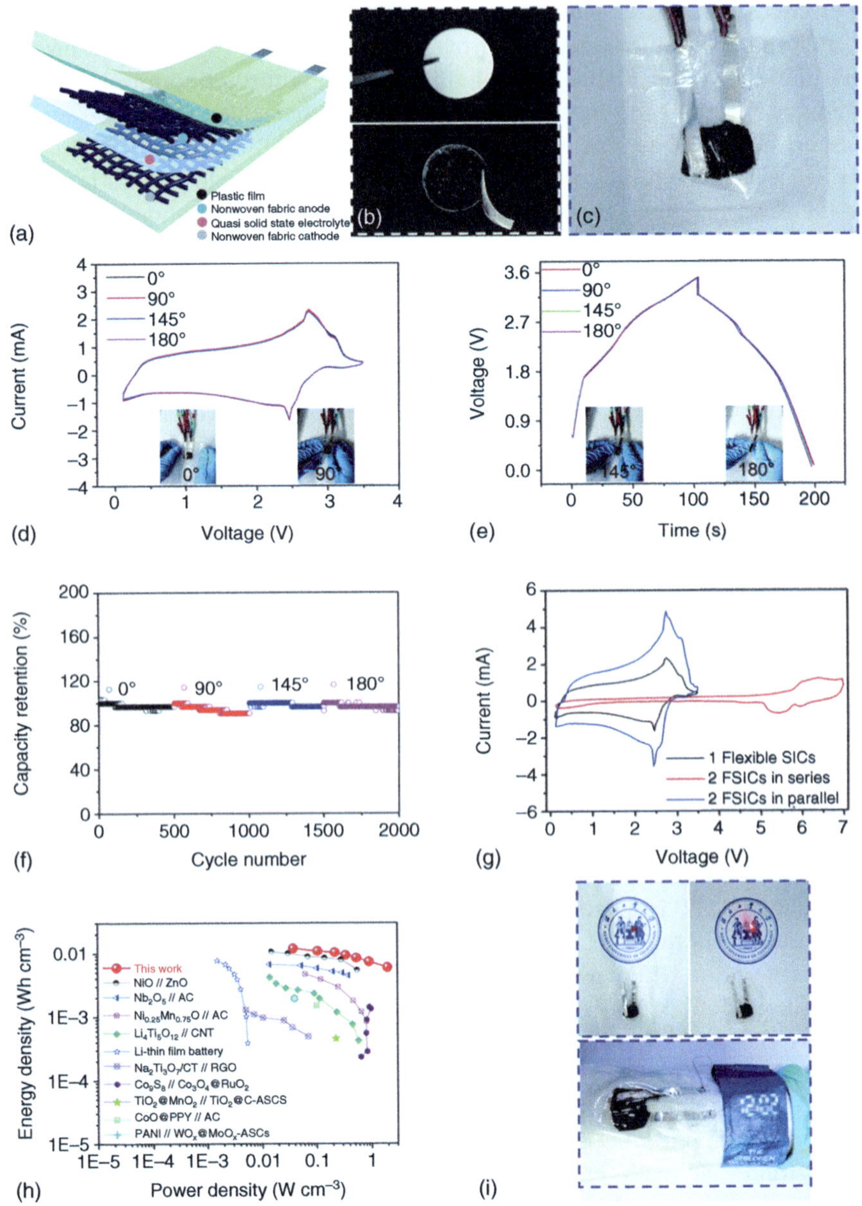

Figure 7.11 (a) Schematic illustration of the configuration of flexible SIC. (b, c) Digital images of the P(VDF-HFP) gel polymer electrolyte and the as-assembled flexible SIC. (d–g) Electrochemical properties in different bending states (d) CV curves, (e) GCD curves (f) Cycle stability, (g) CV curves of two flexible SICs devices, (h) Ragone plots, (i) Digital image of a single flexible SICs device that can light up a red LED, and the digital image of a single flexible SIC device can drive a digital watch (3 V). Source: Liu et al. [56]/from Royal Society of Chemistry.

introducing any environmentally harmful chemicals and can also be scaled up to large-scale production of high-performance stand-alone polyimide-graphene nanocomposite flexible electrodes, making commercialization feasible. Furthermore, the flexible reduced graphene oxide (rGO) films they prepared exhibit a much higher capacity (70 mAh g^{-1}) than commercial activated carbon (44 mAh g^{-1}) when used as a cathode. A flexible SIC was assembled when PI/rGO–rGO was used as the anode and rGO as the cathode; a high energy density of 55.5 Wh kg^{-1} was achieved at a power density of 395 W kg^{-1} and 21.5 Wh kg^{-1} at a power density of 3400 W kg^{-1}, demonstrating excellent electrochemical performance. This work demonstrates an innovative and scalable strategy for the synthesis of next-generation polyimide-graphene composite flexible electrode materials.

7.1.1.3 Flexible Electrodes Based on Carbon Cloth

Carbon cloth, a commercially available woven carbon fiber fabric, is considered an excellent substrate for constructing flexible electrodes due to its simultaneous advantages of high electrical conductivity, porous network, large surface area, and good mechanical flexibility and strength [61]. Therefore, flexible electrodes based on carbon cloth without binder, additives, and current collector can show excellent sodium storage capacity.

Recently, various reactive materials such as $Na_2Ti_3O_7$, MoO_{3-x}, $NiCo_2O_4$, SnS_2, and VO_2, have been reported to be grown or coated directly on the surface of carbon cloth [62–66]. For example, Mo et al. reported the growth of 3D porous $NiCo_2O_4$ nanowire arrays (denoted as NCO@CFC) on flexible carbon cloth by solution method and annealing treatment [67]. Due to the unique layered structure of the composite electrode, the flexible anode exhibited high sodium storage capacity (615 mAh g^{-1} at 50 mA g^{-1}), stable cycling performance (81% retention after 50 cycles), and excellent rate capability (363 mAh g^{-1} at 400 mA g^{-1}). In addition, Balogun et al. synthesized VO_2 nanowires grown on carbon cloth coated with carbon quantum dots (CQDs), which showed high sodium storage capacity (320 mAh g^{-1} at 0.3 C), good rate capability (133 mAh g^{-1} at 60 C) when used as a flexible cathode [68]. For carbon-cloth-based flexible electrodes, carbon cloth provides a stable support structure for the active material and a fast transport pathway for electrons. Meanwhile, the porous nature, loose texture, and open space between adjacent fibers facilitate the penetration of electrolyte, thus increasing the contact area between the active material and electrolyte and providing more reaction sites. In addition, the flexible anode made of carbon cloth usually shows improved cycling stability and rate performance compared to NCO electrodes prepared by conventional slurry coating methods, which indicates that this flexible electrode is a promising candidate for flexible SICs. Dong et al. developed a simple hydrothermal process, combined with annealing treatment to grow $Na_2Ti_3O_7$ nanosheet arrays with strong adhesion on carbon textile [69]. The unique flexible 3D $Na_2Ti_3O_7$ nanosheet array/carbon textile electrode exhibited ultrahigh-rate capability (110 mAh g^{-1} at 3 A g^{-1}) and enhanced cycling performance (70% capacity retention at 1 A g^{-1} for 1000 cycles). The maximum practical bulk energy density and power density reached 1.3 mWh cm^{-3} and 70 mW cm^{-3}, respectively, when thin and flexible SICs assembled with 3D flexible $Na_2Ti_3O_7$ nanosheet arrays/carbon fabrics as anodes and GF films as cathodes. This work demonstrates for the first time the fabrication of flexible SICs as high-performance flexible energy storage devices, which also provides a general strategy to assemble a wide range of flexible configurations.

Based on the same approach, Liu et al. also fabricated TiO_2 nanorods/carbon fiber cloth (CFC) composites by a simple hydrothermal method and subsequent heat treatment [70]. The TiO_2/CFC electrode has a discharge capacity of 148.7 mAh g^{-1} after 2000 cycles at a current density of 1 A g^{-1} and the SIC exhibits good electrochemical properties and excellent flexibility. Several other flexible SICs constructed based on flexible carbon cloth have also been shown to be able to operate at different bending degrees without compromising the electrochemical performance, such as the in situ growth of $CoSe_2$@vertical-oriented graphene (VG) layered structures on carbon cloth by a combination of plasma-enhanced chemical vapor deposition and wet chemistry steps by Xia et al. [53]. When the $CoSe_2$@VG/CC anode was coupled with the AC cathode to assemble a SIC, the electrochemical performance of this flexible SIC was almost unaffected in the bent state. The same growth of active materials on carbon cloth substrates by Gu et al. and Chen et al. also obtained excellent electrochemical properties when constructing flexible SICs together with capacitive cathodes.

7.1.1.4 Flexible Electrodes Based on MXenes

2D transition metal carbon/nitrides, or MXenes, with competing properties, such as metallic conductivity, abundant surface functional groups and hydrophilic properties, and high mechanical modulus have attracted a lot of attention. The rich chemical composition and structure of MXenes and the good flexibility characteristics shown by MXenes-derived films and corresponding composites have attracted the interest of researchers, leading to a lot of research. The physical and mechanical properties of MXenes, and to minimize the detrimental effects of MXenes own defects such as stacking and aggregation, several strategies have been developed including the creation of porous nanostructures and the integration of pillar molecules into the interlayer space of MXene sheets [71, 72]. Gogotsi group reported a method of forming MXenes nanosheets by combining them with sacrificial PMMA spherical templates and self-assembly integrated together to form 3D macroporous MXene films [73]. The 3D macroporous MXene films exhibit good flexibility, high porosity, and high electronic conductivity. Stand-alone macroporous MXene films with three different compositions ($Ti_3C_2T_x$, V_2CT_x, and Mo_2CT_x) can be directly used as binder-free anodes, which provide higher reversible capacity and better rate performance when compared to multilayer MXene films and MXene/CNT composites.

Based on the unique properties of MXene, self-assembly of active materials with it into thin films is also a feasible strategy. Wang's group reported the fabrication of freestanding and flexible $Sb_2S_3/Ti_3C_2T_x$ thin films for high-performance sodium storage anodes by electrostatic self-assembly of negatively charged MXene and positively charged PDDA-Sb_2S_3 nanowires (Figure 7.12) [74]. The $Ti_3C_2T_x$ MXene

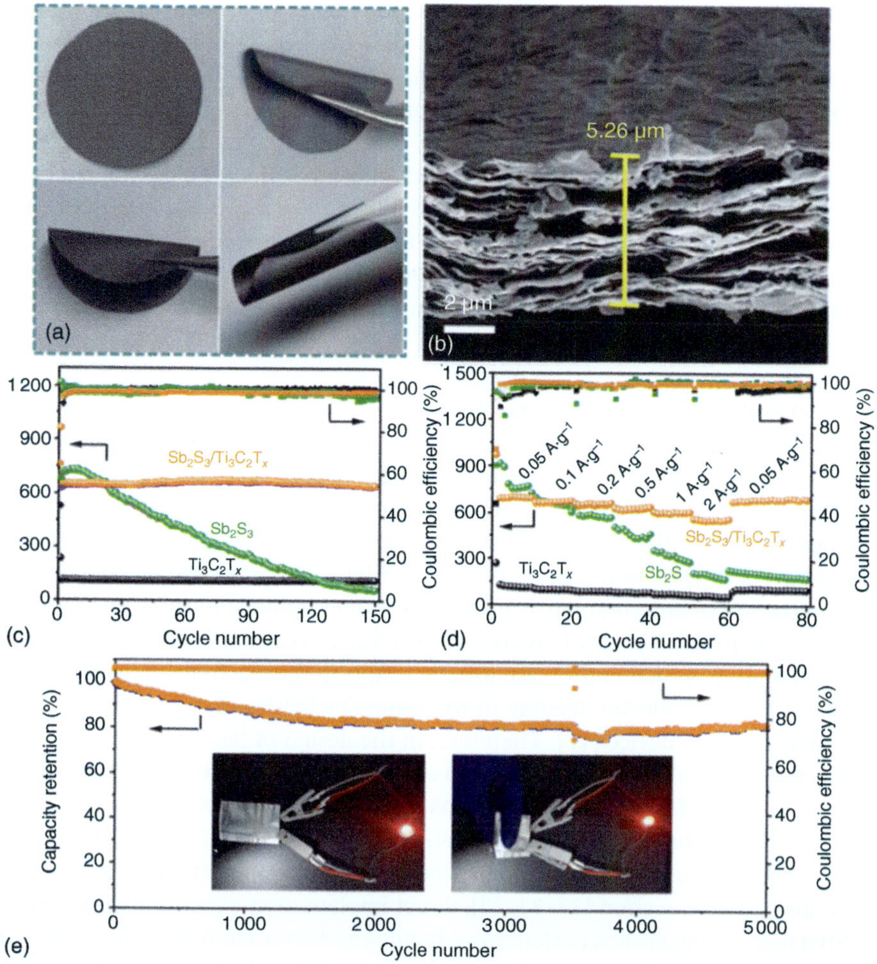

Figure 7.12 (a) Digital photographs of the flexible $Sb_2S_3/Ti_3C_2T_x$ film. (b) Cross-sectional SEM images of the films. (c) Cycling performance at 0.1 A g^{-1}. (d) Rate capability. (e) Cycle performance of the flexible SIC at 1 A g^{-1}; insets: photographs of the SIC that can power a LED indicator under different bending conditions. Source: Yang et al. [74]/Reproduced with permission from Springer Nature.

in this composite electrode improved the charge transfer kinetics, suppressed the volume expansion of Sb_2S_3 and inhibited the dissolution of polysulfide sodium, and the Sb_2S_3 nanowires prevented the repacking of $Ti_3C_2T_x$ nanosheets and promoted the ion diffusion within the thin film electrode. Due to the synergistic effect of 1D Sb_2S_3 nanowires and 2D MXenes, the composite anode exhibited excellent rate performance (553 mAh g^{-1} at 2 A g^{-1}) and cycling stability (465 mA g^{-1} at 1.0 A g^{-1} maintained after 500 cycles) in Na^+ ion storage. In addition, the flexible SICs using $Sb_2S_3/Ti_3C_2T_x$ anodes and AC/rGO cathodes provide excellent energy and power densities as well as outstanding cycling performance with a capacity retention of 82.78% after 5000 cycles at 1 A g^{-1}. The Na^+ ion storage performance of

MXene bonded electrodes was found to be much higher than that of conventional polyvinylidene fluoride bonded electrodes due to the 3D conductive network constructed by MXene nanosheets and the corresponding favorable structural flexibility resulting from the tremendous advantage of 2D MXene in terms of electron conductivity. From this perspective, combining high capacity and of some promising nanomaterials (e.g. TMOs/sulfides/phosphides) with highly conductive MXenes in a well-designed structure or architecture is a promising strategy for the construction of flexible MXene-based integrated electrodes with high electrochemical/mechanical properties [75]. Therefore, a rational MXene electrode structure is important for designing MXene-based flexible electrodes to develop high-performance anodes.

7.1.1.5 Flexible Electrodes Based on Metal Foil

Metal foil as a 2D planar current collector or substrate is considered one of the most important direct choices for constructing electrodes for flexible energy storage devices due to its excellent electrical conductivity, structural designability, and commercial availability [76]. Flexible electrodes using metal foils as current collectors have a stacked multilayer structure similar to conventional electrodes, but without the use of any conductive agents or polymer binders. Flexible energy storage devices require the ability to withstand bending at a certain angle and frequency during operation, which in this case usually results in the electrode material falling out of the current collector, leading to weakening or failure of the electrochemical performance of the device [77]. Therefore, in the design of flexible electrodes, it is usually necessary to meet the requirements of having a certain fatigue strength of the electrode under deformation conditions as well as higher mechanical stability and stronger adhesion between the metal substrate and the active material. Considering the key problems posed by conventional electrode fabrication techniques mentioned in the introduction section [33, 34, 42], direct implantation of metal foil into flexible SICs usually encounters performance failures and even safety hazards; there is a need to develop flexible metal foil-based electrodes. The solution is usually surface treatment (including in situ conversion, hydrothermal treatment, and electrodeposition) on metal foil to align nanostructures and nanoarrays. In contrast to the conventional slurry coating method, neither binders nor conductive additives are used in the preparation of metal foil-based flexible electrodes. So far, various metal foils such as Cu foil, Ti foil, Nb foil, and stainless steel have been used as substrates for the construction of flexible SIB and flexible SIC electrodes. Some active materials such as TiO_2 nanotube arrays, $Na_2Ti_3O_7$ nanoarrays, $Na_2Ti_2O_4(OH)_2$ nanosheets, CuO nanostructures, Cu_2Sb/Cu, Cu_2Se, 3D porous Sn, SnO_2 structures, Cu_3P nanowires, and SnS_2 nano wall arrays have been fabricated as binder-free metal substrate electrodes [37, 77–82].

Titanium foil, with its high electrical conductivity, relatively low cost, and high mechanical and chemical stability, has been used as a substrate for the construction of flexible electrodes [83]. It is often used to directly grow titanium-based compounds in nanoarray structures of flexible electrodes for SIBs and SICs. Among them, TiO_2 is considered as a promising material for sodium storage because of

its network structure with abundant interstitial sites to accommodate sodium ions and convenient ion diffusion, low intercalation potential (0.6 V vs. Na^+/Na), and high theoretical capacity (335 mAh g^{-1}) [15, 84]. To improve the inherent low conductivity and sluggish ion dynamics of TiO_2 materials, encapsulated in carbon fibers by electrospinning as mentioned above, in addition to this, the growth of self-supported TiO_2 nanotube arrays on Ti foil has been proposed as a free-standing and binder-free electrode. For example, Ni et al. synthesized S-TiO_2 nanotube arrays on Ti foil by a simple electrochemical oxidation and sulfidation treatment [77]. This self-supporting nanoarray shows intimate adhesion between S-TiO_2 and Ti foil, which not only inhibits the detachment of S-TiO_2 from the current collector, but also enables fast charge transfer. The ordered nanotube array structure provides a large contact area for effective electrolyte access and an adequate pathway for rapid diffusion of Na^+ ions. Meanwhile, the S doping effectively improves the electronic conductivity, thus enhancing the electrochemical activity of TiO_2 and increasing the electronic conductivity within the nanotubes. These attractive advantages enable the flexible S-TiO_2 anode to have excellent Na^+ ion storage performance, providing ultra-stable cycling performance (136 mAh g^{-1} after 4400 cycles at 136 mA g^{-1}) and ultrahigh-rate capability (167 mAh g^{-1} at 3350 mA g^{-1}).

The use of titanium foil as a substrate for in situ growth or transformation of active materials was likewise realized in Liu's group and Zhi's group, where Liu's group reported a simple and versatile hydrothermal method for direct growth of $NaTi_2(PO_4)_3$ nanowires on Ti foil current collectors, a strategy that uses only a single solute solution and does not involve high-temperature treatment, unlike the previously reported solid-state reaction route and sol–gel treatment [55]. In the absence of conventional binders and additives, these nanowires are grown directly oriented on the collector, providing many convenient electrons transport pathways along the 1D direction. The spacing between the nanowires not only allows sufficient electrolyte penetration to shorten the ion diffusion paths and thus reduce the interfacial resistance/polarization, but also buffers the small volume expansion of the NTP upon Na^+ ion insertion. Furthermore, this 3D nanowires anode has a strong adhesion to the collector due to the spacing between individual nanowires that can effectively release the generated stress, and thus can be more robust to suffer from mechanical deformation. As a result, this NTP anode exhibits an ultrafast charging capability and good cycling stability comparable to that of a pure capacitive AC cathode over a rather wide rate range of 15–150 C (1 C ≈ 133 mA g^{-1}). When utilizing polyvinyl alcohol-sodium perchlorate (PVA-$NaClO_4$) as a hydrogel quasi-solid electrolyte, they assembled a flexible AC/NTP SIC device. It exhibited a high voltage of 2.4 V, good rate performance, high energy and power densities (5.8 and 999 mW cm^{-3}; 62.1 Wh kg^{-1} and 9538 W kg^{-1}), and excellent cycling stability (10 000 cycles). Zhi's group developed a synthetic strategy for the liquid conversion of solid titanium sources to NTP successfully synthesized porous single-crystal $NaTi_2(PO_4)_3$ (PSC-NTP) and coated the surface of PSC-NTP (PSC-NTP@C) with an amorphous carbon layer to enhance its electrical conductivity (Figure 7.13) [85]. In a three-electrode cell with an aqueous electrolyte, PSC-NTP@C exhibited a high initial discharge capacity of 105 mAh g^{-1} at 500 mA g^{-1} and decayed by only

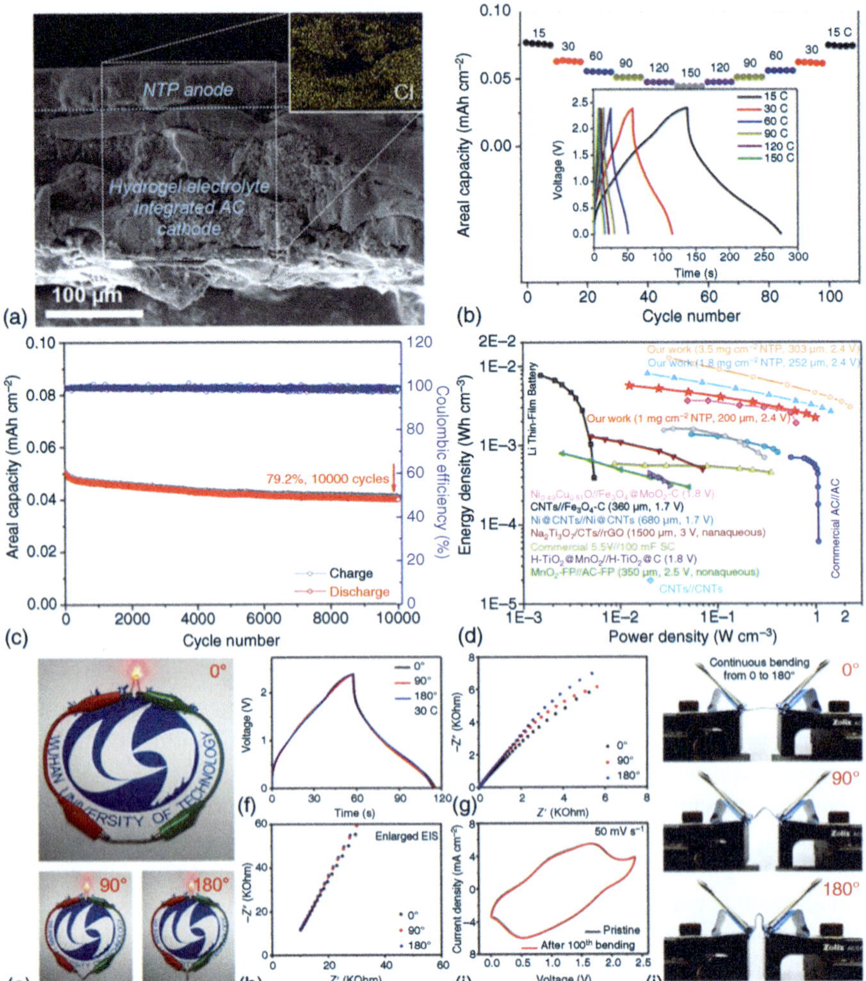

Figure 7.13 AC//NTP quasi-solid-state device. (a) Cross-sectional SEM image, the inset is Cl element distribution. (b) Rate performance, the inset is GCD curves. (c) Cycling performance. (d) Volumetric Ragone plot. (e) Optical images of the device powering LED indicator at different bending angles. (f) GCD curves, (g, h) EIS spectra (0.01 Hz–100 kHz) upon bending. (i) CV curves of the device before and after 100 times continuous bending. (j) Optical images showing the continuous bending equipment. Source: Tan et al. [55]/ Reproduced with permission from John Wiley & Sons.

0.22% in the subsequent 100 cycles. When integrated with ZIF-8-derived N-doped carbon (NC) in a liquid electrolyte, the assembled flexible SIC achieved a good first discharge capacity of 89 mAh g^{-1} at 0.5 mA g^{-1} and maintained a high value of 63 mAh g^{-1} even after 400 cycles.

7.2 Flexible Capacitive Anode and Battery-Type Cathode SICs Cell Configurations

In general, two types of SICs have been reported. The first one is the electrolyte-consuming SIC configuration described above, which works on the principle that both anions and cations in the electrolyte act as charge carriers during the charging/discharging process. The typical anode involves a Faraday Na^+ ion intercalation process, while the cathode is an EDLC-type carbon material. The second type of SIC is the opposite of the first configuration, where the anode undergoes a rapid capacitive process while the cathode is a typical battery-type Na^+ ion insertion material, referred to as a rocking-chair SIC [86]. That is, a hybrid storage mechanism different from that of SIB is involved in the rocking-chair SIC. So far, most of the work has focused on the first type of SICs, where various excellent sodium storage anodes (e.g. $Na_2Ti_3O_7$, Nb_2O_5, TiO_2, $NaTi_2(PO_4)_3$, MoS_2, and $MoSe_2$) are rationally designed to match the fast kinetics of the capacitive AC cathode. However, the low capacity of the AC cathode (typically below 60 mAh g^{-1}), which is known from the equation ($1/C_{SIC} = 1/C_{cathode} + 1/C_{anode}$), leads to a low energy density of the SIC device.

Furthermore, electrolyte-consuming SIC devices are kinetically slow in Na^+ insertion into the anode material since they are coupled with a high-capacity intercalation-based anode and a high-rate EDLC cathode, which can easily lead to kinetic differences between the two electrodes of SICs. The work in Section 7.1 has focused on trying to improve the rate performance of the anode to overcome the kinetic imbalance between the two electrodes in SICs. Although high energy power performance has also been achieved with their efforts, these SICs consume a large amount of electrolyte during charging and may also form some side reactions at high potentials, further leading to lower energy density. In addition, in conventional SICs, the specific capacity of the porous carbon cathode (usually AC) is much lower than that of the battery-type anode. In order to achieve charge balance between the two electrodes, more carbon cathode material is required, which inevitably consumes more electrolyte and also reduces the volumetric capacity of the whole device. To overcome these problems, rocking-chair SIC configurations have been developed in which the cathode involves Na^+ intercalation/de-intercalation, while the anode corresponds to the capacitive desorption/adsorption of Na^+ ions on the electrode surface. For example, Yamada et al. reported a prototype SIC device using an all-aluminum $Na_2Fe_2(SO_4)_3$ cathode and an MXene Ti_2C anode at 2.4 V. Chen et al. demonstrated that by coupling an Al_2O_3-coated $Na_{0.66}Mn_{0.54}Ni_{0.13}Co_{0.13}O_2$ cathode with a commercial CA anode in an organic electrolyte containing Na^+ coupling, a stable SIC device is designed. In addition, $Na_{0.44}MnO_2$ nanorods and $Na_3V_2(PO_4)_3$ (NVP)/N-doped mesoporous carbon nanosheets were also used to construct SICs. The energy density of these SICs was significantly increased, but the corresponding power density strongly depended on the Na^+ intercalation/de-intercalation kinetics of the cathode material. Among various Na-containing compounds, the NVP and fluorophosphate NPF with a Na superion conductor (NASICON)-type structure is considered an important sodium storage cathode material because of its 3D stable crystal structure, high discharge plateau, good thermal stability, and high theoretical energy (≈501 Wh kg^{-1}).

Wang's group designed and fabricated an N-doped mesoporous carbon nanosheet array on a flexible substrate by using an array of metal organic frameworks (MOFs) as a reaction template [87]. The interconnected Co/Zn MOF nanosheets anchored on the flexible CFC surface were first converted into ordered mesoporous carbon nanosheet arrays (mp-CNSs). On this basis they deposited VO_2 and NVP to form interconnected VO_2@mp-CNSs/CFC and NVP@mp-CNSs/CFC for direct use as electrodes without any conductive additives or binders. In these hybrid electrodes, VO_2 or NVP nanoparticles are encapsulated by a highly graphitized carbon shell within the mp-CNSs support. This unique electrode design was shown to offer multiple advantages: (i) suppressing the aggregation of electroactive nanoparticles, (ii) accommodating volume changes during cycling, (iii) maintaining high electrical and ionic conductivity throughout the electrode, and (iv) generating robust flexibility. Thus, the VO_2@mp-CNSs/CFC electrode is shown to have pseudocapacitive properties for Na^+ ion storage, while the battery-type NVP cathode also exhibits ultrahigh-rate charge and discharge properties equivalent to those of a graphene-based SC. Finally, quasi-solid-state hybrid SICs were assembled by simultaneously utilizing flexible VO_2@mp-CNSs/CFC as the anode, NVP@mp-CNSs/CFC as the cathode, and Na^+ ion-conducting gel polymer as the electrolyte. This hybrid SIC device can provide high energy density (≈161 Wh kg^{-1} and 8.83 mWh cm^{-3}) and high power density (24 kW kg^{-1} and 1.32 Wh cm^{-3}), and its performance almost exceeds that of previously reported hybrid devices (Figure 7.14).

Among various sodium energy storage cathode materials, high-voltage NVPF shows great promise because of its high theoretical energy density (501–507 Wh kg^{-1}), which is much higher than its analogue-NVP (394 Wh kg^{-1}). However, the power capability and cycling stability of NVPF is significantly lower than NVP due to its significantly lower electron conductivity. Therefore, the typical strategy is to hybridize NVPF powders with various conductive carbons (e.g. rGO and graphene quantum dots) or ruthenium oxide. On the other hand, it has been found that neatly aligned nanoarrays grown directly on highly conductive substrates have been considered as a new generation of high-performance electrodes due to their special structural advantages. However, so far, NVPF arrays for SICs are still in their infancy due to the complex composition and uncertain directional nucleation. Wang's group used a one-step hydrothermal process to grow large-area, uniform, and well-aligned NVPF arrays on flexible freestanding CNF membranes [88]. Compared to other carbon substrates, the 1D CNF structures can form effective current pathways, shorten ion diffusion lengths, and provide high surface areas, increasing large electrolyte-electrode contact areas. The growth process of NVPF can be controlled by varying the reaction time. This electrode design has unique advantages in the Na^+ extraction/insertion process. First, the open space within the nanoarray promotes electrolyte penetration and Na ion diffusion. In addition,

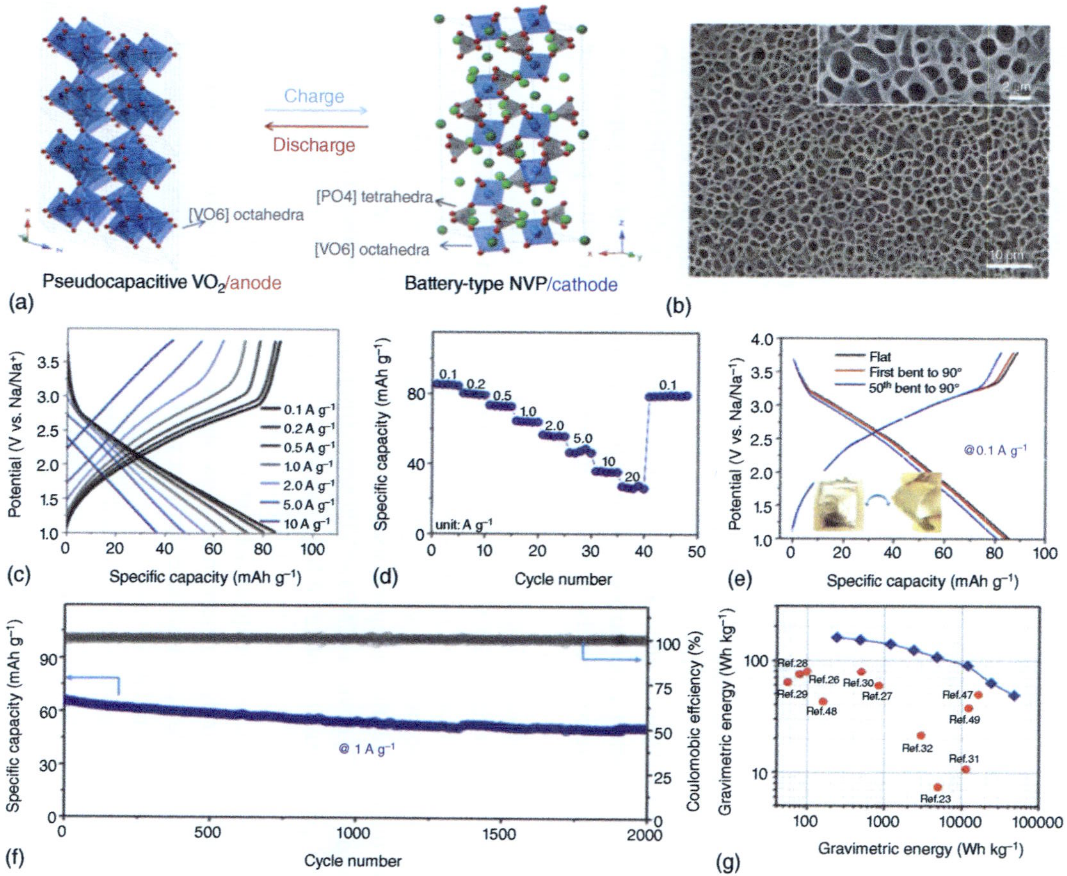

Figure 7.14 SIC full device. (a) Schematic illustration of the SIC device using flexible NVP@mp-CNSs/CFC and VO_2@mp-CNSs/CFC as cathode and anode. (b) SEM images of the P(VDF-HFP) membrane (porous gel polymer). (c) GCD curves and (d) Rate performance of the as-assembled SIC at different current densities. (e) GCD curves of the quasi-solid-state SIC device under flat and bent (90°) status (shown as inset). (f) Cycling performance at 1 A g^{-1} and (g) Ragone plots of the flexible quasi-solid-state SIC. Source: Xu et al. [87]/Reproduced with permission from John Wiley & Sons.

the tight connection between the array and the conductive CNF substrate ensures excellent conductivity and high structural stability of the NVPF electrode. When used as a Na^+ storage cathode (2–4.5 V vs. Na/Na^+), the NVPF array electrode exhibits high-rate performance and good cycling stability. The specific capacity of CNF@NVPF electrode was determined to be about 120 mAh g^{-1} (based on the total mass of NPF and CNF) at a low rate of 0.2 C, which is almost equal to the value of the NPF powder electrode (≈117 mAh g^{-1}). As the rate increases to 20 C, the CNF@NVPF electrode can still maintain a high capacity of 56 mAh g^{-1}, which is higher than that of the NPF powder (14 mAh g^{-1} at 20 C). This indicates that the 3D array electrodes facilitate better rate performance compared to the powder. Cycling stability was further tested at different rates. After 100 cycles at 0.2 C, the specific capacity of CNF@NVPF remained as high as 104 mAh g$^-$1 (87% retention), which is better than the NPF powder electrode (92 mAh g^{-1}, 78% retention). After 500 cycles at 5 C, the specific capacity of CNF@NVPF is still 66 mAh g^{-1}, which greatly exceeds the capacity value of NPF (35 mAh g^{-1}), and they also discuss the importance of high area capacity in terms of practical applications. At a high current density of 0.17 mA cm^{-2}, the CNF@NVPF electrode maintains a high area capacity of 0.78 mAh cm^{-2}, which is much higher than that of the NPF powder (0.09 mAh cm^{-2}). To match the battery-type cathode (CNF@NVPF) in SICs, a capacitor-type carbon anode is also required. Most of the capacitive materials used are commercial activated carbons, but their capacitive performance is relatively low. Therefore, they synthesized PCNFs. The PCNF electrodes have a high specific capacity of 228 mAh g^{-1} at 0.1 A g^{-1} and maintain a high capacity of 135.7 mAh g^{-1} even at 20 A g^{-1} (60% retention with respect to 0.1 A g^{-1}), which demonstrates their excellent rate capability. When combined with a synthetic highly interconnected porous GPE with higher ionic conductivity and greater electrolyte uptake than conventional separators, the fabricated quasi-solid sodium ion hybrid capacitors (SIHC) devices (CNF@NVPF arrays//porous carbon fibers) provide high energy//power densities of 227 Wh kg^{-1} and 4936 W kg^{-1} (Figure 7.15), respectively, more importantly, the use of P(VDF-HFP)-based polymer electrolyte, the energy storage performance of the quasi-solid state SIC is even better than that of the SIC with a liquid electrolyte (commercial glass fiber separator.) Thus, the sodium-ion hybrid capacitor device constructed in their work combines the triple advantage of high performance and good safety.

In addition, his group has proposed a simple encapsulation strategy through CNFs to significantly improve the electrochemical performance of high-voltage NVPF with theoretical energy densities up to 501–507 Wh kg^{-1} [89]. Thanks to the high structural stability of the CNF framework not only the performance of NVPF can be significantly improved but also a self-supporting flexible electrode can be obtained. As a result, the CNF@NVPF electrode achieves excellent storage performance (96% capacity retention at 10 C for 2000 cycles). This cycling stability is significantly better than that of previously reported NVPF materials (Figure 7.16). Based on CNF@NVPF as the cathode and PCNFs as the anode, the assembled SIC full device exhibited an ultrahigh energy density of 182.8 Wh kg^{-1}. Even at

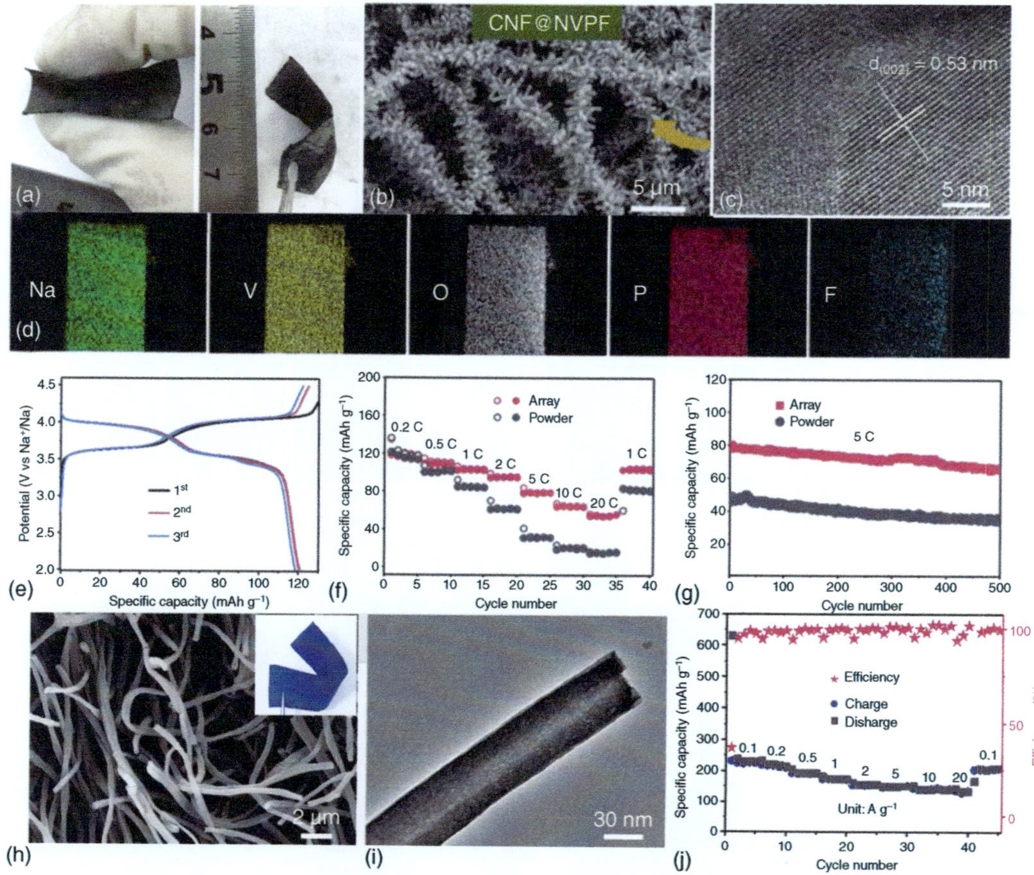

Figure 7.15 (a) Photograpy of carbon nanofiber membrane CNF@NVPF. (b) SEM images. (c) TEM images. (d) Elemental mapping images of individual NVPF. (e) Charge/discharge curves of CNF@NVPF at 0.2 °C. (f) Rate capability for CNF@NVPF and bulk-type NVPF powder electrodes. (g) Cycling stability at 5 °C. The capacitive anode material-porous carbon nanofibers (PCNF). (a) SEM, (h, i) TEM images, and (j) rate capability. Source: Mao et al. [88]/Reproduced with permission from John Wiley & Sons.

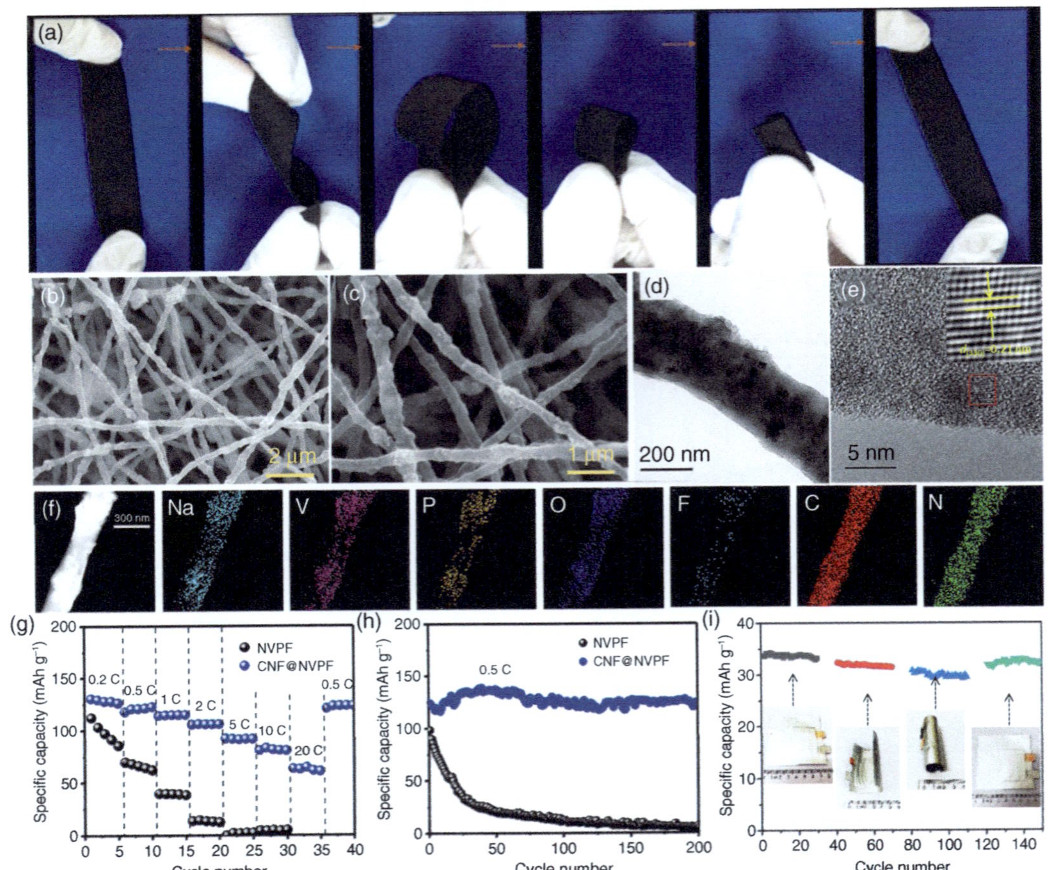

Figure 7.16 (a) Photographs under various bending states of the flexible freestanding CNF@NVPF membrane. (b, c) SEM images. (d, e) TEM images. (f) EDS mapping images. Electrochemical properties of the CNF@NVPF. (g) Rate capability of CNF@NVPF and NVPF powders. (h) Cycling performance. (i) Specific capacities of the pCNF//CNF@NVPF all-flexible pouch cell collected at various bending conditions. Source: Qiu et al. [89]/Reproduced with permission from Elsevier.

the high power of 5898 W kg^{-1}, the corresponding energy value is still as high as 85.3 Wh kg^{-1} as well as a low self-discharge rate.

7.3 Electrolytes in Flexible SICs Devices

To meet the safety, flexibility and versatility requirements of advanced energy storage devices, polymer electrolytes (GPEs) are the best candidates to replace liquid electrolytes because they have the advantages of wide electrochemical window, good thermal stability, and reduced risk of electrolyte solution leakage [90, 91]. In general, GPEs with excellent mechanical flexibility can fit well with the requirements of flexible energy storage devices based on flexible electrodes, moreover that GPEs have higher ionic conductivity than solid electrolytes at room temperature [92]. In GPEs, the liquid electrolyte is immobilized in a polymer matrix, which largely reduces the risk of leakage compared to commercial separators, thus improving the safety of the device. GPEs have attracted increasing attention by combining the advantages of liquid and solid components and by having the dual nature function of electrolyte and separator. The superiority of this combination is reflected in the high ionic conductivity and good interfacial properties of the liquid phase and the good mechanical properties of the solid component. GPEs can be distinguished as heterogeneous (phase-separated) and homogeneous (homogeneous) gels. In general, heterogeneous GPE is the most common and represents a polymeric framework with an interconnected pore system filled with electrolytes. Thus, for heterogeneous GPEs, the transport of electrolyte ions occurs in the swollen gelled or liquid phase. Most GPEs exhibit outstanding ionic conductivity of about 10^{-3} S cm^{-1} at ambient temperature, which can improve the electrochemical performance of GPEs-based energy storage devices [33, 93, 94]. Therefore, GPEs have become one of the most desirable alternatives among various electrolytes for the fabrication of advanced energy storage devices with enhanced safety and flexibility.

In Sections 7.1 and 7.2, flexible electrodes for flexible SICs were highlighted, however, for building flexible SICs, conventional commercial separators are difficult to meet the mechanical flexibility of the devices, and so designing GPEs with mechanical flexibility is also crucial for flexible SICs with excellent electrochemical properties. In this regard, Huang et al. blended fully degradable poly(propylene carbonate) (PPC) with tough poly(vinylidene fluoride) (PVDF) to prepare GPE for high-performance LIBs applications [95]. Electrospun PVDF/PPC polymer electrolyte membranes showed significantly higher ionic conductivity (4.05×10^{-3} S cm^{-1}) than electrospun PVDF membranes (2.11×10^{-1} S cm^{-1}). Zhao et al. developed a self-supported PVDF/P(VC-VAc) (P(VC-VAc) is polyvinyl chloride-vinyl acetate) hybrid GPE could achieve a maximum ionic conductivity of 3.57×10^{-1} S cm^{-1} when the ratio of PVDF/P(VC-VAc) was 7 : 3 [96]. Liu et al. obtained a nonsolvent induced phase separation technique based on PVDF/graphene membrane [97]. The ionic conductivity of GPE increased significantly from 1.85×10^{-3} S cm^{-1} for the pristine PVDF-based GPE to 3.61×10^{-3} S cm^{-1} based on PVDF with 0.002 wt% graphene. Due to the low crystallinity of PVDF-HFP copolymer, it is more frequently used as a polymer matrix for GPE than PVDF. Kim et al. fabricated PVDF-HFP nanofibers by electrospinning technique for GPE of flexible LIBs. PVDF-HFP nanofiber membranes with cross-linked morphology were obtained due to the availability of *N*-methyl-2-pyrrolidone. GPE based on cross-linked PVDF-co-HFP membranes showed an ionic conductivity of 2.3×10^{-3} S cm^{-1} at room temperature, and full cells assembled with PVD/HFP-based GPE showed higher specific capacity than those assembled with commercial separators. Asghar et al. prepared a microporous cellulose/PVDF-HFP membrane by phase change method as host for GPE [98]. The results showed that the GPE displayed a high ionic conductivity of 1.89×10^{-3} S cm^{-1}, a transfer number of 0.89 for Li^+ ions, and a wide electrochemical window of 5.35 V. Kim et al. fabricated a composite GPE based on the combination of PVDF-HFP and nanoscale Al_2O_3, which improved interfacial resistance, ion coordination, and electrochemical stability compared to pure GPE [99].

Currently, inorganic solid electrolytes also play an important role in flexible SICs. The flexible SICs presented above are basically PVDF/HFP-based GPEs, due to the high dielectric constants possessed by PVDF and PVDF-HFP, which show advantages in the ionization of electrolyte salts and concentration of charge carriers. Wang's group prepared a P(VDF-HFP) composite membrane, the surface and cross section of which are shown in Figure 7.17 [87], the composite membrane was gelled by a liquid electrolyte to achieve a GPE with higher ionic conductivity than a commercial separator saturated with a liquid electrolyte. In addition, the sandwich membrane exhibits good mechanical properties and the liquid electrolyte in the GPE is difficult to volatilize, which not only fits the flexible electrode but also greatly improves the safety of the device. Yu's group also constructed GPA using P(VDF-HFP) [100], and when the NTO anode and PSC cathode and the P(VDF-HFP) membrane immersed in the electrolyte solution (the color of the membrane changed from white to transparent, indicating that the organic electrolyte was successfully sandwiched in the pores of the substrate) were assembled into a quasi-solid state SIC, in order to demonstrate the flexibility of the fabricated hybrid ion device, the device was unfolded on a flat surface and rolled with four mandrels of different diameters (as shown in the insets of Figure 7.17i) to evaluate its electrochemical behavior in different bending states. Encouragingly, there is no significant capacity loss under different bending conditions, which indicates its remarkable mechanical robustness and great potential for practical applications. In addition, Figure 7.16k shows that the flexible quasi-solid-state SIC is capable of powering a commercial table lamp and a UT capacitor sign consisting of 50 red LEDs, demonstrating its high energy and high power characteristics.

In many previous reports, it has been well demonstrated that GPEs show advantages in building flexible SICs; however, their effect on the long-term electrochemical performance of SICs compared with conventional glass fiber separators has been rarely reported. Based on this Wang's group using GPEs and commercial glass fiber separators [88], the GCD curves of SICs of both separators are shown in Figure 7.18. The specific capacities of the quasi-solid state SICs were 66.8, 64.8, 62.7, 59.4, 52.3, 47.1, 44.2, and 38.4 mAh g^{-1} at 0.2, 0.5, 1, 2, 5, 8, and 15 C, respectively.

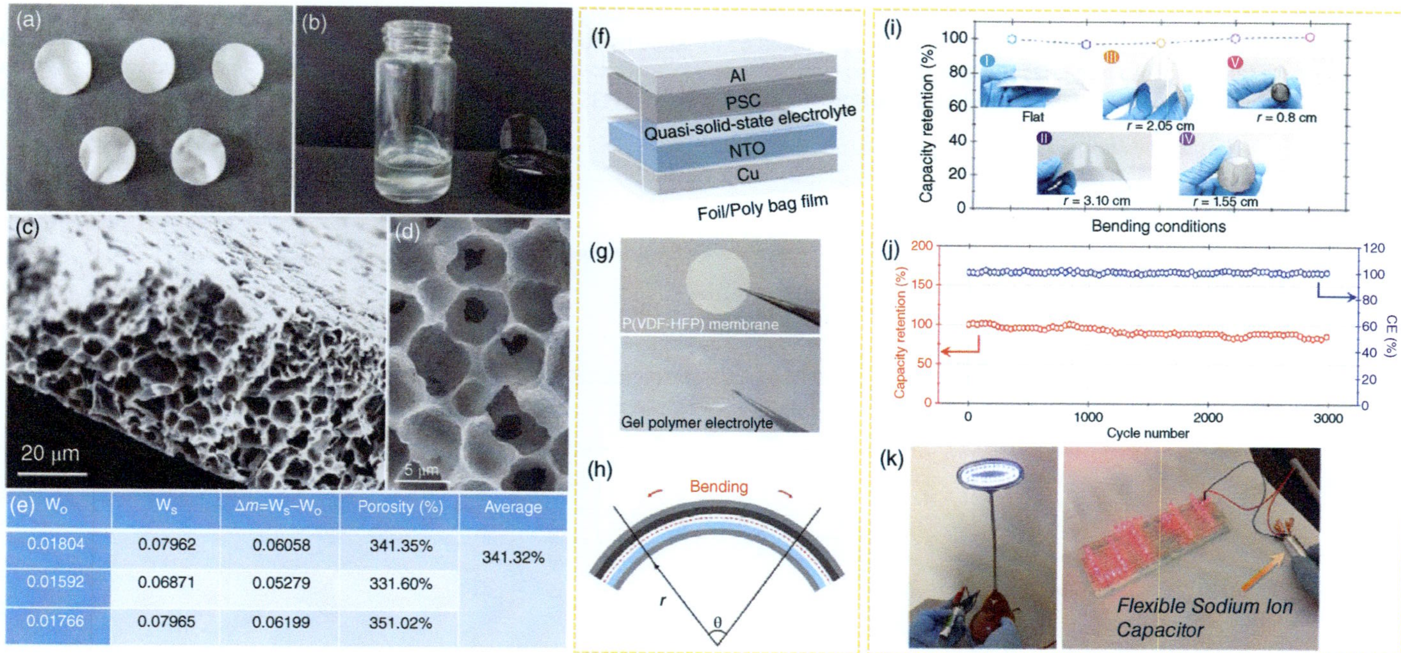

(e) W_o	W_s	$\Delta m = W_s - W_o$	Porosity (%)	Average
0.01804	0.07962	0.06058	341.35%	341.32%
0.01592	0.06871	0.05279	331.60%	
0.01766	0.07965	0.06199	351.02%	

Figure 7.17 (a) Photos of the prepared porous P(VDF-HFP) membrane (b) and its gel polymer electrolyte. (c, d) The cross-sectional SEM image of the P(VDF-HPF) GPE. (e) The mass change of the P(VDF-HFP) membrane before and after adsorbing the $NaClO_4$ (1 M)/PC/FEC (5%) electrolyte. Source: Xu et al. [87]/Reproduced with permission from John Wiley & Sons. (f) Schematic illustration of the structure of the quasi-solid-state SIC. (g) Photos of the prepared porous P(VDF-HFP) membrane and its gel polymer electrolyte. (h) Diagram of the flexible quasi-solid-state SIC under various bending conditions. (i) Stability of capacitive performance of the quasi-solid-state SIC collected at different bending conditions. (j) Cycling stability of the quasi-solid-state SIC. (k) A commercial desk lamp and a UT capacitor logo consisting of 50 red LEDs powered by the assembled quasi-solid-state SIC under rolled state. Source: Reproduced with permission Li et al. [100]. Copyright 2016, The American Chemical Society.

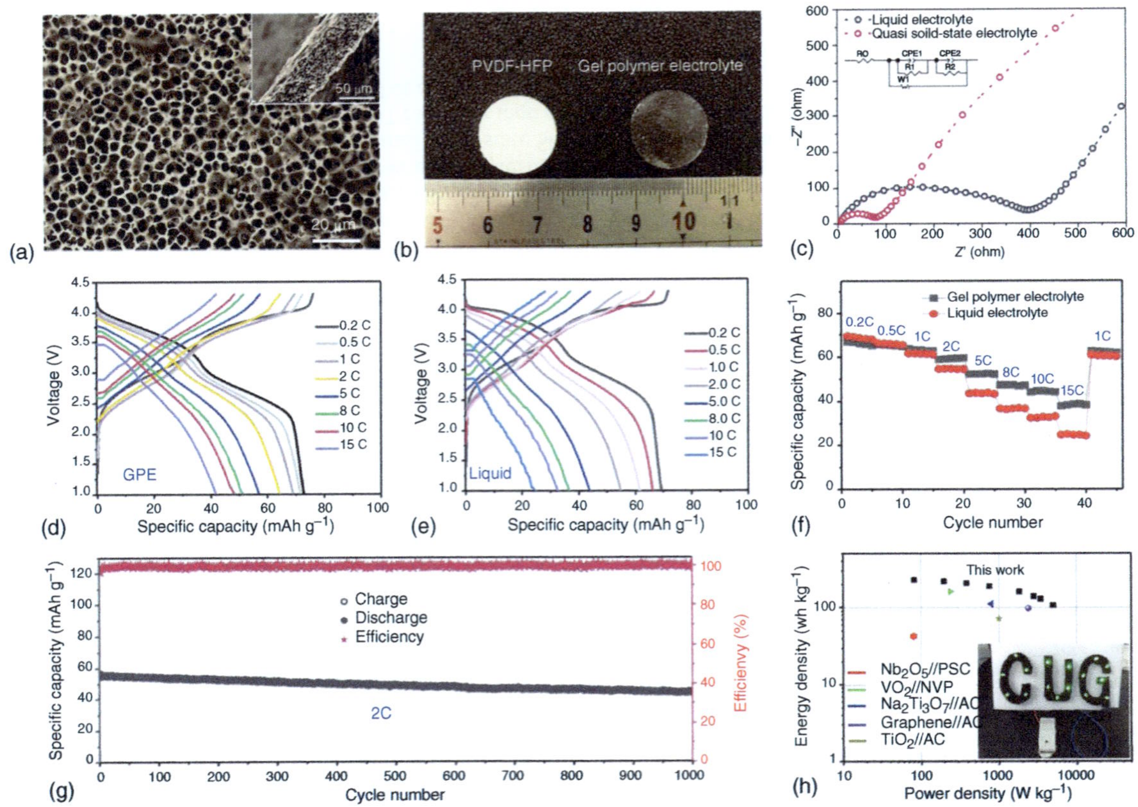

Figure 7.18 Quasi-solid-state sodium-ion hybrid capacitor. (a) SEM images and (b) photograph of P(VDF-HFP). (c) Impedance plots of the CNF@NVPF nanoarray SIHC using liquid electrolyte and quasi-solid-state electrolyte, respectively, GCD curves of sodium-ion hybrid capacitor with (d) gel polymer electrolyte and (e) liquid electrolyte. (f) Rate capability. (g) Cycling stability. (h) 16 green light-emitting diodes (LEDs) are light up by one SIHC device. Source: Mao et al. [88]/Reproduced with permission from John Wiley & Sons.

With a 75-fold increase in current density, the capacity retention is as high as 57.5%. When returned to 1 C, the discharge capacity of the quasi-solid-state SIC remained as high as 61.9 mAh g^{-1}. This rate capability of the quasi-solid-state SIC was significantly better than that of the glass fiber separator SIC (24.4 mAh g^{-1} at 15 C, with 35% retention relative to 0.2 C) (Figure 7.18). This difference can be attributed to the highly continuous porosity of the P(VDF-HFP) membrane, which can lead to lower interfacial resistance and higher sodium ion conductivity. In addition, P(VDF-HFP) is chemically polarized, which can lead to a faster turn-off of ClO_4^- ions in the porous polymer electrolyte. Their study also shows that the resistance value of the polymer electrolyte is much smaller than that of the conventional glass fiber separator. In addition, the quasi-solid SIC demonstrated excellent cycling performance with 80% capacity retention and 99% coulombic efficiency after 1000 cycles at 2 °C. Their SIC can deliver a high energy density of 227 Wh kg^{-1} at 81 W kg^{-1}. At a high-power density of 4936 W kg^{-1}, corresponding to a full charge–discharge of 2.6 minutes, the SIC can still achieve a large energy density of 107 Wh kg^{-1}. Based on the measurement results, the SIC they fabricated exhibits the dual advantages of a high-energy battery and a high-power SC. Furthermore, the ability of the SIC device with PCNF//CNF@NVPF and highly porous P(VDF-HFP)-based GPE to light up a monogram of 16 green LED lamps in parallel further demonstrates its great potential for electrochemical energy storage.

In this chapter, we systematically summarize the recent advances in flexible SICs from three aspects: flexible materials for electrolyte-consuming flexible SICs, flexible materials for rocker-type flexible SICs, and quasi-solid electrolytes. The excellent electrochemical and mechanical properties and low cost of flexible electrodes show great promise for their application in foldable, stretchable, and flexible electronic devices. Different electrode materials have their own advantages and disadvantages. For example, carbon-based materials have good cycling stability, while low capacity is a weakness for practical applications. Quasi-solid electrolytes exhibit a wider window of electrochemical stability, but lower ionic conductivity, and higher ionic conductivity electrolytes typically exhibit lower starting potentials, which limits their application at high voltages. Although flexible SICs exhibit good electrochemical performance and flexibility. However, there are still some problems to be solved, such as low energy and power density, discounted flexibility, and full-cell performance. Nowadays, basic research on electrode materials and encapsulation technologies is very limited, and there is a lack of low-cost mass production technologies and strategies to improve the flexibility of flexible devices are still in the basic research stage, and large-scale synthesis of low-cost, durable flexible electrodes by simple, gentle, and environmentally friendly methods will contribute to the eventual successful commercial application of flexible SICs. Therefore, more effective strategies are needed to improve the performance of flexible SICs. Further improvements can be made in the following areas.

(1) To date, many flexible electrode materials have been reported. However, it is still a great challenge to assemble flexible sodium ion energy storage devices with high energy density and good stability. New active materials should be designed and synthesized with distinct mechanical and electrochemical properties.

(2) In addition to materials, structure is another central factor in achieving the desired flexibility of energy storage devices. As far as the current structural design of flexible LIBs is concerned, typical structural design mechanisms include serpentine patterns, neutral planar designs, and island structures. In addition, some derivative structures inspired by the above structural designs have been reported, such as array structural designs that can enhance the deformability of materials that are inherently less deformable. Likewise, the same structural designs can potentially be applied to flexible SICs.
(3) Currently, most of the work has focused on the material design of half-cells. However, more emphasis should be placed on flexible full-cell structures to achieve electrochemical performance comparable to that of conventional SICs.
(4) General criteria for flexibility assessment are urgently needed. There is a need for a critical evaluation system that comprehensively describes the balance between stability, electrochemical performance, and flexibility, such as bending/tension limits, safety testing, and electrochemical performance under deformation.
(5) Safety and reliability are of paramount importance for flexible sodium ion energy storage devices because of the presence of flammable organic liquid electrolytes and active alkali metals. Safety and reliability are of paramount importance for flexible sodium-ion-based energy storage devices. Inspired by flexible solid-state LIBs, the design of suitable solid-state electrolytes for flexible sodium-ion based devices will bring a bright future for wearable devices.

References

1 Gao, Y., Xie, C., and Zheng, Z. (2020). *Adv. Energy Mater.* 11: 2002838.
2 Xu, T., Du, H., Liu, H. et al. (2021). *Adv. Mater.* 33: 2101368.
3 Zeng, L., Qiu, L., and Cheng, H.-M. (2019). *Energy Storage Mater.* 23: 434.
4 Zhao, Y. and Guo, J. (2020). *InfoMat* 2: 866.
5 Manthiram, A. (2020). *Nat. Commun.* 11: 1550.
6 Yang, C., Xin, S., Mai, L., and You, Y. (2020). *Adv. Energy Mater.* 11: 2000974.
7 Yi, T.-F., Sari, H.M.K., Li, X. et al. (2021). *Nano Energy* 85: 105955.
8 Fang, Y., Luan, D., and Lou, X.W.D. (2020). *Adv. Mater.* 32: 2002976.
9 Liu, Q., Hu, Z., Li, W. et al. (2021). *Energy Environ. Sci.* 14: 158.
10 Dou, X., Hasa, I., Saurel, D. et al. (2019). *Mater. Today* 23: 87.
11 Yang, X. and Rogach, A.L. (2020). *Adv. Energy Mater.* 10: 2000288.
12 Rajagopalan, R., Tang, Y., Jia, C. et al. (2020). *Energy Environ. Sci.* 13: 1568.
13 Shao, H., Wu, Y.C., Lin, Z. et al. (2020). *Chem. Soc. Rev.* 49: 3005.
14 Zhang, Y., Jiang, J., An, Y. et al. (2020). *ChemSusChem* 13: 2522.
15 Cha, G., Mohajernia, S., Nguyen, N.T. et al. (2020). *Adv. Energy Mater.* 10: 1903448.
16 Li, H., Zhang, X., Zhao, Z. et al. (2020). *Energy Storage Mater.* 26: 83.
17 Mishra, K., Yadav, N., and Hashmi, S.A. (2020). *J. Mater. Chem. A* 8: 22507.
18 Han, C., Tong, J., Tang, X. et al. (2020). *ACS Appl. Mater. Interfaces* 12: 10479.
19 Jamesh, M.-I. (2019). *J. Energy Chem.* 32: 15.

20 Cai, P., Zou, K., Deng, X. et al. (2021). *Adv. Energy Mater.* 11: 2003804.
21 Jin, T., Han, Q., and Jiao, L. (2020). *Adv. Mater.* 32: 1806304.
22 Luo, Y., Wang, K., Li, Q. et al. (2020). *Small* 16: 1902719.
23 Liu, Y., Deng, Q., Li, Y. et al. (2021). *ACS Nano* 15: 1121.
24 Zhu, S., Sheng, J., Chen, Y.n. et al. (2020). *Mater. Sci.* 8: 261.
25 Chang, Z., Qiao, Y., Wang, J. et al. (2021). *J. Mater. Chem. A* 9: 4870.
26 Bi, S., Wu, Y., Cao, A. et al. (2020). Mater. Today. *Energy* 18: 100548.
27 Zhang, H., Lv, X., Tian, W. et al. (2021). *Chem. Eng. Sci.* 234: 116453.
28 Liu, X., Ouyang, M., Orzech, M.W. et al. (2020). *Energy Storage Mater.* 30: 329.
29 Zhong, S., Liu, H., Wei, D. et al. (2020). *Chem. Eng. J.* 395: 125054.
30 Li, L., Li, Z., Yoshimura, A. et al. (2019). *Nat. Commun.* 10: 1764.
31 Ru, J., He, T., Chen, B. et al. (2020). *Angew Chem. Int. Ed.* 59: 14621.
32 Yuan, Z., Wang, L., Li, D. et al. (2021). *ACS Nano* 15: 7439.
33 Wang, H., Xu, D., Jia, G. et al. (2020). *Energy Storage Mater.* 25: 114.
34 Xu, D., Wang, H., Qiu, R. et al. (2020). *Energy Storage Mater.* 28: 91.
35 Kong, M., Liu, Y., Zhou, B. et al. (2019). *Small* 15: 1804539.
36 Park, J.-S. and Kang, Y.C. (2019). *Chem. Eng. J.* 373: 227.
37 Pan, L., Huang, H., Zhong, M., and Niederberger, M. (2019). *Energy Storage Mater.* 16: 519.
38 Yang, S., Liu, C., Li, H. et al. (2021). *Energy Storage Mater.* 39: 278.
39 Wang, Y., Wang, K., Zhang, C. et al. (2019). *Small* 15: 1903816.
40 Fan, B.B., Fan, H.N., Chen, X.H. et al. (2021). *ACS Appl. Mater. Interfaces* 13: 19894.
41 Zhang, T., Shi, X., Mao, Z. et al. (2021). *Electrochim. Acta* 399: 139377.
42 Li, Y., Wang, H., Wang, L. et al. (2019). *Small* 15: 1804539.
43 She, L., Li, Q., Zhang, F. et al. (2020). *J. Power Sources* 451: 227744.
44 Wang, D., Liu, H., Shan, Z. et al. (2020). *Energy Storage Mater.* 27: 387.
45 Cao, Y., Ye, Q., Wang, F. et al. (2020). *Adv. Funct. Mater.* 30: 2003733.
46 Xie, F., Zhang, L., Su, D. et al. (2017). *Adv.Mater.* 29: 1700989.
47 Wang, H., Xu, D., Qiu, R. et al. (2021). *Small Struct.* 2: 2000073.
48 Shen, L., Wang, Y., Lv, H. et al. (2018). *Adv. Mater.* 30: 1804378.
49 She, L., Zhang, F., Jia, C. et al. (2021). *Nanoscale* 13: 15781.
50 Liu, Y., Wang, S., Sun, X. et al. (2021). *Energy Environ. Mater.* 0: 1.
51 Wang, G., Wang, S., Sun, X. et al. (2021). *ACS Appl. Energy Mater.* 5: 1114.
52 Chen, W., Zhang, X., Mi, L. et al. (2019). *Adv. Mater.* 31: 1806664.
53 Xia, Z., Sun, H., He, X. et al. (2019). *Nano Energy* 60: 385.
54 Sun, Y., Yang, Y., Shi, X.-L. et al. (2020). *Chem. Eng. J. Adv.* 4: 100053.
55 Tan, J., Zhu, W., Gui, Q. et al. (2021). *Adv. Funct. Mater.* 31: 2101027.
56 Liu, Z., Zhang, X., Liu, C. et al. (2020). *J. Mater. Chem. A* 8: 2777.
57 Gao, J., Zhang, M., Zhang, X. et al. (2022). *Chem.Eng. J.* 433: 133521.
58 Dong, S., Wu, L., Wang, J. et al. (2017). *J. Mater. Chem. A* 5: 5806.
59 Bi, Z., Zhang, Y., Li, X. et al. (2022). *Electrochim. Acta* 411: 140070.
60 Zhao, Q., Yang, D., Zhang, C. et al. (2018). *ACS Appl. Mater. Interfaces* 10: 43730.
61 Liu, X., Xu, W., Zheng, D. et al. (2020). *J. Mater. Chem. A* 8: 17938.

62 Li, Z., Shen, W., Wang, C. et al. (2020). *Correction: J. Mater. Chem. A* 8: 24212.
63 Zhang, H., Lv, Y., Wu, X. et al. (2022). *Chem. Eng. J.* 431: 133233.
64 Chen, S., Zhang, J., Wang, Z. et al. (2021). *Nano Lett.* 21: 5285.
65 Zhuang, H. and Yang, F. (2015). *J. Mater. Chem. A* 3: 19765.
66 Jiang, Y., Song, D., Wu, J. et al. (2019). *ACS Nano* 13: 9100.
67 Balogun, M.S., Luo, Y., Lyu, F. et al. (2016). *ACS Appl. Mater. Interfaces* 8: 9733.
68 Dong, S., Shen, L., Li, H. et al. (2016). *Adv. Funct. Mater.* 26: 3703.
69 Gu, J., Cui, K., Niu, S. et al. (2021). *J. Colloid Interface Sci.* 594: 522.
70 Liu, S., Luo, Z., Tian, G. et al. (2017). *J. Power Sources* 363: 284.
71 Zhang, C., Cui, L., Abdolhosseinzadeh, S., and Heier, J. (2020). *InfoMat* 2: 613.
72 Ming, F., Liang, H., Huang, G. et al. (2021). *Adv. Mater.* 33: 2004039.
73 Zhao, M.Q., Xie, X., Ren, C.E. et al. (2021). *Adv. Mater.* 33: 2004039.
74 Yang, J., Wang, T., Guo, X. et al. (2021). *Nano Res.* https://doi.org/10.1007/s12274-021-3933-7.
75 Zhang, Y.Z., El-Demellawi, J.K., Jiang, Q. et al. (2020). *Chem. Soc. Rev.* 49: 7229.
76 Wang, H.G., Li, W., Liu, D.P. et al. (2017). *Adv. Mater.* 29: 1703012.
77 Ni, J., Fu, S., Wu, C. et al. (2016). *Adv. Mater.* 28: 2259.
78 Zhao, L. and Qu, Z. (2022). *J. Energy Chem.* 71: 108.
79 Kong, L., Tang, C., Peng, H.J. et al. (2020). *SmartMat* 1: 399.
80 Xu, X., Liu, J., Hu, R. et al. (2017). *Chem.* 23: 5198.
81 Zhang, Y., Guo, L., and Yang, S. (2015). *Nanoscale* 7: 14618.
82 Yuan, S., Huang, X.L., Ma, D.L. et al. (2014). *Adv. Mater.* 26: 2273.
83 Lu, X., Li, F., Steimecke, M. et al. (2020). *ChemElectroChem* 7: 737.
84 Rubio, S., Maça, R.R., Aragón, M.J. et al. (2019). *J. Power Sources* 432: 82.
85 Yang, Q., Cui, S., Ge, Y. et al. (2018). *Nano Energy* 50: 623.
86 Gao, L., Chen, G., Zhang, L. et al. (2021). *Electrochim. Acta* 379: 138185.
87 Xu, D., Chao, D., Wang, H. et al. (2018). *Adv. Energy Mater.* 8: 1702769.
88 Mao, Z., Wang, R., He, B. et al. (2019). *Small* 15: 1902466.
89 Qiu, R.Y., Fei, R.X., Guo, J.Z. et al. (2020). *J. Power Sources* 466: 228249.
90 Wang, H., Sheng, L., Yasin, G. et al. (2020). *Energy Storage Mater.* 33: 188.
91 Tang, S., Guo, W., and Fu, Y. (2020). *Adv. Energy Mater.* 10: 1903977.
92 Wu, K., Huang, J., Yi, J. et al. (2020). *Adv. Energy Mater.* 11: 1700892.
93 Zhao, Q., Liu, X., Stalin, S. et al. (2019). *Nat. Energy* 4: 365.
94 Wu, F., Zhang, K., Liu, Y. et al. (2020). *Energy Storage Mater.* 33: 26.
95 Huang, X., Zeng, S., Liu, J. et al. (2015). *J. Phys. Chem. C* 119: 27882.
96 Zhao, M., Zuo, X., Ma, X. et al. (2017). *J. Membr. Sci* 532: 30.
97 Liu, J., Wu, X., He, J. et al. (2017). *Electrochim. Acta* 235: 500.
98 Asghar, M.R., Zhang, Y., Wu, A. et al. (2018). *J. Power Sources* 379: 197.
99 Kim, I., Kim, B.S., Nam, S. et al. (2018). *Materials* 11: 543.
100 Li, H., Peng, L., Zhu, Y. et al. (2016). *Nano Lett.* 16: 5938.

8

Pre-sodiation Technologies

Zirui Song[1], Chang Liu[2], and Xiaobo Ji[1]

[1] *Central South University, College of Chemistry and Chemical Engineering, No. 932, Lushan South Road, Yuelu District, Changsha 410083, China*
[2] *Hunan Institute of Engineering, School of Chemistry and Chemical Engineering, Diligence Road, Xiangtan, Hunan 411104, China*

8.1 Introduction

To solve the aggravating environmental issues as well as meet the demand for increasing energy supply, various electrochemical energy storage (EES) systems have been developed. Secondary metal-ion batteries are identified as one of the most promising EES systems, with the merits of no memory effect, high energy density, and low self-discharge. As the most representative rechargeable metal-ion batteries, lithium-ion batteries (LIBs) have been widely utilized in numerous fields, such as portable electronic devices, electric vehicles, and grid-scale energy storage. Importantly, sodium-ion batteries (SIBs) have received a lot of attention in recent years due to their high abundance of sources in the crust, low-cost, and similar electrochemical (EC) properties to LIBs. It is believed that the electrochemical performances of metal-ion batteries are mainly determined by the electrode materials; thus, tremendous efforts have been put into exploiting high-performance electrode materials in order to develop various new applications with high energy density. However, the related pre-metallation techniques, which are beneficial to achieving enhanced electrochemical performances, have gained scarce attention, seriously hindering the commercialization processes of EES systems.

Pre-metallation chemistry, also known as metal ion pre-doping, is the addition of metal sources to ensure adequate metal content in EES systems. Pre-metallation treatment is an indispensable step, which could be utilized to compensate for irreversible capacity loss, raise the working voltage, and ensure a sufficient concentration of metal ions in the electrolyte. Therefore, it holds the key to boosting the commercialization of EES devices. Furthermore, the pre-metallation technologies could avoid the accident triggered by active metal. It is noted that pre-metallation

Sodium-Ion Capacitors: Mechanisms, Materials, and Technologies, First Edition.
Edited by Guoqiang Zou, Xiaobo Ji, and Hongshuai Hou.

serves multiple roles in the various EES systems. Generally, pre-metallation effects can be divided into two categories, including:

1. Compensating for irreversible capacity loss in secondary metal-ion batteries. To be more specific, for a general metal system, capacity is stored and released based on the reversible insertion–extraction of metal ions in the electrodes, and the required metal ions for anodes originate from the cathodes. During the charging process, which results from the formation of solid electrolyte interface (SEI) film at the anode part, active metal would be consumed, which could decrease the contents of reserved metal spices in the next cyclic process. Taking Si as example, the notoriously low initial Coulombic efficiency (ICE) of around 50% is mainly caused by the formation of SEI film, and the dramatic volume change could consume a lot of metal ions due to the recurrent evolution of the SEI. Notably, pre-metallation technologies can be utilized to compensate for the lost active metal contents, which could further guarantee cycle stability and enhance energy density. Moreover, it has been noticed that performance of full battery system is mainly determined by the metal content in cathode. For example, when P2-type sodium layered oxides are found to be inadequate in sodium sources, pre-sodiation procedure can be ideal to provide sufficient sodium content, which is conducive to building high-performance batteries.
2. Addressing the safety hazard in metal ion batteries, such as Li–S and Na–S batteries. Currently, Li/Na–S batteries are composed of Li/Na metal as anode and sulfur as cathode. The usage of highly active Li/Na metal holds great possibility of engendering safety issues. Although the redox reactions between metal and sulfur can endow the metal–sulfur batteries with high theoretical energy densities, the practical applications of metal–sulfur batteries are seriously restricted by safety concerns. Specifically, during the charge/discharge cycles, the Li/Na dendrites would uncontrollably form and potentially pierce the separator, thereby causing internal short circuit phenomenon, which could bring about a potential accident (e.g. fire disaster). To achieve the commercialization of metal–sulfur batteries, metal-free anodes (such as prelithiated/presodiated materials) could be optimal alternatives, which can fabricate safe metal–sulfur systems.

The above-mentioned advantages of the pre-metallation process have been clearly elaborated, and various methods to implement pre-metallation strategy will be clarified in Sections 8.1–8.3. It is worth mentioning that the development of pre-sodiation has learned a lot from pre-lithiation; therefore, pre-lithiation approaches will be introduced firstly to provide a big picture for the reader to systematically understand pre-metallation techniques.

8.2 Pre-lithiation in Lithium-Ion Batteries

It has been found that different pre-lithiation strategies are based on different mechanisms. Nowadays, the pre-lithiation strategies can be principally summarized as follows: (i) operation with Li metal; (ii) usage of Li-based alternatives; (iii) supply of extra additives. [1].

8.2.1 Operation with Li Metal

Pure Li metal can be operated as pre-lithiation technologies for electrode materials. Currently, three approaches have been frequently utilized: direct contact method (DC), external short circuit (ESC), and electrochemical method (EC) (Figure 8.1) [2].

The DC method is carried out by DC of electrode material with Li metal in the presence of electrolyte. It is a simple and efficient approach for the pre-doping of lithium ions and has been widely applied in laboratories [3]. Si nanowire (SiNW) anode is directly contacted with Li foil in the presence of an electrolyte [4], and the SiNWs can be lithiated when they are touched with Li foil under certain pressure. The two pathways of pre-lithiation are proposed by Liu et al. for the first type, electrons will pass through the peak of the wires when the SiNWs touch Li foil. For another type, electrons could transfer through the contact positions between stainless steel substrate and Li foil, and then up through the bottom of the SiNW, when SiNWs do not contact the Li foil. It is worth mentioning that around 50% of the full capacity of SiNWs could be supplied through this pre-lithiation tactic within 20 minutes. Moreover, a roll-to-roll mechanical pre-alloying method has been noticed, which could achieve the pre-lithiation purpose and is found applicable for large-scale manufacturing [5]. Generally, above-mentioned method is fast and convenient; however, the degree of pre-lithiation can barely be controlled. Therefore, another ameliorated pre-lithiation strategy is developed based on DC method between active material and Li metal [6]. A resistance buffer layer (RBL) between the anode and Li foil has been inserted to regulate the rate and degree of lithiation. Thanks to the porous structure as well as the high electrical conductivity of RBL, the transfer of Li ions and the delivery of electrons could be boost. Furthermore, the

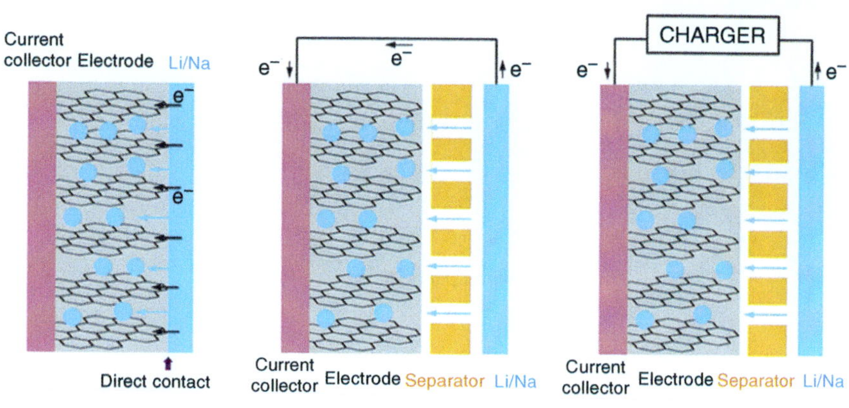

Figure 8.1 Schematic of three different pre-metallation methods: DC, ESC, and ES. Source: Reproduced with permission Kim et al. [2]. Copyright 2014, Royal Society of Chemistry.

soft RBL could ensure close contact between Li foil and anode, thereby bringing out a homogeneous pre-lithiation. Due to the highly reactive nature of Li metal, inert environments ($H_2O < 0.1$ ppm and $O_2 < 0.1$ ppm) are normally required. Interestingly, a trilayer structure of active material/polymer/lithium electrode has been fabricated by Cao et al., the polymer layer could protect lithium against moisture and O_2, and it can gradually be dissolved in the electrolyte, which can ensure the contact between active materials and lithium to achieve uniform pre-lithiation (Figure 8.2a) [7]. Moreover, the degree of pre-lithiation based on this method could be roughly controlled by adjusting the reaction time and the content of Li sources. Nevertheless, the precision degree of pre-lithiation is still very difficult to control. In terms of ESC method, an active electrode and a sacrificial lithium metal electrode are selected, and they are isolated by a porous polymer acting as separator in the electrolyte. A wire is utilized to link the electrodes together to spontaneously drive

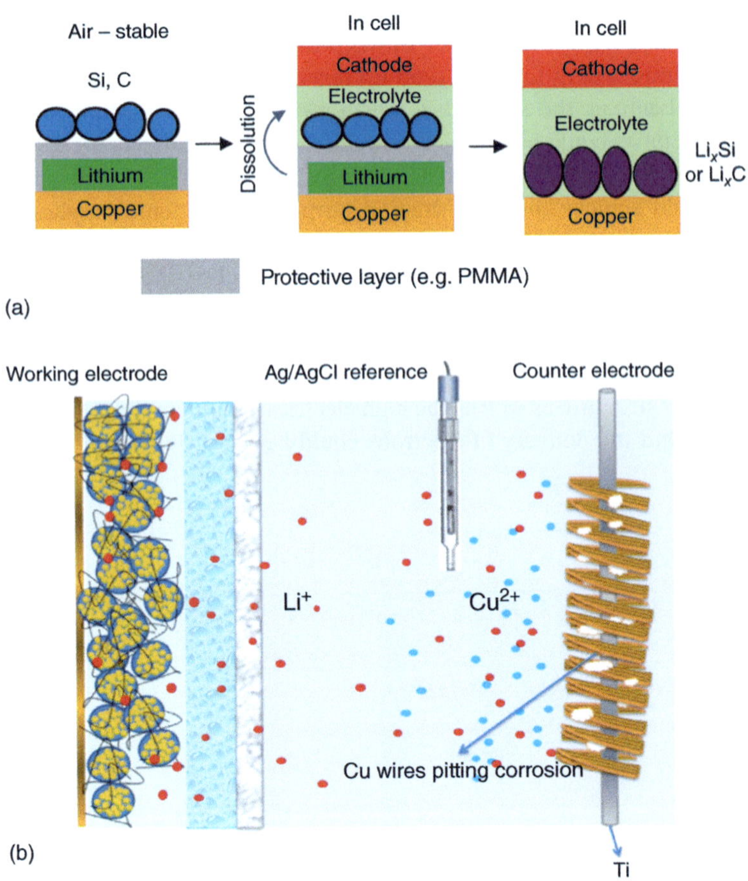

Figure 8.2 (a) Schematic of the procedures to fabricate ambient-air-stable lithiated anode. Source: Reproduced with permission Cao et al. [7]. Copyright 2016, American Chemical Society. (b) Schematic of EC method for pre-lithiation. Source: Zhou et al. [8]/Reproduced with permission from John Wiley & Sons.

Li ions embedded into the electrode. The carbon-coated SiO_x (marked as c-SiO_x) electrode has been pre-lithiated using the ESC method, and it is assembled in a battery configuration with a piece of lithium metal foil and a separator. Spontaneous pre-lithiation process can be initiated by the potential difference between the electrodes [9]. Importantly, unlike the DC method, the degree of pre-lithiation according to ESC strategy can be exquisitely regulated by controlling the shorting time and monitoring the voltage. For example, a high ICE of 94.9% could be delivered for the SiO_x anode, and the structural stability could be maintained. As for the EC method, the pre-lithiation is carried out through cycling in a half-cell at a selected current density, and the lithium metal is chosen as anode while the aiming electrode (cathode) is electronically isolated from Li by the separator [10]. Then, to achieve pre-thiation, several charge/discharge cycles should be implemented within a specific time frame or at a specific potential [11]. In order to tackle the safety issue about the usage of the Li metal in EC process, Zhou et al. prelithiated the Si electrode with a Cu pitting corrosion electrode rather than a Li-donating electrode, which can be identified as Li-metal-free (Figure 8.2b) [8]. This strategy is far safer than the DC method because it avoids short circuiting; additionally, it is green and highly controllable, opening up new avenues for the realization of pre-lithiation. However, the assembling/disassembling procedures of pre-treated half cells in the EC technique would cause an increase in expenditure and complications in manufacturing, which hold great danger for advancing the corresponding commercialization process. Additionally, several innovative strategies based on chemical lithiation have been developed through the operation with Li metal. Li et al. successfully fabricated Li_xSn@polypyrrole via a thermal lithiation process of the Sn nanoparticles with lithium metal at elevated temperature [12]. Meanwhile, three types of Li_xSi ($x = 4.4$, 3.75, and 2.33) have been prepared by Li et al. through high-energy-ball-milling Li metal with Si powder [13]. Importantly, the liquid ammonia is utilized to dissolve the Li metal to accomplish the pre-lithiation process [14]. Moreover, this unique method can even be adopted to over lithiated cathode materials. For instance, a spinel $Li_{1+x}Ni_{0.5}Mn_{1.5}O_4$ cathode materials have been synthesized by lithiating to 5 V, and a relatively high first delithiation capacity could be delivered. Note that the full cell ($Li_{1+x}Ni_{0.5}Mn_{1.5}O_4$ [$x = 0.62$]//Si-graphite) could exhibit a 23% higher reversible capacity than that of the battery paired with $LiNi_{0.5}Mn_{1.5}O_4$ cathode in the first cycle. To conclude, operation of Li metal based on various method (DC, ESC, EC, etc.) could accomplish pre-lithiation. However, all procedures must be implemented in the glove box which could offer an inert condition (H_2O level < 0.1 ppm and O_2 level < 0.1 ppm) in order to tackle the safety concern as the Li metal is extremely sensitive. Therefore, it is of necessity to explore the other pre-lithiation approaches to address the safety issues related to the usage of Li metal.

8.2.2 Usage of Li-Based Alternatives

In this section, usage of Li-based alternatives can be divided into two types: alternatives of Li metal and over-lithiated cathodes. For the first type of pre-lithiation approach, stabilized lithium metal powder (SLMP) has been widely adopted in EES systems to fulfill pre-lithiation [15]. To be more specific, SLMP is composed of Li particles covered by protective Li_2CO_3 layer, and it is dry-air stable. Importantly, the protection layer could be utilized to isolate the moisture (H_2O) and oxygen in the air, which can eliminate the safety hazard while using the Li metal. Generally, SLMP is used in the two following methods to lithiate electrode materials: SLMP can be added into the slurry with active electrode material, and it can also be loaded above the dried electrode sheet. Graphite without SLMP has been cycled in half-cells at 0.1 C (≈37.2 mA g^{-1}), the related initial cycle voltage curves were compared [16]. A much lower voltage (≈0.1 V) of graphite with SLMP in half-cell has been observed, which illustrates the partial lithiation of graphite anode, and the first cycle capacity loss of graphite with SLMP in half-cell is obviously reduced. The corresponding ICE is greatly improved due to the addition of SLMP. Importantly, pre-lithiation of graphite electrode accompanied by SLMP could establish stable SEI layer. It is worth mentioning that the effect of SLMP particles can only be triggered by applying pressure before pre-lithiation, which is aimed to destruct the Li_2CO_3 layer, thereby enabling DC between the Li metal and electrode material [17]. Significantly, various anodes (such as carbon-based, Si-based, and metal oxide materials) of LIBs have been pre-lithiated by the addition of SLMP, so irreversible capacity loss can be eliminated, thus achieving high energy density. Other EES systems have also utilized SLMP particles for the pre-lithiation process. For example, in order to realize the pre-lithiation in the Li–S configuration, the lithium sulfide/microporous carbon (Li_2S/MC) cathode is synthesized by spraying SLMP particles onto the surface of S/MC cathode before the assembly of battery [18]. The formed Li_2S/MC cathode could deliver high Coulombic efficiency as well as long cycling stability. When it comes to the performance of the full cell based on Li_2S/MC//graphite, a stable capacity of around 600 mAh g^{-1} after 150 cycles can be exhibited. However, inhomogeneous distribution and high expense of SLMP are the main drawbacks of this method, so it is very important to exploit other pre-lithiation reagents. Interestingly, another type of lab-made Li metal composites has been successfully prepared by cryomilling Li foil with ionic liquid [P_{4444}]TFSI; this method is stable and easy to handle [19]. After pre-lithiation, the fabricated anodes could deliver excellent electrochemical performance (very low overpotential and ultrahigh area capacity), so this effective tactic can be utilized to compensate for the large irreversible capacity loss. Moreover, the growth of Li-dendrite could be greatly suppressed, shedding light on its practical application. Recently, a fast and controllable pre-lithiation approach was reported by Zhang et al., and a chemical reaction based on Libiphenyl-tetrahydrofuran (Li-Bp-THF) compound as a lithium donor has been developed (Figure 8.3) [20]. Importantly, the SEI layer could be formed during chemical pre-lithiation process, which could enhance the ICE of pre-lithiated hard carbon (HC). When assembled full cell with $LiNi_{1/3}Co_{1/3}Mn_{1/3}O_2$ being cathode, it shows an improved ICE of 90.2%, while the ICE of full cell without pre-lithiation treatment is only 75.5%. Additionally, pre-lithiation of sulfur cathodes could be achieved by utilizing the organolithium reagent lithium naphthalenide (Li^+Naph^-) [21]. In general, Li-containing reagents with lower redox potentials and strong reducing capabilities are thought to be promising alternatives for achieving

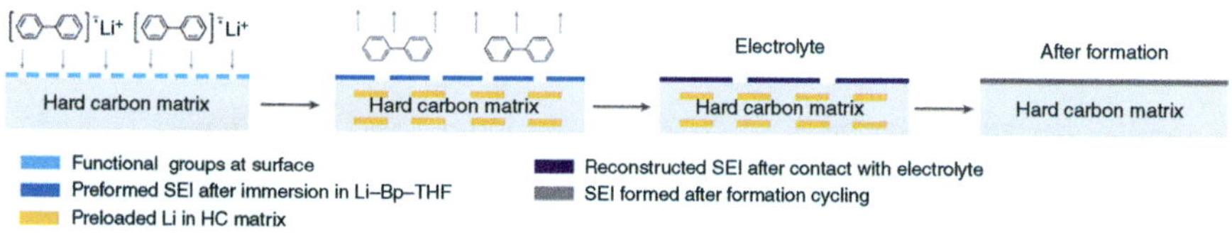

Figure 8.3 Schematic of the pre-lithiation mechanism of Li-Bp-THF. Source: Reproduced with permission Zhang et al. [20]. Copyright 2020, American Chemical Society.

chemical pre-lithiation, and the lithium radical anions of aromatic hydrocarbons have powerful reducing capabilities comparable to lithium metal. In addition, a suitable roll-to-roll design could be conducive to the industrial application of chemical pre-lithiation strategy based on the usage of Li-containing reagents.

As for the over-lithiated cathode strategy, adopting over-lithiated cathode electrodes can also be an effective method to provide additional lithium source in EES systems [22]. Importantly, there is no need for the extra additives in this technique. It should be noted that the selection of over-lithiated/sodiated cathodes is supposed to be thoroughly considered. Specifically, the over-lithiated cathodes should be electrochemically active, which means the redundant lithium can further be released in the first cycle to compensate for the loss of active lithium in anodes. The over-lithiated spinels $Li_{1+x}Mn_2O_4$ and $Li_{1+x}Ni_{0.5}Mn_{1.5}O_4$ have been developed as pre-lithiation reagents [23]. Aravindan et al. utilized the $Li_{1.26}Mn_2O_4$ cathode as pre-lithiation reagent to compensate for the irreversible capacity loss of α-Fe_2O_3 nanorods [24]. Moreover, Gabrielli et al. fabricated blend cathodes (LMNO-B) consisting of $LiNi_{0.5}Mn_{1.5}O_4$ (LMNO-O) and lithium-rich $Li_{1+x}Ni_{0.5}Mn_{1.5}O_4$ (LMNO-R) to make up for the irreversible capacity loss of Si-based anodes [25]. Importantly, LMNO-R as a lithium rich phase could be converted to the standard spinel phase electrochemical activity after all the extra lithium leaving the original structure. The extracted Li ions (indicated as yLi$^+$) during the initial charge/discharge process are depleted by irreversible processes (e.g. the formation of SEI film). Moreover, the remaining Li^+ content (noted as $1-y$Li$^+$) is utilized for the reversibly alloying process of Si anode. In the meantime, LMNO-B electrode containing the exact amount of extra Li, which is required for compensating the irreversible losses of lithium (yLi$^+$), and LMNO-B cathodes can make sure the subsequent reaction of the cathode part. Furthermore, Liu et al. synthesized $Li_3V_2(PO_4)_3$ as cathode to achieve pre-lithiation (Figure 8.4a) [26]. In the initial charge process, the released Li^+ from $Li_3V_2(PO_4)_3$ cathode can be wielded to pre-lithiated HC anode. Subsequently, the subsequent formed $Li_2V_2(PO_4)_3$ cathode can construct a 4 V LIB with the pre-lithiated HC anode, and a maximum energy density of 208.3 Wh kg^{-1} along with a maximum power density of 8291 W kg^{-1} can be manifested with 2000 cycles. Additionally, a "giving comes before receiving" strategy has been proposed by Song et al., and $Li_5V_2(PO_4)_3$ (L_5VP) cathode was prepared to pre-lithiate a HC anode (Figure 8.4c) [28]. Notably, the L_5VP cathode is first synthesized by lithiating $Li_3V_2(PO_4)_3$ (L_3VP) as the "giving" step, and the irreversible capacity loss in HC anode side could be entirely supplied by de-lithiation of L_5VP, which is called

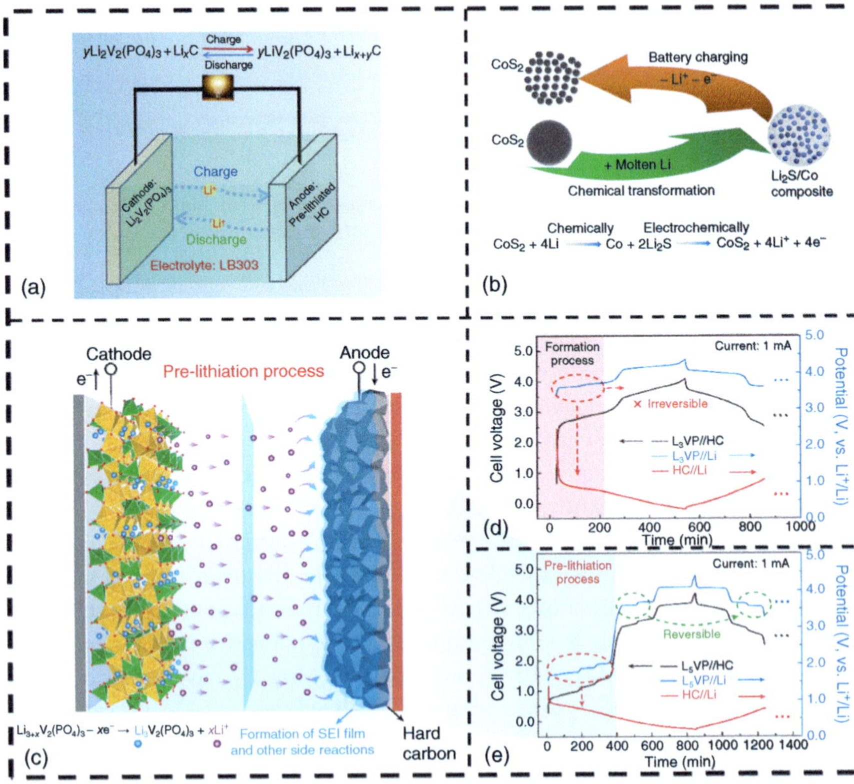

Figure 8.4 (a) Schematic illustration of the full cell ($Li_3V_2(PO_4)_3$//HC). Source: Liu et al. [26]/Reproduced with permission from John Wiley & Sons. (b) Schematic illustration of the chemical synthesis of Li_2S/metal composite as well as the process of lithium extraction during the charge process. Source: Sun et al. [27]/Reproduced with permission from John Wiley & Sons. (c) Schematic illustration of the pre-lithiation process of L_5VP//HC full battery. The chronopotentiometry profiles based on three-electrode tests of (d) L_3VP//HC full battery and (e) L_5VP//HC full battery. Source: (c–e) Reproduced with permission Song et al. [28]. Copyright 2019, Elsevier.

"receiving." It is clearly observed that the two Li ion per formula of L_5VP are extracted below around 3.0 V (vs. Li/Li^+) for the L_5VP//HC full battery, thereby completely compensating for the irreversible Li^+ loss in the HC anode part (Figure 8.4d,e). Moreover, the average potential of HC anode in L_3VP//HC full cell decreased from ≈0.70 to ≈0.45 V (vs. Li/Li^+) in L_5VP//HC full cell. As for the result, the L_5VP//HC pouch full cell delivers a much higher ICE of 96.7% and can exhibit a high energy density of 320 Wh kg^{-1} at room temperature.

8.2.3 Supply of Extra Additives

So far, a great variety of materials have been exploited as extra additives to perform pre-lithiation. For example, Zhao et al. proposed Li_xSi species as the lithiated active material in the anode part, and the Li_xSi–Li_2O core–shell nanoparticles can be used as a great pre-lithiation reagent. The high liberated specific capacity can compensate for the initial cycle capacity loss in the anode part [29]. Notably, a one-step thermal alloying process can be utilized to synthesize the nanoparticles. Multiple merits of Li_xSi nanoparticles for pre-lithiation can be summarized below: (i) Li_xSi can manifest enormous capacity (4200 mAh g^{-1} of Si, 2000 mAh g^{-1} of $Li_{4.4}Si$) at a low potential of around 10 mV vs. Li/Li^+ to pre-lithiate various kinds of anodes; (ii) nano-size Li_xSi particles can be favorable to achieving fast pre-lithiation process of anode materials compared with Li metal foil; (iii) protected by a Li_2O passivation shell, Li_xSi–Li_2O core–shell nanoparticles are stable under dry air condition, which implies the possibility for scale-up manufacturing processes at low cost. The capacity preloaded to Si nanoparticles is 1310 mAh g^{-1}, lower than the theoretical specific capacity, which is caused by the decomposition of the solvent as the Li_xSi possesses strong reducing ability. In this case, Si and graphite anodes are successfully pre-lithiated with Li_xSi nanoparticles, high ICE of 94% and >100% can be achieved, respectively. However, Li_xSi–Li_2O nanoparticles could maintain the capacity for only relatively short durations under the circumstance of low relative humidity. Therefore, it is important to develop pre-lithiation reagents with great stability. Significantly, Zhao et al. prepared a surface modification method to enhance the stability of Li_xSi nanoparticles [30]. Specifically, a continuous and dense coating layer triggered by the reduction of 1-fluorodecane was utilized to protect the Li_xSi. The coating can be treated as an artificial SEI layer, composing of LiF and lithium alkyl carbonate with lengthy hydrophobic carbon chains, could serve as effective passivation layer to contribute to the maintenance in the ambient environment. Moreover, this Li_xSi nanoparticles protected by the artificial SEI could exhibit a high pre-lithiation capacity of 2100 mAh g^{-1}, and negligible capacity decay in dry air after five days was shown. In addition, an excellent capacity of 1600 mAh g^{-1} in humid air (≈10% relative humidity) can be exhibited. Other similar strategies have also been explored. As shown in Figure 8.5a, a one-pot metallurgical pre-lithiation approach was developed by employing the $Li_{22}Z_5$ alloys and $Li_{22}Z_5$–Li_2O composites (Z = Si, Ge, Sn, etc.) [33]. With the merits of high capacity and low chemical potential, $Li_{22}Z_5$ alloys as well as $Li_{22}Z_5$–Li_2O composites can serve as pre-lithiation reagents, and they are able to increase the ICE of anode materials (both graphite and alloy type). Notably, Li_xGe can exhibit the optimal stability under ambient air conditions among all lithiated group IV alloys, which is consistent with the theoretical results of the large binding energy between Li and Ge atoms in $Li_{22}Ge_5$ crystal. Furthermore, owing to the strong binding between O in Li_2O and Li in $Li_{22}Z_5$, the ambient air stability of $Li_{22}Z_5$–Li_2O composites can be ameliorated.

Meanwhile, the cathode additives can also accomplish pre-lithiation effectively [34]. Nevertheless, different from additives applied in anode part, the selection of cathode additives is supposed to be carefully considered, several key traits should be noted to design remarkable pre-lithiation reagents. Firstly, the working potential of pre-lithiation reagents should be appropriate, which should be located in the voltage range between the anode and cathode materials. Secondly, pre-lithiation reagents should be sacrificial, which means the remaining product after decomposition

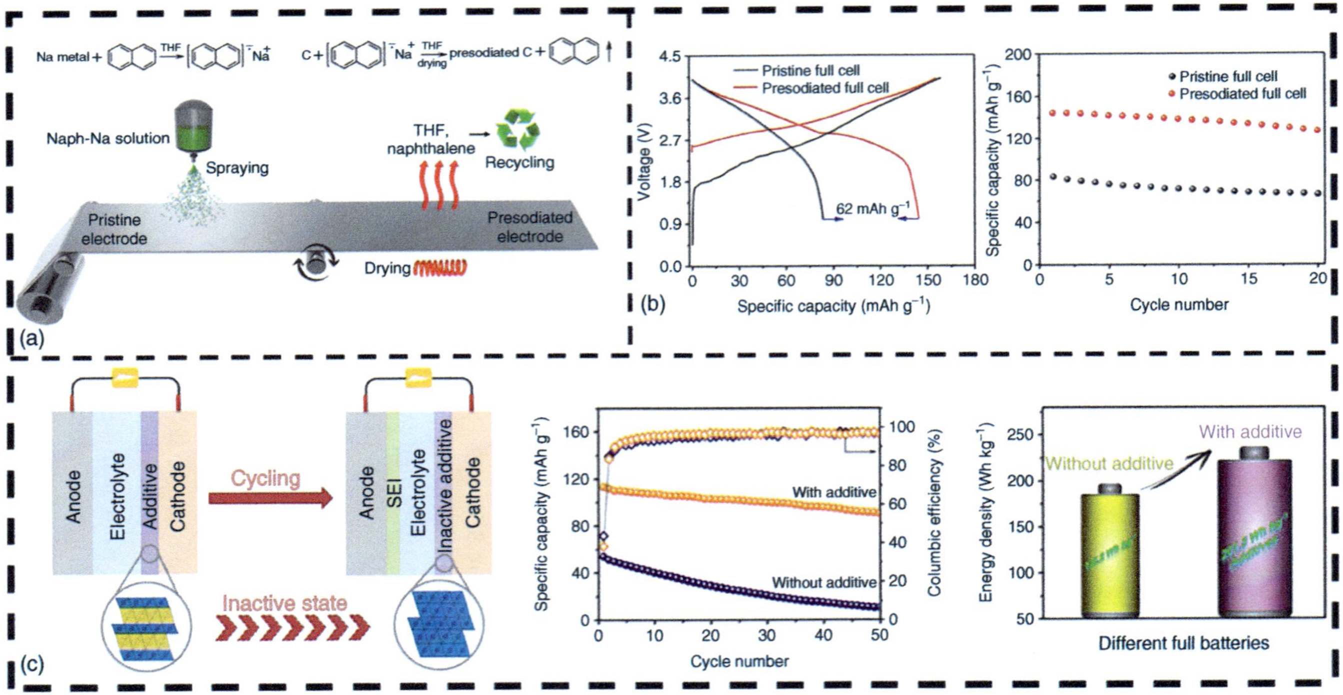

Figure 8.5 (a) Schematic of utilizing a Naph-Na solution to pre-sodiate HC electrodes. (b) Electrochemical performance of sodium-ion full cell ($Na_{0.9}[Cu_{0.22}Fe_{0.30}Mn_{0.48}]O_2$//HC) with pre-sodiation. Source: (a, b) Liu et al. [31]/Reproduced with permission from John Wiley & Sons. (c) Schematic of the mechanism and electrochemical performance of self-sacrifice-assisted traditional method. Source: Reproduced with permission Shen et al. [32]. Copyright 2019, Elsevier.

should be chemically and electrochemically stable in the EES systems; therefore, no harm would be produced to influence the electrochemical performance. Furthermore, the additives should be able to compensate for irreversible capacity loss with a low dosage by having a high specific capacity. Additionally, the operation of additives should be easily to handle, and they are compatible with the existing industrial processing. To meet above-mentioned considerations, various additives have been explored, such as the binary and multinary compound. For example, Li_3N has been utilized as representative binary compound to assist the $LiCoO_2$ cathode, thereby constructing high energy density EES systems [35]. Notably, Li_3N could manifest a theoretical capacity of 2308.5 mAh g^{-1} after decomposition, larger than that of LiN_3 (around 547.4 mAh g^{-1}) as cathode additive. Moreover, Li_3N has better stability than LiN_3, which can be beneficial for scale-up manufacture. It is worth mentioning that the reversible capacity of the blend electrode is obviously enhanced after adding a small amount of Li_3N to $LiCoO_2$ cathode. Moreover, Bian et al. developed Li_2CO_3-passivated Li_3N with high stability [36]; the additive can deliver a large irreversible capacity of 800 mAh g^{-1} after pre-lithiation. It is noted that 2 wt% Li_3N can be used as an additive in a $Li(Li_{0.18}Ni_{0.15}Co_{0.15}Mn_{0.52})O_2$ (LLMO) cathode to have 10% increment of the discharge capacity of a full battery (LLMO//graphite). In addition, another facile strategy was explored to prepare a surface passivated Li_3N material through the reaction of Li metal with N_2 [37]. Significantly, highly crystalline Li_2O and Li_2CO_3 would constitute the passivation layer on the surface, which can enable good environmental stability for Li_3N particles. Remarkably, various cathodes (such as $LiCoO_2$, $LiN_{i0.6}Co_{0.2}Mn_{0.2}O_2$, and $LiFePO_4$) can deliver larger capacities with the usage of Li_3N additives in the initial charging process compared with the corresponding pristine electrodes.

It was discovered that Li_2S could also be used as the pre-lithiation agent. Zhan et al. utilized Li_2S to supply enough lithium ions to compensate for the loss of active lithium source in the initial cycle and improve the electrochemical performance of LIBs [38]. A high irreversible capacity of 1084 mAh g^{-1} can be delivered by mixing Li_2S in the core–shell structured pre-lithiation material with Ketjenblack and poly(vinylpyrrolidone). After evaluating the pre-lithiation effect of Li_2S, a full cell ($LiFePO_4(Li_2S)$/Si-C) is constructed. As a result, compared with the pristine $LiFePO_4$/Si-C full cell, the energy densities of the $LiFePO_4(Li_2S)$/Si-C full cell have been remarkably increased ≈13.4%, ≈26.7%, ≈65.0%, and ≈110.2% after the first, 10th, 100th, and 200th cycle, respectively. Notably, the unique characteristics of conversion reactions can be beneficial to delivering high capacity, which could be donated for anodes. Based on this strategy, M/Li_2O composite was explored as a new kind of cathode pre-lithiation additive [39]. It is discovered that M/Li_2O nanocomposites hold more than four times the theoretical specific capacity of current cathodes (724 mAh g^{-1} for Co/Li_2O with a molar ratio of 3/4, 799 mAh g^{-1} for Fe/Li_2O with a molar ratio of 2/3, and 935 mAh g^{-1} for Mn/Li_2O with a molar ratio of 1/2). To be more specific, with the addition of 4.8% Co/Li_2O, the $LiFePO_4$ electrode manifests 11% higher overall capacity than that of the pristine $LiFePO_4$ electrode with no additive. Owing to this strategy, the specific capacity and energy density of a $LiFePO_4$/graphite full cell have been remarkably enhanced.

Interestingly, the Li_2S/Co nanocomposite has been synthesized to carry out pre-lithiation. The detailed mechanism is illustrated in Figure 8.4b, lithium source in the composites can be electrochemically released during the first charging process through an inverse electrochemical conversion reaction [27]. A high specific capacity of 670 mAh g^{-1} can be delivered by Li_2S/Co nanocomposites, which could efficiently compensate for the lithium loss of anodes. Moreover, the multicomponent $Fe/LiF/Li_2O$ nanocomposites was prepared as cathode pre-lithiation additives, which could release a great capacity of 550 mAh g^{-1} based on the multielectron inverse conversion reaction during the first charging process [40]. It is proved that the $Fe/LiF/Li_2O$ nanocomposite can act as an additive to work with various cathodes (such as $LiCoO_2$, $LiFePO_4$, and $LiNi_{1-x-y}Co_xMn_yO_2$) and exhibit outstanding compensation effect. Furthermore, a conformal Li_2O/Co nanoshell on $LiCoO_2$ particles was in situ prepared via chemical formation method, the pre-lithiation reagent could be utilized as a built-in material with high capacity to compensate for the initial lithium loss [41]. With the addition of 1.5 wt% Li_2O/Co, a capacity of 15 mAh g^{-1} can be increased for the $LiCoO_2$ compared with the pristine $LiCoO_2$ cathode. As for the multinary compound, various materials (such as $Li_{0.65}Ni_{1.35}O_2$ [42], Li_2CuO_2 [43], Li_2MoO_3 [43b], Li_5FeO_4 [44], and Li_6CoO_4 [45]) have been widely exploited for the pre-lithiation process in EES systems. However, it should be noted that the de-lithiated products are not electrochemically active and the remained part in the cathode would greatly impair the electrochemical performance of EES systems in the practical manufacture [46]. Herein, original strategy based on organic sacrificial salts has been intensively developed. With the merits of abundant resources, low-cost, and environmental friendliness, sacrificial organic salts are conducive to promoting commercialization of EES systems. In summary, the rational design of organic salts has provided a new avenue for pre-lithiation techniques, which can greatly avoid the complicated assembly process in EES systems.

8.3 Pre-sodiation in Sodium-Ion Batteries

As the anodes in sodium-ion batteries can also be adopted in the most common sodium-ion capacitors, the pre-sodiaition methods in battery configuration can be conducive to inspiring the relative application in Na ion capacitors. Herein, the pre-sodiation strategies in Na ion batteries have been summarized in this section. Nowadays, development of pre-sodiation techniques is still in the early stage, based on different mechanisms, pre-sodiation strategies can be basically categorized as follows: (i) operation with Na metal; (ii) usage of Na-based alternatives; and (iii) supply of extra additives.

8.3.1 Operation with Na Metal

Similar to DC method applied in pre-lithiation, Moeez et al. developed a pre-sodiation method to pre-sodiate $Na_{0.67}Fe_{0.5}Mn_{0.5}O_2$ and HC electrodes, the increased ICE, improved cyclability along with enhanced capacity can be delivered. Moreover, the formation of a thick cathode electrolyte interface (CEI) film caused by the pre-sodiation process can contribute to preventing the dissolution of the active materials and corresponding deposition on the anode side, thus improving cycle stability [47a]. In addition, Liu et al. develop a strategy of metallurgically prealloying Na to reduce the open-circuit voltage (OCV) and suppress the gas release [5b]. Encouragingly, when the alloy foil is used as the anode in $Na_3V_2(PO_4)_2F_3$(NVPF)//Na–Sn full cells, the ICE is significantly improved from 25% to 75%, and from the 2nd cycle to the 100th cycle, the average Coulombic efficiency can be maintained up to 99.44%. After 100 cycles, the cathode-weight normalized capacity is reduced to a still-acceptable value of 81.4 mAh g^{-1} (NVPF) from 112.5 mAh g^{-1} (NVPF).

8.3.2 Usage of Na-Based Alternatives

Normally, the usage of Na-based alternatives can also be divided into two types: alternatives of Na metal and over-sodiated cathode. For the former type, Liu et al. developed a simple and high efficiency pre-sodiation strategy at electrode-level based on this strategy, it can be realized through spraying a sodium naphthalene (Naph-Na) solution (Figure 8.5a) [31]. Specifically, Naph-Na could react with oxygen-containing functional groups, thus forming Na-containing functional groups during pre-sodiation process. Besides, a SEI layer can be gradually built on the surface of the pre-sodiated active carbon electrode with the existence of liquid electrolytes. Owing to this pre-sodiation strategy, a large irreversible capacity of 60 mAh g^{-1} is supplied, and the ICE of obtained HC anode can get an increase of 20%. Notably, the energy density of full cell ($Na_{0.9}[Cu_{0.22}Fe_{0.30}Mn_{0.48}]O_2$//HC) could be elevated to 240 Wh kg^{-1} after pre-sodiation treatment (Figure 8.5b). Furthermore, this simple pre-sodiation approach has been exploited to be applied in other anode materials for SIBs, such as metal oxides-based and alloy-based anodes [47].

As for the over-sodiated method, Mirza et al. prepared the $Na_4V_2(PO_4)_3$ (marked as Na_4VP) cathode to pre-sodiate the HC anode [48]. Firstly, $Na_3V_2(PO_4)_3$ (Na_3VP) was converted to Na_4VP cathode through a sodiation step. Then, Na_4VP could offer the extra one sodium ion by a de-sodiation step below 2.2 V (vs. Na^+/Na) to entirely compensate for the irreversible capacity loss of the anode, and the formed Na_3VP can directly be used as the cathode for the following cycles. As for the electrochemical result, the full cell (Na_4V//HC) achieved an excellent specific energy of 265 Wh kg^{-1}, 76% higher than that of the full cell without pre-sodiation (Na_3VP//HC). Moreover, great rate performance (62.32 mAh g^{-1} at 10 C) as well as cycling stability (around 80.0% capacity retention after 100 cycles at 1 C) of full cell (Na_4V//HC) can be delivered.

8.3.3 Supply of Extra Additives

Generally, P_2-$Na_{0.67}[Fe_{0.5}Mn_{0.5}]O_2$ cathode is identified with the issue of sodium deficiency, which could be addressed by utilizing the additive of NaN_3 [49]. It is worth mentioning that 60% increase of the reversible capacity in the full battery configuration with P_2-$Na_{0.67}[Fe_{0.5}Mn_{0.5}]O_2$ cathode could be obtained with 10% addition of NaN_3 in the cathode composites. Meanwhile, Na_2S as the pre-sodiation reagent has been successfully deployed in the Na–S configuration, and the electrochemical performances of Na-S batteries (Na_2S/C//HC) could be enhanced [50]. Interestingly, nanoscale mixtures of metal (M) and binary compound as pre-lithiation reagents fabricated through a facile cathode method has been developed, which can manifest high pre-lithiation efficiency and good compatibility [51]. Especially, there is an irreversible phase transition for $NaCrO_2$ material in the high voltage region, which could deliver a high irreversible capacity of up to 230 mAh g^{-1}. Owing to this irreversible phase transition mechanism, $NaCrO_2$ was proposed as an effective self-sacrificial additive to compensate for the loss of sodium ions (Figure 8.5c) [32]. Importantly, the cycle stability of the HC//$Na_3V_2O_2(PO_4)_2F$ full cell is obviously improved with the assistance of sacrificial regents, and the energy density can be significantly increased to 201.5 Wh kg^{-1}, which is 1.65 times that of the pristine cell without additives. Moreover, Shanmukaraj et al. successfully prepared $Na_2C_4O_4$ as organic additive in cathodes, which can settle the issues of irreversible capacity loss of anodes in SIBs [52]. The cathode ($Na_3(VO)_2(PO_4)_2F$ with $Na_2C_4O_4$) can show a highly stable capacity of 135 mAh g^{-1} as well as stable cycling performance after pre-sodiation.

8.4 Pre-sodiation in Sodium-Ion Capacitors

Sodium-ion capacitors, as the combination of SIBs (large energy density) and supercapacitors (high power density), were built for next-generation energy storage systems. Note that full sodium ion capacitors (SICs) could not exhibit any available performance without pre-sodiation techniques as adequate Na^+ need to be provided to compensate for the irreversible consumption due to the formation of SEI as well as side reactions [1a, 53]. More importantly, including the benefits mentioned in metal-ion batteries, pre-sodiation approaches can also widen the working voltage window of full capacitors and minimizing the electrolyte consumption during long term cycling in SICs. Normally, SICs are constituted of a battery-type electrode as anode and a capacitor-type electrode as cathode. However, porous carbon materials (activated carbon) demonstrate the physical adsorption/desorption behavior of anion and could not offer metal sources for anodes, which means that the need of Na sources for anodes can only be provided from the electrolyte, this would cause the loss of electrolyte, intrinsically influencing the long-term stability. Significantly, the pre-sodiated anodes can deliver flat and low potentials curves, which could enable the wide working voltage range and increase energy densities of SICs [54]. Normally, pre-sodiation strategies applied in SICs can be divided into the following categories: electrochemical method, alternatives method, and sacrificial additives method.

8.4.1 Electrochemical Method

Generally, EC method is carried out by assembling and disassembling the half-cells, in which the targeted electrode and sodium metal serve as working electrode and counter electrode, respectively. The whole pre-sodiation process must be executed in extreme inert atmosphere ($O_2 < 0.1$ ppm and $H_2O < 0.1$ ppm) as the consequence of high activity of sodium metal. Notably, the pre-sodiation degree of selected material could be easily controlled through the regulation of cutoff potential. Owing to the trait of simplicity, most pre-sodiation experiment has been done by EC strategy. Early in 2012, Kuratani et al. successfully pre-sodiated HC, the anodes were galvanostatically pre-sodiated to 80% of the full capacity of carbons. To be more specific, sodium ions were galvanostatically pre-doped into HC at 30 °C in the half-cells with Na metal as the counter electrode. The reduction cutoff potential was set up to be 0.0 V vs. Na/Na^+, and then the anodes were oxidized with the cutoff potential of 1.5 V vs. Na/Na^+ at the same current density. The whole pre-sodiation would be carried out for three cycles, after that, the state of charge of the anodes was adjusted to be 80%. It is discovered that the potential of the anodes after pre-sodiation was around 0.0 V vs. Na/Na^+, which could bring the higher output potential difference of the electrodes in full SICs. The assembled full SICs (AC as cathode), which could deliver a capacity retention of 91% after 1000 cycles [55]. Yang et al. utilized EC approach to pre-sodiate $NiCo_2O_4$ as anode, the full SICs (AC//$NiCo_2O_4$) showed an energy density of 60 Wh kg^{-1} at 10 000 W kg^{-1} (Figure 8.6a), and 85.0% of the initial capacitance can be retained after 2000 cycles at the current density of 3 A g^{-1}, much better than the performance of the SICs without pre-sodiation [56]. For another example, the SICs (AC//$Li_4Ti_5O_{12}$/reduced graphene oxide) fabricated by Kim et al. delivered a maximum energy density of 83 Wh kg^{-1} at the power density of 25 W kg^{-1}, and it could maintain an energy density of 28 Wh kg^{-1} at 5 kW kg^{-1}. Moreover, the full SICs exhibited excellent cycling stability after 30 000 cycles (Figure 8.6b) [57].

8.4.2 Alternatives Method

Due to the trait of more mild and uniform reaction in liquid phase, alternative method utilizing the compound of sodium-dissolved solution to accomplish the mission of pre-sodiation was developed. Brady et al. proposed Na-biphenyl to pre-sodiate $Ti_3C_2T_x$ MXene in the solution of diglyme. An Ar-filled glovebox with strict inert environment ($O_2 < 0.1$ ppm, $H_2O < 0.1$ ppm) was utilized for the chemical sodiation process to produce sodiated MXene ($Na_yTi_3C_2T_x$). Specifically, Na metal was mixed with biphenyl, and then the added electrode materials were added to the solution to be pre-sodiated. After that, the mixture of reagents was added to diglyme, the resulting solution was found deep blue–green, and the color cleared up as the sodium ions were intercalated. Based on the theoretical simulation, multiple sodiation sites and formation of a 2D sodium domain structure are observed (Figure 8.6c). It is found that at distances of r between 3 and 8 Å, only minor signal variations of atom–atom pair correlations appeared, which

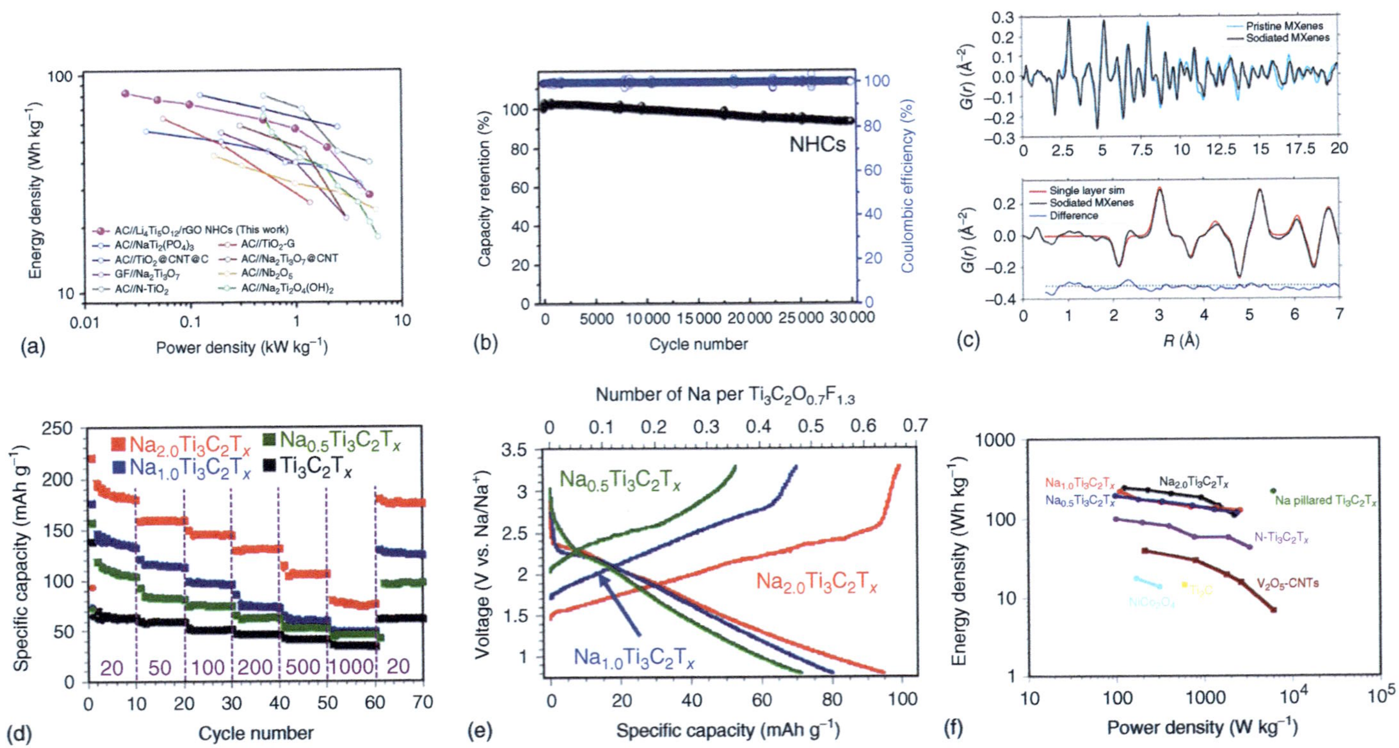

Figure 8.6 Ragone plots (a) of assembled SICs [56]. Cycle performance (b) of assembled SICs. Source: (a, b) Kim et al. [57]/With permission of ELSEVIER. (c) Neutron PDFs and difference PDF analysis. (d) Rate performance of $Na_2Ti_3C_2T_x$, $Na_1Ti_3C_2T_x$, $Na_{0.5}Ti_3C_2T_x$, and $Ti_3C_2T_x$ at different current densities. (e) The first galvanostatic charge–discharge cycles. (f) Ragone plots of series of $Na_2Ti_3C_2T_x$, $Na_1Ti_3C_2T_x$, and $Na_{0.5}Ti_3C_2T_x$ in SICs. Source: (c–f) Brady et al. [58]/With permission of American Chemical Society.

indicate a disordered Na^+ distribution and/or the presence of multiple sodiation sites at interfaces. Variations in PDF peak intensities between pristine MXenes and MXenes after pre-sodiation are detectable in the long-*r* regions. This can be caused by the following important factors: (i) The increase of spacing or gap between adjacent two sheets after the uptake of sodium ions. (ii) The extra contribution from Na-to-MXene atom–atom correlations. (iii) The change of layer stacking order and the deformation of in-plane lattice. According to the results in Figure 8.6d and Figure 8.6e, with the merit of large amount of pre-doped sodium ions in $Na_2Ti_3C_2T_x$, it can show a larger first cycle de-sodiation capacity than $Na_1Ti_3C_2T_x$ and $Na_{0.5}Ti_3C_2T_x$. Moreover, a remarkable specific energy of 249 Wh kg^{-1} was exhibited at specific power of 124.5 W kg^{-1} when assembled the SICs with a free-standing activated carbon film as cathode and $Na_2Ti_3C_2T_x$ as anode, respectively (Figure 8.6f) [58]. However, restricted by the volatility, and flammability of solvents, further development of this method needs to be explored in terms of addressing the safety concern.

8.4.3 Sacrificial Additives Method

It should be noted that the sacrificial cathode additives method could be handled simply in the existing fabrication conditions, where degree-controlled in situ pre-sodiation is possible, shedding light on scale-up manufactures. Currently, the sacrificial sources could be divided as inorganic compound and organic compound. Na_2S [59], $NaNH_2$ [60], $NaBH_4$ [61], Na_2CO_3 [62], and $Na_2C_2O_4$ [63] were demonstrated as inorganic sacrificing additives to perform pre-sodiation. With a high theoretical capacity of 687 mAh g^{-1}, the full SICs (Maxsorb//Na-Sn_4P_3) utilizing Na_2S as pre-sodiation reagent demonstrates a high specific energy of 48 Wh kg^{-1} at specific power of 1 kW kg^{-1}. Jezowski et al. prepared $NaNH_2$ and $NaBH_4$ to assist the AC//HC-Na SICs and AC//Sn_4P_3 SICs delivering the energy density of 35 Wh kg^{-1} at 1 kW kg^{-1} and 33 Wh kg^{-1} at 2 kW kg^{-1}, respectively (Figure 8.7a). Pouch-type SICs (AC//HC) with Na_2CO_3 as the sacrificial additive exhibited a superior performance of 103 Wh kg^{-1} at 13 kW kg^{-1}, and it can power 300 (LED) bulbs. Interestingly, in terms of massive dosage issue of additives, Song et al. deployed interfacial bonding effect (Figure 8.7b,c) to successfully decrease the practical decomposition potential of $Na_2C_2O_4$ to 3.95 V, which could liberate sufficient sodium ions to compensate for the irreversible loss. Inspired by the fascinating physicochemical properties of carbon dots with various functional groups, strategy of building interfacial bonding (Na—O—C) between 3D conductive network and $Na_2C_2O_4$ through high-energy ball milling was carried out. Based on the theoretical calculations, it was revealed that strong interfacial bonds can contribute to superior conductivity as the band gap can be completely vanished, which indicates that coupling effect between sodium, oxygen, and carbon atoms can efficiently improve electronic conductivity of sodium oxalate. Compared to the weak van der Waals force, the strong charge transportation between $Na_2C_2O_4$ and carbon dots could facilitate the electrons' migration from sodium oxalate to carbon dots, thereby improving the conductivity of $Na_2C_2O_4$ electrode. Notably, a unique double coating method (Figure 8.7d) was utilized to

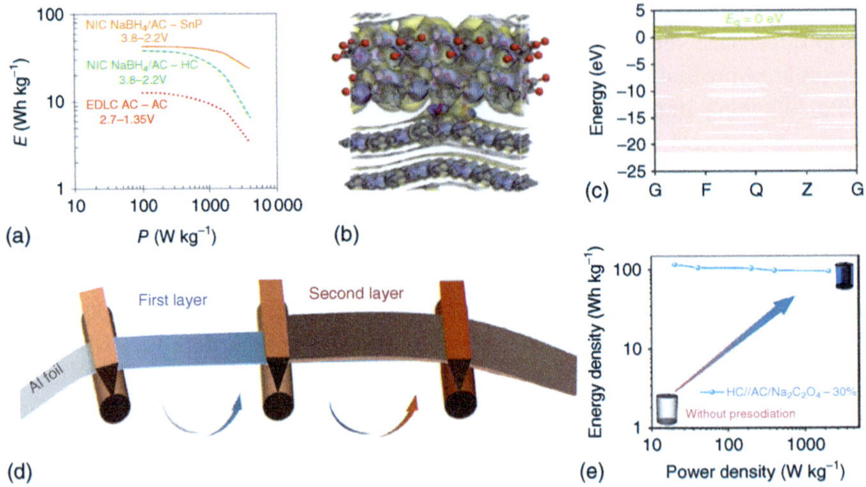

Figure 8.7 (a) Ragone plots of the SICs ($NaBH_4$/AC//hard carbon and $NaBH_4$/AC//tin phosphide). Source: Jezowski et al. [61]/Walter de Gruyter GmbH/Licensed under CC BY 4.0. Schematic illustrating (b) and band structure (c) of interfacial bonding strategy. (d) Original double-coating approach. (e) Ragone plots of the AC//HC SICs. Source: (b–e) Ref. [63]/Springer Nature/Licensed under CC BY 4.0.

fabricate the electrode compounds. Firstly, the activated carbon electrodes were composed of AC, conductive carbon, and PVDF with a mass ratio of 8 : 1 : 1. Then individual $Na_2C_2O_4$ slurries were coated on the top of activated carbon electrodes. This coating method could avoid the structural collapse of electrode material caused by the gas evolution as the sacrificial reagent decomposes. To verify the efficiency of sodium oxalate as sacrificial additive for in situ pre-sodiation, sodium oxalate was added to cathodes in full SICs with various materials as anodes. Note that full SICs without pre-sodiation could barely exhibit available performance, with a dosage of 30 wt% of $Na_2C_2O_4$ compared to the mass of cathode, the SICs (AC//HC) ameliorated by the unique double-coating method could manifest an excellent energy density of 118.2 Wh kg^{-1} at 20 W kg^{-1} (Figure 8.7e).

Importantly, organic sacrificing sources have been aroused extensively exploration as the consequence of low-cost, structural diversity, and environmental friendliness. The SICs (AC-$Na_2C_2O_4$//Sn_4P_3) constructed by Pan et al. can display a high specific energy of 44 Wh kg^{-1} at 1 W kg^{-1} (Figure 8.8A) [64]. In addition, with the merit of the formation of Na_2CO_3 decomposed from $C_4O_4^{2-}$, the full SICs exhibited a preeminent capacitance retention of 94% after 11 000 cycles (Figure 8.8B). It is well known that organic carbonyl compound can be counted as a promising family of cation-insertion material; cations can be reversibly inserted and released based on the redox reaction. Nevertheless, the electrochemical instability at the high-voltage region limits the application. Unlike inorganic materials, organic materials could enrich available utilization for pre-sodiation strategy with the diverse structure. An organic sodium salt ($Na_2C_6O_6$) is successfully proposed by Zou et al. as the alternative sodium source to accomplish the in situ pre-sodiation process. In this

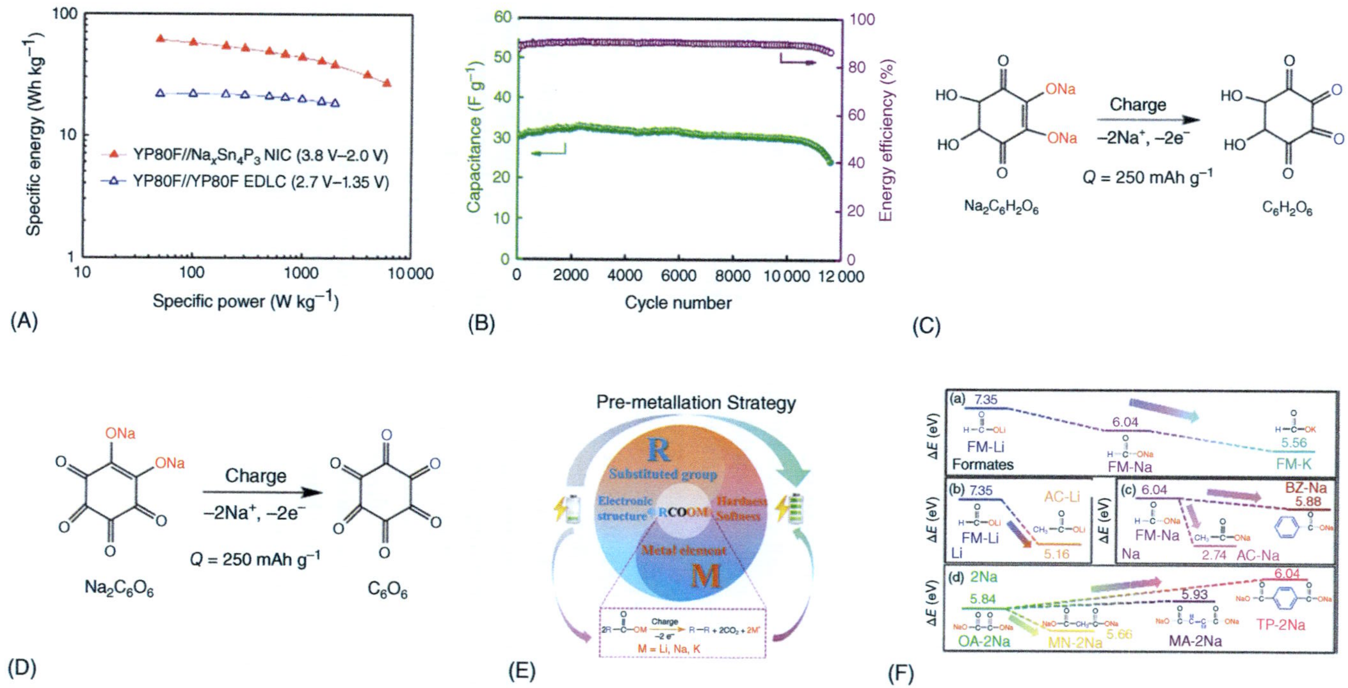

Figure 8.8 Ragone plots (A) and cycle performance (B) of YP80F//$Na_xSn_4P_3$ SICs. Source: (A, B) Pan et al. [64]/ELSEVIER/Licensed under CC BY 4.0. (C) Schematic de-sodiation process of $Na_2C_6O_6$. (D) Schematic de-sodiation process of $Na_2C_6H_2O_6$. Source: (C, D) Zou et al. [65]/Reproduced with permission from John Wiley & Sons. (E) Schematic illustration of modulated bonding energy. (F) Bonding energy of O—M in different systems. Source: (E, F) Zou et al. [66]/Reproduced with permission from John Wiley & Sons.

work, traditional slurry method was utilized: The composite cathode electrodes were prepared by mixing $Na_2C_6O_6$, activated carbon, conductive carbon, and PVDF under various ratios. As for the pre-sodiation process, it is found that introducing organic sodium salts (disodium rhodizonate and tetrahydroxy-1,4-benzoquinone disodium) into the cathode can serve as the sodium sources to compensate for the irreversible Na^+ loss (Figure 8.8C,D) [65], owing to the irreversible electrochemical characteristic, the anodes (including carbon and TiO_2) have been successfully pre-sodiated, and the corresponding SICs could deliver excellent electrochemical performance. Interestingly, a molecularly compensated pre-metallation approach has been employed by Zou et al. to precisely regulate the decomposition potential of sacrificial additives through manipulating the bonding energy of oxygen-metal moiety (Figure 8.8E) [66]. Sacrificial cathode additive method could ensure adequate metal sources for enhanced performance of advanced energy storage devices. Nevertheless, the precise regulation and rational design of decomposition potential of additives have been a great issue. It is worth mentioning that pre-metallation strategy based on molecular regulation has been successfully applied through the reaction of Kolbe electrolysis. The electrochemical oxidation potential of metal carboxylate could be adjusted by the bonding energy between oxygen and metal. The theoretically results further verified the electron-donating effect as well as low charge density of cations can reduce the interaction of M—O bonding, thereby decreasing the decomposition potential of sacrificial additives (Figure 8.8F). Additionally, the low oxidation potential is recognized as a vital factor for sacrificial cathode additives, as it can perfectly avoid the decomposition of electrolyte, thus ensuring the long cycling performance of EES devices. However, recent substitution method would increase the redundant molecular weight of additive as the relative functional groups cannot offer a contribution to capacity, thereby limiting the output capacity and energy density of the whole configuration. To solve this hazard, in-depth understanding about the function of introduced group and the oxidation potential should be explored. Importantly, Zou et al. discovered the electronic effect and regiochemistry of substitution in pre-sodiation process. Under the guidance of density functional theory calculations, it is disclosed that the highest occupied molecular orbital level can be utilized to predict and regulate the electrochemical oxidation potential of organic sacrificial additives, and higher energy of highest occupied molecular orbital could bring about a lower oxidation potential. Moreover, the strong electron-donating effect of induced by the p–π conjugation structure is conducive to donating the electron to the –COONa unit, which could elevate of the level of highest occupied molecular orbital, thus reducing the oxidation potential. In addition, the different substituent group sites (such as ortho-substitution and para-substitution) with different electron-donating features can increase the electron cloud density of the aromatic backbone, and the orientation of electric charges toward the phenylic carbon atom coupled with the –COONa can be altered by the resonance effect, which can reduce the practical oxidation potential of additives. In summary, electronic effect as well as regiochemistry of the functionality triggered by electron-donating substituent, p–π conjugation, and optimized regiochemistry have been utilized to elevate the highest occupied molecular orbital level of pre-sodiation additives, thereby dramatically reducing the corresponding decomposition potential as well as cutting down the additional molecular weight [67].

References

1 (a) Zou, K., Deng, W., Cai, P. et al. (2021). *Adv. Funct. Mater.* 31: 2005581. (b) Zhan, R., Wang, X., Chen, Z. et al. (2021). *Adv. Energy Mater.* 11: 2101565. (c) Song, Z., Zhang, G., Deng, X. et al. (2022). *Adv. Funct. Mater.* 32: 2205453.

2 Kim, M., Xu, F., Lee, J.H. et al. (2014). *J. Mater. Chem. A* 2: 10029.

3 Sivakkumar, S.R. and Pandolfo, A.G. (2012). *Electrochim. Acta* 65: 280.

4 Liu, N., Hu, L., McDowell, M.T. et al. (2011). *ACS Nano* 5: 6487.

5 (a) Xu, H., Li, S., Zhang, C. et al. (2019). *Energy Environ. Sci.* 12: 2991. (b) Liu, W., Chen, X., Zhang, C. et al. (2019). *ACS Appl. Mater. Interfaces* 11: 23207.

6 Meng, Q., Li, G., Yue, J. et al. (2019). *ACS Appl. Mater. Interfaces* 11: 32062.

7 Cao, Z., Xu, P., Zhai, H. et al. (2016). *Nano Lett.* 16: 7235.

8 Zhou, H., Wang, X., and Chen, D. (2015). *ChemSusChem* 8: 2737.

9 Kim, H.J., Choi, S., Lee, S.J. et al. (2016). *Nano Lett.* 16: 282.

10 (a) Xia, Q., Yang, H., Wang, M. et al. (2017). *Adv. Energy Mater.* 7: 1701336. (b) Ma, H., Geng, H., Yao, B. et al. (2019). *ACS Nano* 13: 9161. (c) Yang, B., Chen, J., Lei, S. et al. (2018). *Adv. Energy Mater.* 8: 1702409.

11 (a) Han, C., Tong, J., Tang, X. et al. (2020). *ACS Appl. Mater. Interfaces* 12: 10479. (b) Zou, K., Cai, P., Liu, C. et al. (2019). *J. Mater. Chem. A* 7: 13540. (c) Zou, K., Cai, P., Wang, B. et al. (2020). *Nano-Micro Lett.* 12: 121.

12 Li, S., Wang, C., Yu, J. et al. (2019). *Energy Storage Mater.* 20: 7.

13 Li, X., Kersey-Bronec, F.E., Ke, J. et al. (2017). *ACS Appl. Mater. Interfaces* 9: 16071.

14 Dose, W.M., Blauwkamp, J., Piernas-Muñoz, M.J. et al. (2019). *ACS Appl. Energy Mater.* 2: 5019.

15 (a) Weinberger, M. and Wohlfahrt-Mehrens, M. (2016). *Electrochim. Acta* 191: 124. (b) Wang, L., Fu, Y., Battaglia, V.S., and Liu, G. (2013). *RSC Adv.* 3: 15022. (c) Cao, W.J., Shih, J., Zheng, J.P., and Doung, T. (2014). *J. Power Sources* 257: 388.

16 Wang, Z., Fu, Y., Zhang, Z. et al. (2014). *J. Power Sources* 260: 57.

17 Forney, M.W., Ganter, M.J., Staub, J.W. et al. (2013). *Nano Lett.* 13: 4158.

18 Zheng, S., Chen, Y., Xu, Y. et al. (2013). *ACS Nano* 7: 10995.

19 Pu, K., Qu, X., Zhang, X. et al. (2019). *Adv. Sci.* 6: 1901776.

20 Zhang, X., Qu, H., Ji, W. et al. (2020). *ACS Appl. Mater. Interfaces* 12: 11589.

21 Wu, Y., Momma, T., Ahn, S. et al. (2017). *J. Power Sources* 366: 65.

22 Aravindan, V., Satish, R., Jayaraman, S., and Madhavi, S. (2018). *ACS Appl. Energy Mater.* 1: 5198.

23 (a) Moorhead-Rosenberg, Z., Allcorn, E., and Manthiram, A. (2014). *Chem. Mater.* 26: 5905. (b) Mancini, M., Axmann, P., Gabrielli, G. et al. (2016). *ChemSusChem* 9: 1843. (c) Aravindan, V., Arun, N., Shubha, N. et al. (2016). *Electrochim. Acta* 215: 647.

24 Aravindan, V., Nan, S., Keppeler, M., and Madhavi, S. (2016). *Electrochim. Acta* 208: 225.

25 Gabrielli, G., Marinaro, M., Mancini, M. et al. (2017). *J. Power Sources* 351: 35.

26 Liu, Y., Yang, B., Dong, X. et al. (2017). *Angew. Chem. Int. Ed.* 56: 16606.

27 Sun, Y., Lee, H.-W., Seh, Z.W. et al. (2016). *Adv. Energy Mater.* 6: 1600154.

28 Song, Z., Feng, K., Zhang, H. et al. (2019). *Nano Energy* 66: 104175.

29 Zhao, J., Lu, Z., Liu, N. et al. (2014). *Nat. Commun.* 5: 5088.

30 Zhao, J., Lu, Z., Wang, H. et al. (2015). *J. Am. Chem. Soc.* 137: 8372.

31 Liu, X., Tan, Y., Liu, T. et al. (2019). *Adv. Funct. Mater.* 29: 1903795.

32 Shen, B., Zhan, R., Dai, C. et al. (2019). *J. Colloid Interface Sci.* 553: 524.

33 Zhao, J., Sun, J., Pei, A. et al. (2018). *Energy Storage Mater.* 10: 275.

34 Shanmukaraj, D., Grugeon, S., Laruelle, S. et al. (2010). *Electrochem. Commun.* 12: 1344.

35 Park, K., Yu, B.-C., and Goodenough, J.B. (2016). *Adv. Energy Mater.* 6: 1502534.

36 Bian, X., Pang, Q., Wei, Y. et al. (2018). *Chem. Eur. J.* 24: 13815.

37 Sun, Y., Li, Y., Sun, J. et al. (2017). *Energy Storage Mater.* 6: 119.

38 Zhan, Y., Yu, H., Ben, L. et al. (2018). *J. Mater. Chem. A* 6: 6206.

39 Sun, Y., Lee, H.-W., Seh, Z.W. et al. (2016). *Nat. Energy* 1: 15008.

40 Du, J., Wang, W., Sheng Eng, A.Y. et al. (2020). *Nano Lett.* 20: 546.

41 Liu, X., Tan, Y., Wang, W. et al. (2020). *Nano Lett.* 20: 4558.

42 Jeżowski, P., Fic, K., Crosnier, O. et al. (2016). *Electrochim. Acta* 206: 440.

43 (a) Zhang, S.S. (2017). *J. Power Sources* 343: 322. (b) Park, M.-S., Lim, Y.-G., Kim, J.-H. et al. (2011). *Adv. Energy Mater.* 1: 1002.

44 Park, M.-S., Lim, Y.-G., Hwang, S.M. et al. (2014). *ChemSusChem* 7: 3138.

45 Lim, Y.-G., Kim, D., Lim, J.-M. et al. (2015). *J. Mater. Chem. A* 3: 12377.

46 (a) Park, M.-S., Lim, Y.-G., Park, J.-W. et al. (2013). *J. Phys. Chem. C* 117: 11471. (b) Sun, X., Zhang, X., Zhang, H. et al. (2014). *J. Power Sources* 270: 318. (c) Jeżowski, P., Fic, K., Crosnier, O. et al. (2016). *J. Mater. Chem. A* 4: 12609. (d) Anothumakkool, B., Wiemers-Meyer, S., Guyomard, D. et al. (2019). *Adv. Energy Mater.* 9: 1900078.

47 (a) Moeez, I., Jung, H.-G., Lim, H.-D., and Chung, K.Y. (2019). *ACS Appl. Mater. Interfaces* 11: 41394. (b) Liu, M., Zhang, J., Guo, S. et al. (2020). *ACS Appl. Mater. Interfaces* 12: 17620. (c) Self, E.C., Delnick, F.M., Ruther, R.E. et al. (2019). *ACS Energy Lett.* 4: 2593. (d) Shen, Y., Zhang, J., Pu, Y. et al. (2019). *ACS Energy Lett.* 4: 1717. (e) Wang, G., Li, F., Liu, D. et al. (2020). *Energy Storage Mater.* 24: 147.

48 Mirza, S., Song, Z., Zhang, H. et al. (2020). *J. Mater. Chem. A* 8: 23368.

49 Martinez De Ilarduya, J., Otaegui, L., López del Amo, J.M. et al. (2017). *J. Power Sources* 337: 197.

50 Bloi, L.M., Pampel, J., Dörfler, S. et al. (2020). *Adv. Energy Mater.* 10: 1903245.

51 Fu, R., Wu, Y., Fan, C. et al. (2019). *ChemSusChem* 12: 3377.

52 Shanmukaraj, D., Kretschmer, K., Sahu, T. et al. (2018). *ChemSusChem* 11: 3286.

53 Cai, P., Zou, K., Deng, X. et al. (2021). *Adv. Energy Mater.* 11: 2003804.

54 (a) Dewar, D. and Glushenkov, A.M. (2021). *Energy Environ. Sci.* 14: 1380. (b) Song, Z., Zou, K., Xiao, X. et al. (2021). *Chem. Eur. J.* 27: 16082.

55 Kuratani, K., Yao, M., Senoh, H. et al. (2012). *Electrochim. Acta* 76: 320.

56 Yang, D.F., Sun, X.M., Lim, K. et al. (2017). *J. Power Sources* 362: 358.

57 Kim, M.S., Roh, H.K., Jeong, J.H. et al. (2019). *J. Power Sources* 409: 48.

58 Brady, A., Liang, K., Vuong, V.Q. et al. (2021). *ACS Nano* 15: 2994.

59 Pan, X.X., Chojnacka, A., Jezowski, P., and Beguin, F. (2019). *Electrochim. Acta* 318: 471.

60 Jezowski, P., Chojnacka, A., Pan, X., and Beguin, F. (2021). *Electrochim. Acta* 375: 8.

61 Jezowski, P., Crosnier, O., and Brousse, T. (2021). *Open Chem.* 19: 432.

62 Sun, C., Zhang, X., Li, C. et al. (2021). *J. Power Sources* 515: 230628.

63 (a) Song, Z., Zhang, G., Deng, X. et al. (2022). *Nano-Micro Lett.* 14: 53. (b) Sun, C., Zhang, X., Li, C. et al. (2022). *J. Energy Chem.* 64: 442.

64 Pan, X., Chojnacka, A., and Beguin, F. (2021). *Energy Storage Mater.* 40: 22.

65 Zou, K.Y., Cai, P., Tian, Y. et al. (2020). *Small Methods* 4: 8.

66 Zou, K., Song, Z., Gao, X. et al. (2021). *Angew. Chem. Int. Ed. Engl.* 60: 17070–17079.

67 Zou, K., Song, Z., Liu, H. et al. (2021). *J. Phys. Chem. Lett.* 12: 11968.

9

Conclusions and Future Perspective

Kangyu Zou, Wentao Deng, Hongshuai Hou, Guoqiang Zou, and Xiaobo Ji

Central South University, College of Chemistry and Chemical Engineering, No. 932, Lushan South Road, Yuelu District, Changsha 410083, China

Since the first proposed sodium-ion capacitors (SIC) device proposed in 2012, many researches on SICs have been reported and the development of SIC is accelerating. To date, the development of SIC is still in its initial stage, and the major reported literatures are emphatically divided into four key aspects: definitions and mechanisms, configurations, electrode materials, and key technologies, which are summarized in Figure 9.1.

9.1 Definitions and Mechanisms

The battery-type and capacitor-type electrochemical storage behaviors are synergistically proceeded in SIC system, which could manifest advantages of both traditional supercapacitors (SCs) and sodium ion battery (SIBs), achieving the combined merit of high energy density and large power density. Notably, the fundamental definitions of different electrochemical storage mechanisms should be clearly distinguished. Generally, it is well known that capacitor-type electrode intrinsically exhibits the electrical double-layer capacitor (EDLC) behavior. Furthermore, EDLC mechanism is governed by the electrostatic accumulation of the electrode/electrolyte without involving the faradaic reaction, which only shows a highly reversible physical adsorption/desorption process. On the other hand, battery-type materials mainly consist of intercalation/conversion/alloying reactions of anode and intercalation reactions of cathode, which are essentially controlled by the Na^+ faradaic reaction. In order to pursue the excellent electrode materials equipped with the peculiarities of both large capacity and high rate capability, pseudocapacitive mechanism has been employed for modifying the electrode materials. With the fast development of pseudocapacitance, the types of pseudocapacitance and pseudocapacitive materials should be better comprehended. It is noticed that pseudocapacitance materials and battery-type materials can be accurately distinguished by electrochemical characteristics (including cyclic voltammetry [CV], galvanostatic charge-discharge

Sodium-Ion Capacitors: Mechanisms, Materials, and Technologies, First Edition.
Edited by Guoqiang Zou, Xiaobo Ji, and Hongshuai Hou.

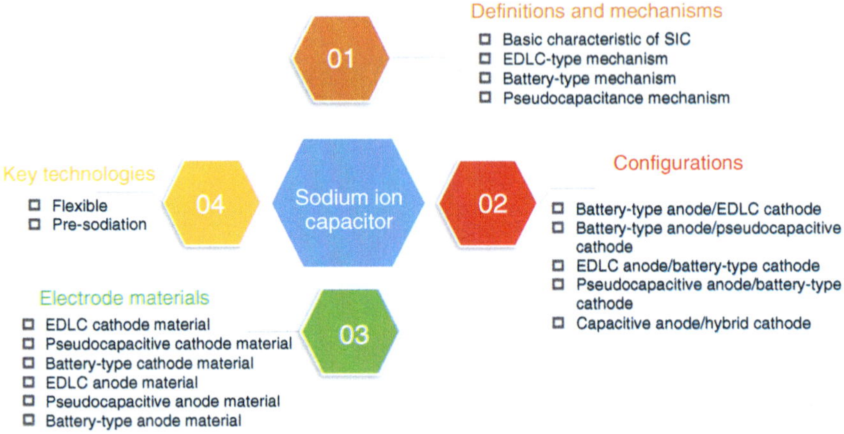

Figure 9.1 Summary of the key fields of SICs.

[GCD], and electrochemical impedance spectroscopy [EIS] curves) and quantitative kinetic analysis. In principle, under the circumstance of pseudocapacitance material, the b can be approximately 1; when $b \approx 1$, it could result from EDLC effect, not necessarily pseudocapacitance effects.

9.2 Configurations

Originally, the supercapacitor device could be classified into the symmetric supercapacitor or asymmetric supercapacitors, according to the energy storage mechanisms of both electrodes. In addition, the SIC, as asymmetric capacitor, possesses the different energy storage mechanisms, which manifest the collaboration of the two types of charge storage features (faradaic and non-faradaic behaviors). The classic SIC configuration is composed of battery-type anode and capacitive-type cathode. Note that, triggered by the development of different electrode materials as device components, the concept of SIC configuration has been gradually expanded. Moreover, some novel SIC configurations have been explored, which are still in the primary stage. Specially, "battery-type anode/pseudocapacitive-type cathode," "capacitive-type anode/battery-type cathode," "pseudocapacitive-type anode," and "battery-type cathode" configurations of SIC systems have been distinguished. It is noticed that the evolution of SIC configuration could occur, resulting from the exploration of new electrode materials, which is aimed at pursuing high energy/power traits. Interestingly, the definition of "sodium-ion pseudocapacitor" has been proposed by involving the reversibly pseudocapacitive anions (such as PF_6^-) into the cathode's intercalation/deintercalation reaction. Moreover, a novel concept named sodium metal hybrid capacitor has been successfully proposed. Specifically, sodium metal hybrid capacitors could exhibit the ultrahigh energy density, which is attributed to the relatively large theoretical capacity (1166 mA h g^{-1}) and (−2.71 V vs. standard hydrogen electrode) of metallic sodium. However, some

SIC configurations are still controversial, and the strict differences in configuration should be carefully distinguished.

9.3 Electrode Materials

In order to pursue the high-performance SICs, various materials have been explored as the appropriate electrode to fabricate SIC devices. In terms of the electrode materials, the improvement of Na^+ diffusion capability is crucial for battery-type electrode to match the fast kinetics of the capacitive-type electrode. Thus, the nanostructure engineering, heteroatom doping, surface coating, and hybrid composite design can bring out the extra pseudocapacitive characteristic, which ensure the quick Na^+ transfer. Moreover, another emphatically considered key property is the relatively low potential accompanied with high capacity, which could enable the enlarged working potential of SIC for improved energy density. Meanwhile, it is important to investigate the electrochemical reaction mechanism of the battery-type materials, potentially boosting the Na^+ storage ability. For example, the sodium storage mechanism of hard carbon (HC) is controversial, and the structural model of HC is still unclear. Moreover, the Na^+ insertion/de-insertion behavior and phase transition mechanism of Na-containing battery-type cathode in SIC system should also be studied. Notably, the volume expansion issue of battery-type material can be ameliorated, considerably decreasing the cycling durability of integrated SIC system. For capacitive-type electrode, triggered by the limited adsorption/desorption behavior, the carbon cathode possesses the unsatisfied specific capacitance, which leads to the troublesome difficulty of electrode manufacturing due to the mass matching. Therefore, pore size regulation, hierarchical porous architecture design, and functional group modification have been employed to boost the electrochemical performance of carbon cathode. Importantly, the EDLC mechanism for accommodating the anions in SIC must be systematically analyzed. In short, fast kinetics of battery-type electrode and large specific capacity of capacitive-type electrode are regarded as vital element in the construction of traditional SICs. Moreover, the exploration of novel electrode material can propel the development of SIC system.

9.4 Key Technologies

Nowadays, flexible and pre-sodiation are two important key technologies of SIC devices. Flexible feature can endow the energy storage device remarkable potentials in wearable electronics market, conformable monitoring skin sensors to implantable medical devices. However, developing reliable and flexible electrodes is still a great challenge to meet the requirements of practical applications. Note that the flexible SIC device is in its early stages, and only a few substrates (such as carbon fiber film) have been directly utilized. It is well known that different substrates have their respective merits and disadvantages. Although carbon substrates (including graphene and carbon cloth) manifest the good electrical conductivity, their large-scale application is seriously limited by the expenditure and the manufacturing technology. Meanwhile, metal substrates are beneficial for fabricating the nanoarray architecture, but the heavy weight and rigidity are fatal issues in flexible devices. Therefore, designing and optimizing flexible electrodes can be realized by employing new substrates and novel structures, which could not only improve the electrochemical performance, but also simplify the fabrication procedure. On the other hand, gel-polymer electrolyte is another critical component in flexible SIC device. And, the low ionic conductivity and poor oxidation resistivity seriously restrict the development of gel-polymer electrolyte, which should be obtained more attentions.

Pre-sodiation techniques can provide the extra sodium sources needed to compensate the irreversible capacity loss triggered by the formation of solid electrolyte interphase (SEI) film. Specially, pre-sodiation strategy is capable of enlarging the working voltage and increasing the Na^+ concentration in the electrolyte for SIC system. Most importantly, the implementation of pre-sodiation can ensure the high security accompanied with simplified assembly process, greatly propelling the commercialization of SIC devices. Nowadays, the reported pre-sodiation methods are mainly classified into three types: operating with sodium metal, employing alternatives of sodium metal, and designing sacrificial sodium sources. Note that the different pre-sodiation methods possess their own mechanisms. Although significant progresses of pre-sodiation have been made, the development of pre-sodiation is still slow due to the lack of in-depth mechanism research in pre-sodiation technique. Moreover, the precise regulation of pre-sodiation degree and the interface chemistry of electrodes are two key directions in pre-sodiation, which are often ignored.

9.5 Future Perspective

Nowadays, although many related works about SICs have been reported, the development of SICs is still in its infancy stage. Moreover, it is a great challenge to exploit potential applications and achieve the commercialization of SICs. Thus, as shown in Figure 9.2, several following aspects must be considered for the future advancements:

(1) Optimizing the characteristics parameters of SICs.
In order to better propel the practical applications of SICs, the key parameters need to be systematically studied, which are ignored in the theoretical research. For example, it is of significance to design the cell voltage range achieve satisfactory electrochemical performance. The energy density of SIC is determined by the cell voltage range, which could be greatly increased by the increase of working voltage. However, when the working voltage is increased, some issues with the electrolyte decomposition and Li electroplating can occur, which seriously decreases the ratability and cycle life of SICs. Therefore, it is necessary that the optimization of cell voltage acquire the outstanding integrated electrochemical performance of SICs.

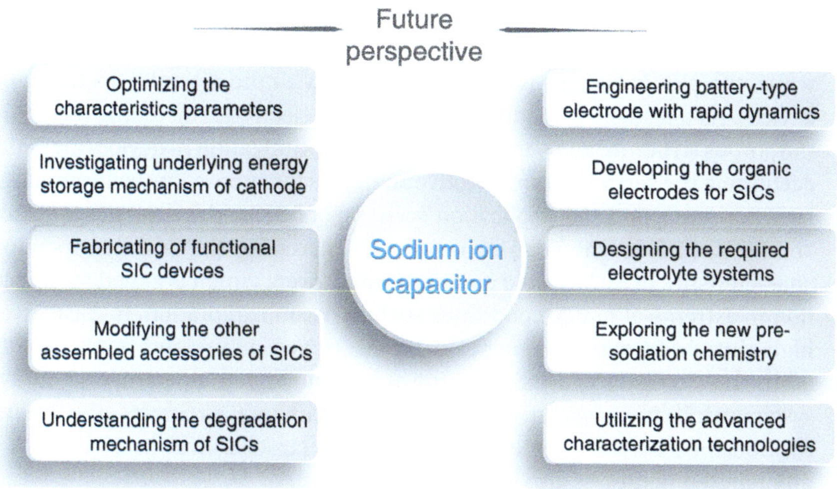

Figure 9.2 Summary of the important aspects of SICs in the future perspective.

Triggered by the distinct energy storage mechanisms of cathode and anode, the high capacity and low cycling durability of anode are not comparable to cathode. Thus, it is difficult to improve the energy/power densities and cycling lifespan in the unbalanced system. Moreover, the energy density of the SIC can be inevitably decreased due to the overweight electrode materials. Thus, regulating the anode/cathode mass ratio is a general method to maximize the utilization of the active materials and further balance the electrochemical performance of SIC. However, due to the different material system and configuration of SIC, the fundamental principles about the anode/cathode mass ratio are very hard to unify.

On the other hand, the self-discharge behavior and high–low temperature character are important parameters for the development of practical SICs, which should receive more and more attention. The self-discharge behavior is known as the spontaneous voltage decay of charged energy storage systems with time in the absence of an external load. Particularly, due to the surface adsorption/desorption mechanism of carbon cathode part, the self-discharge phenomenon is very severe. Moreover, the charged EDLC cathode is in a state of high Gibbs free energy, which will spontaneously decline under the thermodynamic "driving force," leading to the energy loss. Unfortunately, the self-discharge process in SIC systems has been rarely explored. Therefore, mitigating the self-discharge phenomenon is beneficial for the development of advanced SICs. In view of this, the self-discharge mechanisms, influence factors, and suppressing strategies for carbon cathode in supercapacitors should be carefully comprehended, which provides more feasibilities in studying the self-discharge phenomenon in SIC system. Notably, the self-discharge mechanisms in supercapacitors could be divided into three categories: charge redistribution, Faradic reaction, and ohm leakage.

Furthermore, the charge redistribution and Faradic reaction are determined by the intrinsic structure and chemical components of carbon matrix; meanwhile, the ohmic leakage is primarily related with the assembly technique of devices. As a result, the modification of carbon electrode (including regulating the porous structure and surface groups) and optimization of device (including decorating the separator) are effective strategies for suppressing self-discharge. In order to expand the application scenarios of SICs, the diversity and variability of climate should be considered. Therefore, the wide temperature characteristics of SICs should be suitably improved, which is aimed for utilization in extreme climates/conditions such as ocean exploration, tropical areas, high-altitude drones, and polar expeditions. Hence, an in-depth analysis of the failure mechanisms of SICs operating at low and high temperatures should be systematically investigated. Moreover, the fundamental structure–effectiveness relationships between materials and their electrochemical performances at different temperatures also should be uncovered.

(2) Engineering battery-type electrode with rapid dynamics
Because of the sluggish reaction dynamics of battery-type electrodes, the electrochemical performance of SICs is seriously restricted by the kinetic mismatch triggered by the differences in electrochemical storage mechanisms of anode/cathode. Therefore, engineering fast-kinetics battery-type materials is a vital strategy to alleviate the kinetics mismatching. Nowadays, various strategies, including nanostructure design, porosity modification, hetero-interface construction, and heteroatom doping, have been applied to boost the kinetics of battery-type electrodes by introducing the enhanced pseudocapacitive behavior, ultimately achieving both high energy and high power densities. It is crucial to comprehensively analyze pseudocapacitive mechanisms and pseudocapacitive processes to well understand the relationship between "structure and rate" of electrode materials for designing high-performance SICs. Moreover, exploiting new-type anodes equipped with suitable channels for Na^+ diffusion, which is beneficial to accelerate the charge transfer, thus improving the electrochemical performance of SICs.

(3) Investigating underlying energy storage mechanism of cathode
Although the carbon cathodes manifest the quick adsorption/desorption behavior with capacitive feature, they deliver the unsatisfactory specific capacities, which potentially increase the matching difficulty of electrode materials. Hence, it is important to thoroughly explore the energy storage mechanism in Na-based electrolytes, which provides a feasible direction for designing carbon cathodes for advanced SICs. Intrinsically, the optimization of porous characteristics for carbon cathode in electrolyte is regarded as the pivotal approach to improve the capacitive behavior. Therefore, the appropriate pore size and large Brunner-Emmett-Teller (BET) specific area are fundamental factors for accumulating ions at the electrode/electrolyte surface. Unfortunately, it is very hard to precisely regulate the internal structure and porosity of carbon materials, hindering the in-depth investigations in capacitive behavior, which

should be urgently resolved. Moreover, bringing out the pseudocapacitance by heteroatom functionalization and morphology design is also an available solution to fabricate the high-capacity carbon cathode. Meanwhile, the use of advanced pseudocapacitive cathode to replace traditional carbon materials is considered as a great potential tactic for enhanced electrochemical performance of SICs.

(4) Developing the organic electrodes for SICs
The merits of renewable, low-cost, environment-friendly, and organic materials as highly promising electrode materials have attracted lot of attention in electrochemical energy storage systems. However, the feasible applications of organic electrodes in SICs have been rarely studied. Therefore, the various organic electrodes with redox active (such as quinone, aromatic carboxylate, conjugated anhydride, and polymer) should be explored in SICs. Significantly, the exploration of relationship between the molecular, structure, and electrochemical performance can better elucidate the reaction mechanism of organic compounds. Note that the dissolution issue of the organic materials in conventional organic electrolytes as a main bottleneck should be effectively resolved, which can ensure the excellent cyclability of all-organic SIC. As a result, salinization, the combination with carbon substrate and regulating electrolyte are regarded as the most straightforward method of preventing the dissolution of organic electrode materials. In addition, in the future, increasing the tap density of the organic materials and improving the volumetric energy density of SICs can give a good chance of propelling the development.

(5) Fabricating of functional SIC devices
By virtue of flexible and solid-state features, the practical applications of SIC can be greatly accelerated. In terms of flexible SIC system, it is noteworthy that searching new substrates and constructing novel structures are powerful approaches to obtain the high-performance flexible electrodes. Moreover, the heavy weight and rigidity of electrode are critical elements for possible practical applications in flexible/bendable electronics. Note that the research about all solid-state SIC devices is in its infancy. Although solid electrolytes could ensure the safety of SIC, the low ion conductivity of solid-state electrolytes and incompatibility between electrode material and electrolyte are the major challenges. It is expected that the functional SIC devices, as an important technology, follow the trend toward large-scale energy storage. Furthermore, the 3D printing, as a burgeoning technology, provides more possibilities to construct customized architectures in a facile and versatile manner for SIC devices.

(6) Designing the required electrolyte systems
The field of electrolytes in rechargeable secondary batteries has been explored, but the investigation of electrolytes in SIC is at an early stage. Significantly, it is necessary to establish a basic guideline for designing the required electrolyte system, which can efficiently boost the electrochemical performance of SICs. Through the optimization of components, the desired electrolyte can meet the requirements of large dielectric constant and low viscosity for high ionic

conductivity, low melting point, high boiling point, and high flash point for safety. It is noticed that the Na^+ solvation structure can be regulated by the concentration, anion, and solvent, which could potentially affect the Na^+ diffusion dynamics in the SEI film. Different from the SIB, the capacitive behavior associated with adsorption/desorption anion should also be emphatically considered in SIC system. Moreover, introducing the electrolyte additives (including the film-forming additive, flame-retardant additives, and overcharge protection additives) can maximize the electrochemical performance of SICs. Typically, film-forming additives are mainly focused on protecting the electrode and improving the interphase physical properties (such as ionic conductivity, viscosity, and wettability). Flame-retardant additives (such as organic phosphorous compounds) possess the strong ability for scavenging radicals and terminating reactions, which ensures the high security of SICs.

(7) Modifying the other assembled accessories of SICs
It is well known that the other assembled accessories of SICs like current collector, separator, conductive additive, and binder play the vital roles in the integrated electrochemical performance, especially in large-scale field. These related researches can afford more possibilities for constructing high-performance SICs.

(8) Exploring the new pre-sodiation chemistry
Because of the high activity of Na metal, new pre-sodiation technologies must be developed to avoid potential accidents. Moreover, the facile and scalable features should be mainly considered, reducing technological barriers to accommodate practical application of SICs. Notably, although various pre-sodiation methods have been explored, the pre-sodiation technique is empirical and lack of theoretical guidance. For instance, the reaction potential of pre-sodiation additive is difficult to be precisely adjusted. In addition, the modulation of pre-sodiation degree is still a great challenge. Therefore, the in-depth mechanisms of different/pre-sodiation methods should be systematically investigated, promoting better underlying understanding of pre-sodiation behaviors, which provides an available direction for pre-sodiation chemistry of advanced SIC systems.

(9) Understanding the degradation mechanism of SICs
The long-term stability of SICs is mainly decided by the deterioration situation of individual components of SICs. And the fast degradation behavior would cause the undesired performance and short lifetime of SICs. In fact, both the degradation mechanisms and interplay between different subcomponents are quite complicated, which includes the chemical and mechanical deterioration. Generally, chemical degradation processes in SIC systems can be mostly related to the electrolyte decomposition and reduction, SEI formation, binder decomposition, active material dissolution, gas evolution, and so on. Moreover, long-term chemical degradation could bring out the passive influences through the enhanced electrical resistance, side products for attacking or passivating the active materials, and the consumption of electrolyte. It is noticed

that notorious parasitic reactions are not the sole reason for degradation. Meanwhile, the mechanical degradation triggered by the volume changes and stress variation of electrode during the charge/discharge can lead to the cracking of active materials, electrode pulverization, and loss of contact between electrode and current collector. More importantly, the intrinsic features of various degradation mechanisms in SIC should be urgently elucidated at the nanoscale, which is required by the employment of novel in situ and ex situ characterization techniques.

(10) Utilizing the advanced characterization technologies
As an emerging energy storage system, many fields of SIC are still unclear. It is very necessary to investigate thoroughly the energy storage mechanism of electrode and interface chemistry of electrode/electrolyte by the employment of advanced characterization technologies. For example, synchrotron X-ray radiation techniques can explore the evolution of material chemistry, providing key information for electrochemical energy storage mechanisms and SEI films, such as X-ray photoelectron spectroscopy, X-ray absorption spectroscopy, and nuclear magnetic resonance spectroscopy. Moreover, in addition to conventional characterization techniques, some novel techniques can be utilized in the investigation of SICs. Specially, time-of-flight secondary ion mass spectrometer (TOF-SIMS) can be used for evaluating the chemical distribution at nm-level for electrode, solid electrolyte, and interface. Atom probe tomography can record the atomic-level chemical resolution for elemental species. It has been noticed that thermal analysis characterizations can analyze the generated gas to identify thermal decomposition products. Significantly, in situ and operando measurement techniques can intuitively monitor the complicated electrochemical process in SIC system, which can capture the precise information to deeply understand the evolved changes during the charge/discharge behaviors.

(11) Combing the computational results.
With the developments in theoretical calculation and computer technology, the incorporation of calculations and experiments is a preferable practice in understanding and developing new material and electrolyte chemistries for SIC systems. For instance, density functional theory (DFT) provides in-depth insights into essential physicochemical properties such as activation energy of ion diffusion, thermodynamic stability, spatial distribution of electrons, average voltage, and spectroscopic assignment. Furthermore, molecular dynamics (MD) simulation is beneficial for analyzing the dynamic behavior (including solvated structures, migration mechanisms, and electrolyte decomposition pathways) in liquid electrolyte. Nowadays, the artificial intelligence/machine learning (AI/ML) science is regarded as an accelerator for the design and optimization of the electrochemical energy storage systems at all scales, from material design and advanced electrochemical characterization to large-scale manufacture, which can hopefully provide more possibilities for overcoming

the major troublesome challenges by dealing with a massive number of variables and large quantity of data.

In brief, the research on the SIC system is immature, and many pivotal directions are still required to perform big breakthroughs, such as modified electrode materials, functionalized electrolytes, and advanced device configurations. Moreover, this book summarizes the various aspects of SIC in a broad perspective from mechanism to full-cell configuration, from cathode to anode, from flexible device to pre-sodiation technology, and from history to future, which convincingly offers in-depth understandings and directional principles of SIC system. It is believed that the fabrication of high-performance SIC devices will be realized in the near future, which is attributed to the acceleration in the development of SIC triggered by the more and more fundamental studies.

Index

Sodium-Ion Capacitors: Mechanisms, Materials, and Technologies, First Edition.
Edited by Guoqiang Zou, Xiaobo Ji, and Hongshuai Hou.